THE COTTON PLANTATION SOUTH
SINCE THE CIVIL WAR

CREATING THE NORTH AMERICAN LANDSCAPE

Gregory Conniff
Bonnie Loyd
Edward K. Muller
David Schuyler
Consulting Editors

George F. Thompson, Series Founder and Director

Published in cooperation with the
Center for American Places, Harrisonburg, Virginia

"THE
COTTON PLANTATION SOUTH
SINCE THE CIVIL WAR"

✿ ✿ ✿

(Charles Shelton)

CHARLES S. AIKEN

THE JOHNS HOPKINS UNIVERSITY PRESS
BALTIMORE AND LONDON

The Johns Hopkins University Press
2715 North Charles Street
Baltimore, Maryland 21218–4319
The Johns Hopkins Press Ltd., London

Parts of Chapters 1, 2, and 12 are from "New Settlement Patterns of Rural Blacks in the American South," which was published in *Geographical Review* 75 (October 1985); portions of Chapter 12 are from "Race as a Factor in Municipal Underbounding," published in *Annals of the Association of American Geographers* 77 (December 1987), and "A New Type of Black Ghetto in the Plantation South," published in *Annals of the Association of American Geographers* 80 (June 1990); and the first part of Chapter 4 is from "The Decline of Sharecropping in the Lower Mississippi River Valley," which appeared in *Geoscience and Man*, edited by Sam B. Hilliard (1978). I thank the American Geographical Society, the Association of American Geographers, and the School of Geoscience at Louisiana State University for allowing use of these works.

Library of Congress Cataloging-in-Publication Data
will be found at the end of this book.

A catalog record for this book is available from the British Library.

ISBN 0-8018-5679-5

To the memory of my father and mother

"Dont you see?" he cried. "Dont you see? This whole land,
the whole South is cursed, and all of us who derive from it,
whom it ever suckled, white and black both,
lie under the curse? Granted that my people brought the curse
onto the land: maybe for that reason their descendants alone
can—not resist it, not combat it—maybe just endure and
outlast it until the curse is lifted. Then your peoples' turn
will come because we have forfeited ours. But not now.
Not yet. Dont you see?"

". . . You're wrong. The curse you whites brought
into this land has been lifted. It has been voided and
discharged. We are seeing a new era, an era dedicated,
as our founders intended it, to freedom, liberty and equality
for all, to which this country will be the new Canaan."

William Faulkner, *Go Down, Moses*

Contents

Preface

❦ ❦ ❦

Mention of the American South conjures vivid images, which are stronger than for any of the nation's realms with the exception of, perhaps, the American West. Across mental landscapes of the South are a host of pictures, negative ones for many persons, which have been created and reinforced by both the popular media and scholarly works. Central to the images of the South are the plantation and a host of related features, ranging from the abuses of African American slavery to objects of mythical places. The plantation is also among the most misunderstood institutions of American history. The demise of the plantation has been pronounced many times by different scholars, but the large industrial farms survive as significant parts of, not just the South's, but the nation's agriculture.

Central to the survival of the Southern plantation, and also to the many pronouncements of its death, are changes in the geography of the large farms. During the century following the Civil War, the spatial arrangement of the Southern plantation changed twice. The emergence of tenancy after the Civil War produced major geographical alterations in the plantation, and the demise of traditional tenancy, which commenced in the Great Depression of the 1930s, was accompanied by a second spatial reorganization of the great farms. Although the plantation endures in the South, it does not survive in all the regions where it once flourished. In addition to spatial changes in plantations which made them seem to vanish, the regional demise of plantation agriculture also contributed to the premature pronouncements of the death of the agricultural system. Large expanses of the South where plantations once flourished are now essentially void of them. In other regions, however, the great farm exists in its modern geographical forms.

From the time of their importation into continental North America in the seventeenth century, blacks were intimately associated with the plantation. In 1900, after almost three centuries on the North American mainland, approximately 90 percent of the blacks, a population that had grown to nearly nine million, were still in the South. And in the South virtually all blacks were still in the plantation regions. Then began the great twentieth-century intersectional migration of blacks from the rural South to the urban North and West. By 1970, 40 percent of the nation's blacks were in the North, 7 percent in the West, and 53 percent in the South.

But what happened to the blacks who remained in the plantation regions? Most of the recent scholarship on contemporary African Americans has focused on large metropolises. Little attention has been shown to the African Americans who reside in the plantation South, their historical homeland. At the end of the twentieth century, several million of the nation's blacks still lived in the countryside, small towns, and cities of the plantation regions. This population also underwent a profound redistribution. Not only did a rural-to-urban migration occur, but a microscale redistribution of blacks also took place within the countryside. Few blacks, however, remained on plantations, and less than 5 percent of the black labor force was even employed in agriculture.

This study has two principal themes. One is to trace the geographical changes in plantation agriculture between 1865 and 1970. The other is to evaluate the relationship of African Americans to the plantation during the thirteen decades of sweeping economic, social, and political changes from the first Reconstruction through the second. Although Southern agriculture is sometimes depicted as lethargic, almost static, from the 1880s into the 1940s, the changes, especially the landscape changes, were profound. The plantation landscape of 1865 had been so completely reordered by 1910 that few structures from the antebellum period survived. Another sweeping of the Southern plantation landscape began in the 1930s, accompanied by the severing of blacks from the institution to which they had been intimately bound since their importation to the North American continent. The new landscapes with their new settlement patterns are ones in which many blacks live just beyond the borders of contemporary and former plantations.

The current geography of the plantation regions is one that was created, not only by changes in the plantation system, but also by the civil rights movement and the "War on Poverty," the second Reconstruction. The latter chapters, which interpret these events, are more detailed than the earlier ones, which analyze the New South plantation, its alteration, and its demise. My interpretation of the New South plantation is in the context of the voluminous body of literature treating alterations in Southern agriculture from the first Reconstruction through the Great Depression of the 1930s. Despite the relatively large numbers of studies on the civil rights movement and the War on Poverty, none comprehensively examines the impacts of these events in the plantation regions.

Certain of the study's topics, such as school desegregation, apply to the plantation South as a whole. To control the length and complexity of the study, I focus on cotton, historically and geographically the plantation South's most important crop. The study employs the cotton plantation regions of Georgia, Alabama, and Mississippi as primary examples. Not only are these three states part of the core of the plantation South, but they also are among the Southern states with the largest relative black populations and have significant numbers of counties and municipalities in which blacks are the majority population. Many of the critical events of the civil rights movement, including Mississippi Freedom Summer and Martin Luther King Jr.'s Selma Campaign, occurred in these states. By 1990, Georgia, Alabama, and Mississippi, together with Louisiana, led the nation in the number of black elected officials, most of whom held offices in the plantation regions. I do not pursue the story of plantation agriculture beyond 1970, because by that time almost all of the nation's blacks were dissociated from the plantation system.

The manuscript was written during a period in which the terminology used for the United States' largest minority group was changing. *African American* appeared to be superseding *black* as the preferred title. Earlier, in the 1960s, *black* replaced *Negro* and *colored*. Throughout most of the book, I employ *black*. *Black* fits the context of the period after 1966, when the term came into widespread use as a nomen that implied racial pride and alluded to political power. Original terms are retained in quotations, and the newer *African American* is employed in certain places in the text.

Acknowledgments

❧ ❧ ❧

The research for this study was conducted during two periods. In the 1960s, I studied the transformation of the Southern plantation from the labor-intensive New South form to that of the modern mechanized neoplantation. I also explored the origin and disintegration of the Southern Piedmont as a major cotton region. This research was at the University of Georgia under the direction of Merle Prunty Jr., a demanding scholar, wise counselor, and good friend. During my graduate-student days, Charles Wynes opened to me the literature on the history of the New South, and he and Louis DeVorsey taught me the value of historical manuscripts and careful analysis and documentation.

The second phase of the research was from 1985 to 1992. In the early 1980s, I began to wonder what had happened to the blacks who had been so integral a part of the plantation system, had been disassociated from it, but had remained in the plantation regions. What began as a study of local population redistribution and changes in settlement patterns soon led me to an analysis of the civil rights movement and the War on Poverty in the plantation regions. In turn, these topics directed study of the new politics and the efforts to find jobs to replace those lost in agriculture.

I was assisted in my research by many persons from strangers I met on back roads who told me of the local area, to farmers who shared confidential financial details, to civil rights activists who candidly discussed successes and failures, to skilled librarians and archivists who searched until they found that for which I had only a vague reference. I am especially grateful to Dabney Wellford, who gave me access to the file cabinets of the library of the National Cotton Council in Memphis when I began my initial research. At the time, the newspaper and magazine articles and technical reports constituted one of the largest collections on mechanization of cotton production and its consequences. Danny Bellinger at the King Center in Atlanta patiently searched until he located copies of all the Southern Christian Leadership Conference, Student Nonviolent Coordinating Committee, and Mississippi Freedom Democratic Party papers that I requested. I thank William Creech at the National Archives in Washington, who spent hours reading Office of Economic Opportunity documents that I sought and took an enlightened attitude toward what should be available to researchers. Numerous officials of the U.S. Department of Agriculture, the Department of Housing and Urban Development, and the Justice Department graciously provided me with their time and copies of documents which I requested.

David C. Barrow III provided me with copies of historical maps of Oglethorpe County, Georgia, and showed me the location of the plantation at Philomath which was once owned by his family. Larry Farmer not only explained to me the successes and problems of community and economic development in the Yazoo Delta but also shared some of his childhood experiences in Panola County, Mississippi. The death of my father, Claude Aiken, while I was writing the manuscript revealed the extent to which I was dependent on him for many of the social, economic, and technological details of the agrarian South. He lived through most of the twentieth century and made the transition from labor-intensive to mechanized cotton production. He had quite a diverse knowledge, including the changes in cotton production caused by the boll weevil, the operation of a cotton gin, and the dimensions of the rooms and construction techniques of tenant houses. My father also was the first to call my attention to the new settlement pattern of blacks which began to emerge after 1960.

A grant from the National Science Foundation (No. SES8420508) at the time the Geography and Regional Science Division was administered by Ronald Abler funded the critical research on recent population redistribution and new settlement patterns of blacks in the plantation regions. Lisa Roberts ably assisted with the project and spent endless hours entering and analyzing the census data that revealed the spatial patterns. I also thank the University of Tennessee, Knoxville, and C. W. Minkel, associate vice chancellor and dean of the Graduate School, for three Faculty Research Grants that aided in vital parts of the research. The Southern Regional Education Board provided a grant for research at the King Center in Atlanta, and a grant from the Association of American Geographers funded a final trip to the National Archives in Washington to obtain information from Office of Economic Opportunity documents that were declassified for my use.

I am grateful to John Fraser Hart, Peirce Lewis, and Donald Meinig, who encouraged me to quit chopping my research up into articles and to produce a comprehensive book. I particularly thank John Hudson, who not only encouraged me in this project but read the first draft of the manuscript and offered a number of helpful suggestions. George Thompson of the Center for American Places solicited the manuscript, and I especially appreciate his comprehension of the complexity of the rural South. George not only encouraged me to produce the book that I wanted to write but waited patiently through several deadlines for delivery of the manuscript.

Sidney Jumper and the Department of Geography at the University of Tennessee, Knoxville, provided annual course reductions for research and also gave me an additional course reduction during a spring semester. The department's Cartography Laboratory, under the direction of Will Fontanez, drafted the maps. In addition to Will, I thank Charles Lafon, who standardized the final computer versions, and Louise Matthews and Brian Debartolo, who compiled many of the initial maps. Becky Fontanez helped with the tables and various other parts of the manuscript. The manuscript was improved by the editing of Grace Buonocore, who remembered the minutest details and corrected inconsistencies.

My wife, Mary Ann, who is my inspiration, read two entire drafts and made suggestions as to syntax as well as found spelling errors. Charles and John patiently endured their father's seeming obsession with an eight-year project that took part, but I hope not all, of the time that he could spend with them.

CHAPTER ONE

Overview of the Southern Plantation

✣ ✣ ✣

But we have not come to praise the plantation, and certainly not to bury it. . . . It is idle to expect its early demise in the "black belt" of the United States.

Ulrich Bonnell Phillips, "Plantations with Slave Labor and Free"

Since World War II . . . winds of change were said to have swept through the Third World. . . . It is ironic that although plantations were one of the main sources of conditions that generated the winds of change, they have managed to withstand the tempest.

George Beckford, *Persistent Poverty*

THE PLANTATION IS OF THE PAST, AND YET IT IS OF THE FUTURE

A few years before his death in 1932, Ulrich Bonnell Phillips, one of the prominent historians of the American South during the first half of the twentieth century, journeyed from his academic haven at the University of Michigan across the new farmlands of California's Central Valley. Among the places he visited was a 9,000-acre "ranch" on which the principal agricultural endeavor was not the herding of livestock but the cultivation of sugar beets. Hoe gangs of "Hindus, Sicilians, Mexicans, and men of yet other stocks" toiled down mile-long furrows of vast alluvial fields. Phillips immediately recognized that, despite the place and the absence of Southern stereotypes, the great American "ranch" was actually a plantation.[1]

Although racial rhetoric and lapses into nostalgic agrarianism mar and detract from his work, Phillips had particular geographical insights into plantation agriculture which some later scholars fail to discern. The Southern plantation was not destroyed by the fall of slavery, and slavery was not the critical element that defined the plantation. Moreover, the plantation is a world agricultural system, and as a plantation area the American South is not unique. The southeastern United States is simply on the northern fringe of the world's tropical and subtropical areas into which European colonial powers extended the industrial agricultural system as a means of obtaining dependable, inexpensive quantities of sugar, tea, spices, and other products that could not be grown in the middle latitudes. By the early twentieth century in continental North America the plantation had evolved well beyond the simple colonial model to a more economically and technologically sophisticated factory stage that utilized new motorized machinery and required even greater economies of scale. In 1903, Phillips wrote that the "most successful grain farms in the West are really plantations, where great gangs of men and machines work under a single direction."[2] Essentially the spatial production units of American agriculture were becoming more the embodiment of the despised plantation than the cherished family farm. Whether large, highly capitalized farms grew sugar beets in California, wheat in Colorado, or cotton in Georgia, they were plantations.

Industrial-type farms that specialized in the mass production of a commercial crop originated several centuries before the term *plantation* was used. The concept of Western plantation agriculture evolved in the Mediterranean Basin, which Europeans partially conquered in the twelfth century during the Crusades. For Europeans, sugarcane, which they found the Muslims growing, not only was a less expensive source of sugar than honey but also a more versatile one. Great fortunes were to be made through the production of large quantities of sugar for exportation to western Europe. The commercial production of sugar began near Tyre and spread to other parts of the Mediterranean Basin. From the thirteenth into the fifteenth century, Cyprus was the principal center. The feudal estate, modified for intensive commercial crop production, was the spatial model for the industrial farm system that evolved. Under traditional feudalism, lords of manors collected payments from peasants who lived in villages and worked the land, but they did not closely supervise the farming. Because sugarcane estates in the Mediterranean realm were for intense production of a commercial crop, they were carefully managed. Families of laborers housed in villages not only tended fields but were integrated into all aspects of sugar production, including the operation of the sugar mills and the packaging of the commodity for export. Lack of sufficient labor eventually led to the purchase of slaves for agricultural workers. The initial slaves were captives from the Mediterranean borderlands, but by the mid-fourteenth century blacks from sub-Saharan Africa had begun to supplement them.[3]

From the Mediterranean realm the new industrial agricultural system was spread by Portugal and additional emerging European colonial powers to other parts of the world, including the Caribbean islands and the South and North American continents. The term *plantation* originated in sixteenth-century England during the reign of the Tutors as the name for a new settlement of people. The first colonial plantation consisted of Scots and English who were settled in Ireland. The second plantation, Virginia, was on the North American continent. Gradually, the term acquired other meanings. In Virginia, an area cleared for crops became known as a plantation, and from this usage the term also became a synonym for *farm*, a meaning that persisted in the American South into the early nineteenth century.[4] Increasingly, however, *plantation* was associated primarily with large commercial farms, a usage that was well established by the American Civil War.

In 1860, the plantation regions of the South were the northern fringe of a plantation America culture sphere, an area of dispersed districts, which extended from southeastern Brazil northward into Central America, the Caribbean, and the southeastern United States.[5] In the United States, the plantation regions were in a crescent-shaped zone that spread southward and westward from Maryland into eastern Texas (map 1.1). Three seaboard nuclei—coastal Virginia and Maryland, coastal South Carolina, and the lower Mississippi Valley in Louisiana—were the embryonic areas where plantation agriculture began during the colonial period. Each nucleus emphasized a different crop: tobacco in Virginia-Maryland, rice in South Carolina, and sugarcane in Louisiana. The crop that became the principal one of the plantation South, cotton, was not commercially significant until after the American Revolution. Increasing demand for cotton caused by the Industrial Revolution and invention of the Eli Whitney principle of removing seed from the type of cotton which could be grown in the inland South stimulated production. Both the plantation system and cotton spread into the interior of the South after 1800.[6] Until the 1830s, the South's plantations were confined to coastal areas and areas along navigable rivers. The invention and adoption of the railroad permitted the expansion of plantation agriculture beyond waterways, and the South's earliest railroads were constructed to and through plantation regions.

Since the agricultural system emerged in the Mediterranean Basin centuries ago, several characteristics have distinguished plantations from other types of farms.[7] First, plantation agriculture requires high capitalization compared with most other types of farming. For new crops and in new regions, the potential profit that can be realized from a planting venture is so large that even speculative capital is invested. Second, plantation agricul-

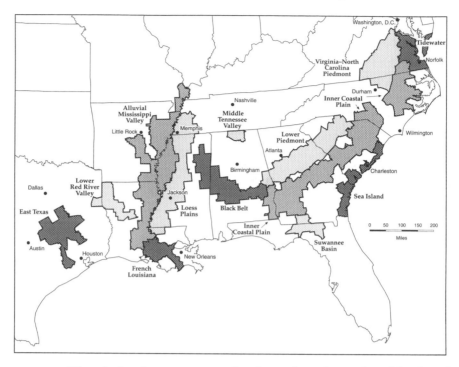

MAP 1.1. Historical and contemporary Southern plantation regions (developed employing U.S. Bureau of the Census 1916, [1948]; Woofter et al. 1936; Holley, Winston, and Woofter 1940; Welch 1943; Prunty 1955; Anderson 1973, 1982; Hart 1982; Hilliard 1984)

ture is significantly focused at both the farm and regional scales. Although subsistence crops might be grown to help sustain the labor force, only one commercial crop is usually emphasized. As in other types of industrial ventures, specialization contributes to efficiency. Skills required for planting, harvesting, and processing a crop are competently learned and perfected. In addition, the marketing of one crop is more cost-effective than the selling of two or more. Third and fourth, both the size of the landholding and the labor force are large enough to achieve economies of scale. Historically, neither a few acres nor three or four workers could produce enough of a staple crop to support a processing facility, such as a sugar mill or cotton gin, which expedited competitive marketing. Size of landholding alone, however, is not what distinguishes a plantation. A few intensely farmed plantations of 100 or 200 acres have existed, but most have contained 300 acres or more and have been larger than the legendary American family farm. The labor force, historically, was composed of entire families, not just heads of households. Counting children, more than one hundred persons could easily make up the labor force on a plantation of a few hundred acres. Approximately one laborer was required for every acre of sugarcane in the Mediterranean Basin.[8] Although slavery quickly became associated with the new industrial farms, a large labor force, rather than a slave labor force, is what distinguishes the plantation.

The fifth characteristic of a plantation is one that differentiates it from the traditional feudal estate. Plantations require careful management throughout the year. The managerial demand on large plantations is so intricate that it consists of a hierarchy that extends down from the owner to manager to foremen over field tracts and company businesses. The sixth characteristic is that a plantation has a unique geographical form that spatially distinguishes it from other types of farms. A nucleated settlement complex has traditionally been an overt element of the geography. The most important building in the complex is the one from which management disseminates. Although the storied big house historically served this purpose on many of the South's plantations, the headquarters often is an office building or is located in a related business such as a store. Larger plantations have a facility to process the crop, primarily to reduce bulk, before the commodity is shipped to market. Rice and sugar mills, cotton gins, and tobacco barns became striking features of settlement complexes on Southern plantations. The buildings that house the primary power supply, exclusive of human, are also components of the settlement complex. Historically, on Southern plantations these buildings were mule and horse barns, but today they are large tractor and machinery stations. Houses for workers are also major components of the settlement complex. Usually the dwellings are built in a row along one or both sides of a road near the headquarters. On pre–Civil War Southern plantations a group of slave dwellings was termed the *slave quarter* or *quarters* and a single dwelling termed a *quarter house*.

EVOLUTION OF THE SOUTHERN PLANTATION

Primarily on the basis of the differences in agriculture, a Lowland South with its emphasis on plantation crops evolved distinct from an Upland South dominated by white yeoman farmers. The Lowland South had larger landholdings and a larger black population and emphasized the great staples—rice, tobacco, sugarcane, and cotton. Not only did the Upland South have smaller landholdings and fewer blacks, but grains, especially corn and wheat, and livestock, primarily hogs, cattle, horses, and mules, were the basis of its husbandry.

Plantations reached their numerical apex and greatest geographical extent in the American South early in the twentieth century. Ironically, at the very time the Southern plantation was at the height of its numerical and spatial importance, the idea prevailed that the great farms had vanished from the South. As with certain other American myths, this one persisted among the educated elite. In colleges and universities across the United States during the 1920s and 1930s, students learned from American history textbooks, including ones written by eminent historians Charles A. and Mary R. Beard and Samuel E. Morison and Henry Steele Commager, that the South's plantations were destroyed by the Civil War and an agrarian revolution in which the large estates were replaced by small farms.[9]

The myth of the demise of the Southern plantation developed for three principal reasons. First, it arose because personal accounts of changes in plantation agriculture were misinterpreted. For planter families the Civil War was a total one that brought catastrophic economic, social, and political changes. Their world was turned upside down. The period of adjustment to a new order, including reorganization of plantation agriculture, extended for more than two decades. In the uncertain atmosphere of the period it was not unusual for planters and their children to write of the ruin of the plantation system. Statements such as "that the old plantation system of farming must now be generally abandoned throughout the South is too obvious to require argument"[10] addressed the passing of what was, but not the emergence of the new agriculture that replaced it. Descriptions of the ruin of the old plantation system by post–Civil War Southern scholars especially gave credence to the idea of the demise of the Southern plantation. Virginia historian Philip A. Bruce wrote in 1900, "Thirty-five years after Appomattox, the ruin of the economic and social system which prevailed in the age of slavery and of the large plantation is complete."[11] Taken out of the context of the scope of his work, even particular statements by Phillips in his early works leave the impression that the Southern plantation was destroyed by the war and the fall of African American slavery. Restoration was his theme in his first major plantation study. "When the plantation comes to be re-established predominantly in the fertile parts of the South, it will bring order out of the existing chaos," wrote Phillips in 1903.[12]

Failure of scholars to view the Southern plantation in geographical context is a second reason for the myth that the plantation disappeared from the South following the Civil War. Because of its close bond with slavery, the pre–Civil War Southern plantation was unique to United States agriculture, but it was not exceptional in the scope of world agriculture. In basic spatial components, the cotton plantation in Mississippi differed little from the sugar plantation in the Caribbean or the coffee plantation in Brazil. Some recent scholars are so taken with the uniqueness of slavery in the American experience that they have interpreted it to be the principal feature that defined the Southern plantation. Writing of the demise of the plantation after the Civil War, Ransom and Sutch state: "One of the most dramatic and far-reaching developments of the postemancipation era was the decline of the plantation system of agriculture and its replacement by tenant farming. The reorganization of the antebellum plantation into smaller tenancies, each operated by a single family, was both swift and thorough. We know that by 1880 the plantation system had ceased to exist."[13]

Ransom and Sutch are led to this interpretation primarily by their failure to take a holistic view. They acknowledge that after the Civil War plantations were not broken up and land ownership remained concentrated. However, they qualify their interpretation and add to the confusion over interpretation of the postbellum South by attempting to make a "distinction between the plantation as a system of labor organization and the plantation as a large landholding."[14] But slavery was not the critical factor that distinguished a plantation. Even Phillips comprehended that the plantation "was less dependent upon slavery than slavery was upon it; and the plantation regime has persisted on a considerable scale . . . in spite of the destruction of slavery."[15] Even during the pre–Civil War era, Southern plantations worked entirely or in part by free laborers existed among those tilled by slaves.[16]

The primary reason that the myth of the disintegration of the Southern plantation arose is that in the post–Civil War agricultural censuses the number of Southern farms increased substantially while the average farm size decreased dramatically.[17] Between 1860 and 1880 the number of farms in the South more than doubled, increasing from 549,109 to 1,252,249, and the average farm size declined from 365 to 157 acres.[18] The growth in number of farms and the decrease in average farm size, in large part, were due to the increase of sharecroppers and other types of tenant farmers on plantations and the United States Census Office's enumeration of each tenant unit as a farm. In their interpretations of census data, journalists and scholars often mistakenly equated *farm* with *landholding*. A 1901 article in the *Boston Transcript* concluded that the "revolutionary increase in small farms" indicates that "the great plantations of some states" have been "almost entirely eliminated, as in the black counties of Mississippi."[19] But the terms are not synonymous; they have distinct geographical meanings. A *landholding* is the cadastre and *farm* the agricultural operation unit. Plantation-size landholdings persisted in the South following the Civil

War, but most were subdivided into several tenant farms. Unlike the North, where a tenant farmer usually was an independent renter, in the South many tenants were closely supervised laborers on large landholdings with centralized management.

Between 1865 and 1970, the geography of the Southern plantation was altered twice. A different spatial arrangement of the plantation typified each of the South's three historical eras: the Old South (1600s–1865), the New South (1880–1940), and the Modern South (since 1970). During the Old South era, which extended from the introduction of the plantation system into continental North America in the 1600s until the end of the Civil War, a nucleated settlement pattern in which the slave quarter was prominent was a primary spatial feature. After a relatively brief transition period following the war, the New South plantation had begun to emerge by 1880. Tenant farming produced significant alterations in the spatial form of the Southern plantation. Although there were several geographical differences between the Old South and New South plantations, the disappearance of the slave quarter and the dispersal of dwellings onto tenant units were primary. The Southern plantation entered the third period of geographical change during the Great Depression of the 1930s. As the emergence of tenants altered the plantation after the Civil War, the demise of traditional tenancy began to change it less than three-quarters of a century later. The period of transformation extended over the three decades that encompassed the New Deal, the Second World War, and the civil rights movement. The Modern South plantation, what Prunty termed the *neoplantation*, was spatially dominant by 1970. The neoplantation resembles the Old South plantation in its nucleated settlement pattern. However, machines and chemicals are substituted for laborers; and the workers, who live in the clusters of dwellings, are paid weekly wages.[20]

THE OLD SOUTH PLANTATION

Three geographical aspects of African American slavery are significant: slaves were concentrated in particular regions of the South; within regions, they were concentrated on a relatively small number of landholdings; and on the landholdings, they were housed in nuclei of dwellings. In 1860, most of the South's slave population was confined to distinct regions spread across the plantation crescent (map 1.2). The major regions were the Virginia Tidewater, the Sea Island coast of South Carolina and Georgia, the lower Piedmont of South Carolina and Georgia, the Black Belt across central Alabama into northeastern Mississippi, the Loess Plains from

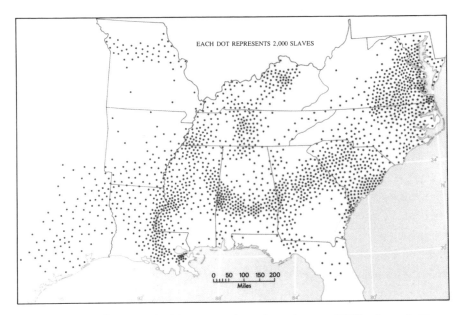

MAP 1.2. Distribution of African American slaves in 1860 (Hilliard 1984). Courtesy of Louisiana State University Press

southwestern Tennessee southward across Mississippi into northeastern Louisiana, and the alluvial Mississippi Valley in southern Louisiana. Because definitions of *plantation* are not quantitatively specific, scholars have used different figures for the number of slaves a person in the antebellum South would have had to have owned to qualify as a planter. Stampp conservatively set the figure at twenty. He found that in 1860 only one-fourth of the families in the South owned slaves, and only 12 percent of the slaveholders had twenty or more.[21]

Insights into the degree of concentration of slave ownership and regional variations in it are provided by the 1860 census schedules. St. Helena Parish, Beaufort District, South Carolina, was part of the Sea Island rice-cotton region, one of the first plantation areas in the South; Oglethorpe County, Georgia, was in the Piedmont cotton region at the northern edge of the Lowland South plantation country; the southern division of Macon County, Alabama, was in the Alabama-Mississippi Black Belt, which by 1860 was a mature inland cotton region; and Tunica County, Mississippi, was in the Yazoo Basin, part of the emerging alluvial Mississippi Valley cotton region. In St. Helena Parish, planters composed 51.4 percent of the slaveholders and owned 88.7 percent of the blacks, and in Tunica County they constituted 45.0 percent of the owners and held 87.3 percent of the slaves. The sample areas in the Piedmont and the Black Belt cotton regions indicate that the degree of concentration of ownership was not as great but surpassed 60 percent. In the southern division of Macon County, Alabama, 34.5 percent of the slaveholders belonged to the planter class and owned 71.7 percent of the slaves; in Oglethorpe County planters accounted for 24.2 percent of the masters and held 63.5 percent of the slaves.[22]

The plantation headquarters was usually the house in which the owner or an overseer resided (map 1.3). The expression "big house," which frequently was used on plantations to refer to this building, implied authority as well as size of structure. One or more barns housed horses used for riding and for pulling carriages and other conveyances and work stock for the fields. The settlement complex also contained a facility for processing the principal crop of the plantation. On cotton plantations, which by 1860 had become the predominant type, a gin house was found on all but the smallest. The gin house was a two-story building (fig. 1.1). It housed the gin, the machine that removed the seed from the fiber, on the second floor and the drive gear, to which mules or oxen were hitched, on the ground floor. Some of the gins were water-powered, and by the 1850s a few steam-powered ones had begun to emerge. The cotton press, a tall vertical structure with a packing screw with two long appendages to which animals were hitched, sat next to the gin house.[23] Additional buildings in the settlement

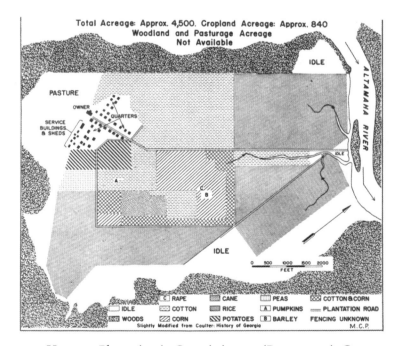

MAP 1.3. Hopeton Plantation in Georgia in 1827 (Prunty 1955). Courtesy of the American Geographical Society

FIG. 1.1. Old South–era cotton gin. A planter created such a facility by purchasing a gin stand, the running gear, and the press, often from different manufacturers or their agents, and building the gin house using a carpenter's pattern book. In 1860, the typical "gin," the device that removed the seed and became known as the gin stand, was located on the second floor. The size and capacity of the gin were measured in terms of the number of circular "saws" that revolved between slots through which lint, but not seed, could pass. The most popular gins of 1860 contained sixty saws and could remove the seed from enough lint in a day to make three to six 400-pound bales. The speed of such a gin was related to the daily capability of the plantation mules and slaves as much as to its size. Six mules and eight to fifteen adults were required to keep such an operation running continuously. Most of the labor was used to haul seed cotton to the gin stand and to move the lint from the gin house to the press using baskets (Tompkins 1903).

complex included one or more smokehouses for preserving and storing meat, a provisions house in which other types of food were stocked, a blacksmith shop, and sometimes grist- and sawmills.

Together, the buildings of even a relatively small plantation formed an imposing complex set among the fields and woodlands (fig. 1.2). Numerous travelers through the antebellum South, ranging from prominent scholars to anonymous journalists, described the nucleated settlement pattern of the plantation landscape in countless references. Throughout the writings of Frederick Law Olmsted, who left one of the most detailed accounts of the antebellum Southern landscape, are frequent references to "quarters," "negro settlement," "little hamlet," "little villages," and "range of negro houses."[24] Detailed maps of antebellum plantations show the headquarters; slave quarters, which usually are depicted as one or two rows of houses; and other buildings that composed the settlement complex (map 1.4). Former slaves in their remembrances of bondage made numerous references to the nucleated settlement pattern. In the narratives of former slaves collected during the Great Depression by the Federal Writers' Project of the Works Progress Administration are an impressive number of statements by elderly blacks who recalled living on large plantations with nuclei of buildings. There is a tendency to dismiss the narratives as unreliable recollections of the elderly in which interviews with the blacks were manipulated and edited by the whites who staffed the Federal Writers' Project. Recollections—such as the one by Lina Hunter, who remembered that on the Billups plantation in Oglethorpe County, Georgia, "back of de big house . . . was just rows and rows of slave cabins"—may be inaccurate, but they are not untrue.[25]

Small slaveholders clustered their two or three slave dwellings around the owner's residence. Olmsted described the farmstead of a small north Mississippi slaveholder which consisted of the owner's house, "a neat building of logs," next to "three negro cabins; one before the house and two behind it."[26] Large slaveholders arranged the dwellings for blacks in definite patterns. Small planters usually constructed the dwellings in a row, which frequently was behind or to one side of the master's residence. The largest planters built their numerous cabins on "streets" to form what appeared to be villages. The most common form was a single street with houses on both sides.[27] From the beginning of slavery in North America, most dwellings of blacks were poorly constructed. Small log houses with dirt floors, faulty

FIG. 1.2. A row of preserved slave houses on the Sam Davis plantation in Tennessee. Courtesy of the Sam Davis Foundation

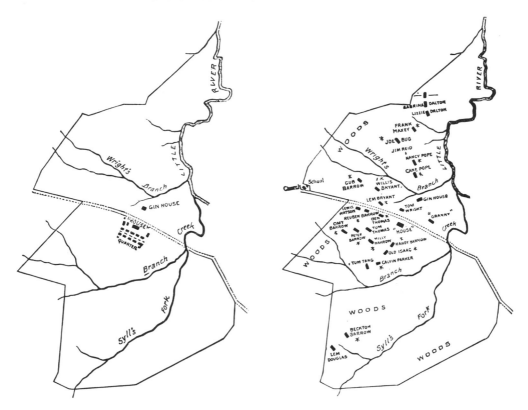

MAP 1.4. The Barrow family's Sylls Fork Plantation, Oglethorpe County, Georgia, 1860 and 1881 (Barrow 1881)

chimneys, and poor ventilation were regarded as fire traps and harbingers of disease.[28] Articles and letters in newspapers and journals subscribed to by planters in the 1840s and 1850s recommended that such houses be abandoned in favor of ones that were above ground and better ventilated and had brick or stone chimneys. In part the recommendations about ventilation were based on the prevalent belief that "bad air" caused disease. Air was thought to be contaminated not only by contact with human excrement, garbage, and general human filth, including unwashed clothing and bedding, but also by decaying wood and vegetation. Another half century would pass before the actual causes of malaria, typhoid, cholera, yellow fever, hookworm, pellagra, and other diseases that plagued the South began to be discovered.[29]

By the end of the Old South era, details for construction of suitable slave houses and hamlets had become rather standardized. Slave dwellings had evolved through several generations, and both single-room and multiroom slave houses were constructed. Although dwellings with more than two rooms were built, most slave houses were single and double pens. Houses were one story with gabled roofs. They were built 1 to 3 feet off the ground with the sills resting on rocks or bricks and were constructed of chinked logs or of planks. Doubles with adjacent pens were more common than double pens with a central hallway or "dogtrot." Double-pen dwellings either had chimneys and fireplaces at each end or were designed as a "saddlebag," which had a central chimney that served both rooms. Each room of a double-pen slave house had a door to the outside. Usually, there was no connecting door between the two rooms, for each housed one slave or one household. On some plantations, men, women, and children who were not parts of family units were housed in a long building divided into three or more compartments. The most common house recommended by planters was a single pen of logs or boards, 16 by 18 feet.[30]

The dwellings of slaves varied greatly from the simple, rustic ideal described in publications of the 1850s and preserved in historical exhibits of the present. At one extreme was a small group of slaves who enjoyed better housing. House servants on large plantations, and all of the chattel of idealistic masters, usually lived in multiroom frame or brick dwellings with plastered walls.[31] At the other extreme were miserable hovels. On the eve of the Civil War, a substantial number of the blacks in the South lived in unchinked, windowless, single-pen log shacks with dirt floors and clay and stick chimneys. But whatever the condition of the slave houses, across the plantation regions most of the dwellings were clustered near the big house in hamlets that dotted the landscape.

EMERGENCE OF THE NEW SOUTH PLANTATION

When the Southern plantation peaked in numerical importance early in the twentieth century, almost all traces of the nucleated settlement pattern that characterized the Old South era had disappeared except in the Louisiana sugar plantation region and what remained of the Sea Island rice region.[32] Throughout the cotton and tobacco plantation regions of the Lowland South, most rural blacks lived in dwellings scattered across the landscape. In many instances, the whites who were employed during the depression of the 1930s by the Federal Writers' Project to interview former slaves selected blacks with whom they were well acquainted. In the descriptions of the settlement patterns on pre–Civil War plantations, it is evident that the elderly blacks were attempting to convey to the younger white interviewers impressions of something that they knew was unfamiliar to them. In trying to characterize the settlement pattern on the Gillum plantation near Prairieville in the Alabama Black Belt, Hanna Jones pointed to the city street in front of her house and exclaimed, "De quarters looked jus' lak dis street dat I lives on now." And Josephine Cox explained to her interviewer that the Bogard plantation in Marshall County in the Mississippi Loess Plains was such a "big place" that "the street between the quarters was longer than this street in front of your house."[33] Describing Macon County, Alabama, in 1934, Charles S. Johnson observed, "Here live almost hidden in the fields, or at wide intervals along the country road, over 11,000 Negroes." And Edgar Thompson in a 1940 description of the New South plantation landscape wrote: "The plantation which the visitor in the South sees today, is generally an estate, a group of little farms, cultivated on shares. Dilapidated cabins are sprinkled over the estate, one to each tract or farm. . . . This general and typical spacial pattern of the present-day Southern plantation, [is] a spacial pattern quite different from that of slavery."[34]

Although blacks were euphoric over freedom and had great expectations about deliverance from bondage, most quickly realized that the new situation in which they found themselves did not differ substantially from that of slavery. The words that a Union army officer spoke to a group of South Carolina freedmen at the end of the Civil War were prophetic for the majority of rural blacks. "You are free but you must know that the only difference you can feel yet, between slavery and freedom, is that neither you nor your children can be bought or sold. You may have a harder time . . . than ever before; it will be the price you pay for your freedom."[35] In 1910, nearly half a century after emancipation, 89 percent of the nation's blacks were still in the South, most still a rural people concentrated in the plantation regions (map 1.5).

With emancipation, planters were confronted with the problem of devising a new plantation labor system whereby the former slaves could again be employed. What is especially significant about the new organization that developed is that blacks were able to assert themselves and extract limited concessions from the planters. The concessions, in turn, influenced the geographical form of the plantation.

One method by which former slaves could be reintegrated within plantation agriculture was to continue the work system of slavery which employed gangs of blacks under the direction of the owner, a white overseer, or a black driver. The gang system was designed for maximum efficiency with a set work pace and daily tasks. Whitelaw Reid, who traveled through the Natchez district in 1866, found that the large plantations were operated much as they had been before the Civil War with the former slaves working

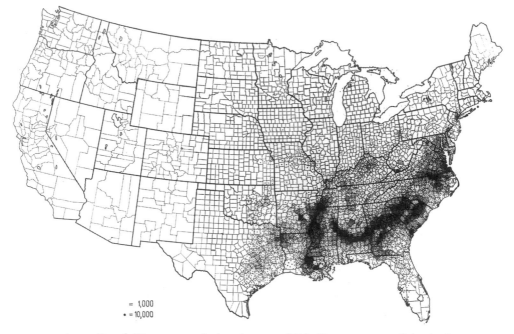

= 1,000
• = 10,000

MAP 1.5. Rural Negro population in 1910 (U.S. Department of Agriculture, 1919)

in large, closely supervised gangs.[36] A variation of the gang was the squad system. Squads contained smaller numbers of workers and usually were composed of extended families or a group of persons who preferred to work together.[37] Although the squad system offered the laborers more freedom than the gang system, squads also were closely supervised and frequently were commanded by blacks who were counterparts to the drivers of slavery. On David Barrow's Sylls Fork Plantation in Oglethorpe County on the lower Georgia Piedmont, the squad system was introduced immediately after the war. Each of the two squads was under the control of a driver called a "foreman."[38] Although some planters paid former slaves who worked in gangs and squads cash wages, the shortage of currency in the war-torn South resulted in the widespread practice of paying with a share of the crop.

Another method of employing former slaves on plantations was to engage them as tenant farmers. Several types of farm tenancy existed in the South and the nation during the post–Civil War period. The types were distinguished on the bases of what the renter and the landlord provided in the production of a crop and the method by which the landlord was paid. Some tenants paid a pre-agreed amount of money or agricultural commodities. Others paid a part of the crop, with the amount dependent on what the landlord and the tenant provided.

Resistance of the former slaves to the new work systems that employed them in gangs and squads was immediate. The freedmen did not want the slavelike controls that continued to be placed on them, especially the abuses of overseers and drivers. In addition, for several years following the Civil War the belief prevailed among the former slaves that the federal government would institute land reform. Plantations would be confiscated and subdivided into small farms that would be given to them. The desire for more personal freedom coupled with the belief that they should have their own farms led the former slaves to rebel against work systems that they viewed as new forms of bondage. Among the few weapons available to them was refusal to work in gangs and squads and work slowdowns by pretense of illness or lack of comprehension.

Brooks in his detailed study of post–Civil War alterations in the plantation system, which employed interviews with planters who contended with the transition from slave to free labor, emphasized that the former slaves, not the planters, were the ones who preferred tenant farming.[39] Recent studies of the post–Civil War period also stress that widespread emergence of tenant farming was largely due, not to a deliberate strategy by planters, but to the insistence of former slaves.[40] One argument is that tenant farming emerged from labor market bargains struck between planters and former slaves.[41] To the extent that tenancy offered the former slaves more personal freedom, it was a compromise. But tenant farming, especially sharecropping, quickly became a repressive labor system in which legal devices, informal agreements, paternalism, and occasional violence were used to control workers. For the planters, the advantages of tenancy were a more dependable labor supply and labor forces composed of entire households rather than confined, potentially, to just the heads of households. Planters, however, lost the efficiency of the gang and squad systems and total control of all farm operations. For the former slaves the overt benefit of tenancy was the right to work their own small farms. Field work was no longer under the hour-by-hour supervision of drivers, overseers, or landowners.

The new tenure arrangements under which plantations began to be farmed produced significant changes in the internal spatial organization. Not only were the large landholdings fragmented into many small tenant units, but profound alterations occurred in the settlement pattern, especially in the location of houses. Former slaves began to abandon the slave quarters and to erect crude dwellings on their tenant plots. A South Carolinian wrote in 1877, "The negroes used to be kept . . . in cabins clustered together near the residence of their owner or overseer. Since their liberation they have shown quite a tendency to desert the relics of their former subjection." He continued by observing, "Many of the 'negro quarters' . . . on isolated and *very* rural plantations are yet inhabited, but a large number all over the country are tenantless and going to ruin." Robert Somers, who traveled across the South five years after the conclusion of the Civil War, described the changes in the settlement patterns of blacks on plantations in the Tennessee Valley of northern Alabama: "It was usual in slavery times to concentrate the 'quarters,' and the cribs, and the mule stables, near the homestead. But under the free contract, by which the negro field-hand has become a sharer of the crop . . . it is found convenient to spread him . . . more about, near his work."[42]

One of the most detailed contemporary analyses of the settlement changes that accompanied the development of the tenant system on Southern plantations is David C. Barrow Jr.'s study of the alterations on his father's 2,365-acre Sylls Fork Plantation in Oglethorpe County on the lower Georgia Piedmont.[43] Not only did Barrow describe the changes in labor arrangements between 1865 and 1881, but he also mapped the alteration in the settlement pattern (map 1.4). The problems that arose over close supervision under the squad system, especially the continued use of a driver, together with abuse of the plantation mules and disputes over division of the crops, led to the adoption of the tenant system and partitioning of the plantation into small farm units. One by one the houses in the old slave quarters were abandoned, and dwellings were erected on the tenant farms.[44]

Significantly, the efforts to create the new dispersed settlement pattern, like the demand for tenant farms, originated among the former slaves, not among the planters. Some planters, such as James Gillespie of the Natchez district in Mississippi, opposed the efforts of the former slaves to reject the nucleated settlement pattern of slavery.[45] Like the tenant system, the abandonment of the quarters eroded the close supervision that planters could exercise over the freedmen. Among the former slaves the desire to leave the hamlet was so keen that even well-constructed dwellings in the quarters were renounced for shacks in the fields and woods. Some former slaves interviewed in the 1930s remembered their houses in the quarters as superior to their contemporary dwellings.[46] Hurricane, the Natchez district plantation of Joseph Davis, brother of Jefferson Davis, had quarters that surpassed the ideal in slave housing. Each well-built cottage had two large rooms with plastered walls. Yet the quarters on Hurricane were abandoned even though the plantation had been purchased from the Davis family by former slaves. Ben Montgomery, the leader of the freedmen who owned the plantation, was killed in 1874 when the wall of a cottage that was being razed fell on him.[47]

From the viewpoints of planters and some outside observers, including Somers, making the dwellings more convenient to fields was among the primary reasons for alteration of the settlement pattern. Barrow gave the desire of blacks for more "elbow-room" as the basis.[48] But reasons of the former slaves for the relocation of residences from the quarters ran much deeper. Because as tenants blacks found themselves still caught in the plantation system, they had to find ways to express their newfound freedom within it. The slave quarters, like the squad system and drivers, were associated with the close supervision of bondage. The dispersed houses, like the tenant units, were geographical expressions of freedom. The more astute observers of the post–Civil War era saw this meaning in the new spatial arrangement of the dwellings of blacks. Edward King, who traveled extensively across the South in the 1870s, wrote concerning the emerging settlement pattern: "The thing which struck me as most astonishing here, in the cotton lands [of the lower Mississippi Valley], as on the rice plantations of South Carolina, was the absolute subjection of the negro. . . . The only really encouraging sign in their social life was the tendency to create for themselves homes, and now and then to cultivate the land about them." More than sixty years later Thompson observed that although the dilapidated cabins that were sprinkled across the plantation seemed "to reach toward the Big House for protection," they also seemed to be "demanding independence." Merle Prunty Jr. argued in his study of the Southern plantation that "personal freedom" was the paramount factor in the dispersal of houses.[49]

Two additional objects that also were overt symbols of the new freedom within the plantation system, churches and schools, emerged. As slaves, it was not uncommon for blacks to accompany their masters to church, and church sanctuaries had sections in the balcony or at the back reserved for blacks. The withdrawal of blacks from white churches into their own congregations and denominations, especially on larger plantations, began during the antebellum period, and it rapidly accelerated during Reconstruction. With emancipation also came a fervent desire among the former slaves to learn to read and write. Because education was forbidden to them as slaves, being able to read and write was among the paramount expressions of freedom.[50] In addition, the belief prevailed among former slaves, a conviction that was passed to successive generations of blacks, that education was fundamental to economic and social advancement. Booker T.

Washington, who was a product of the postemancipation education frenzy, recalled: "Few people who were not right in the midst of the scenes can form any exact idea of the intense desire which the people of my race showed for an education. . . . It was a whole race trying to go to school. Few were too young, and none too old, to make the attempt to learn."[51]

Small, poorly constructed church buildings, which frequently also served as schools, rapidly began to appear on the reordered landscape of the New South, often on plantation property. Some planters gave or sold former slaves lots on which to erect the buildings. David Barrow gave the tenants on Sylls Fork a corner tract on the road for a church and a school.[52] Despite their miserable physical conditions, the churches and the schools were of chief importance, for from Reconstruction through the civil rights movement, most of the black leadership and most of the social interaction among blacks in the plantation regions were within these two institutions.

RESUBJECTION OF BLACKS

When the former slaves commenced their journey into freedom, they had few material possessions and no formal education. Their two major assets were their labor and potential political power. Black labor, however, had a limited market, for during most of the New South era little demand for it existed outside the Southern plantation regions. In the plantation South, however, cheap black labor remained a critical component upon which the agricultural system was based. Through "bargaining" their labor, former slaves achieved an illusory independence as tenant farmers within the plantation system. Increasingly, though, planters compromised labor as a commodity with which blacks could bargain.

The concession to tenant farming was loaded in favor of the planters, and with the passage of time the control of planters was reasserted and became more intense. What in the 1880s might have seemed to have been a significant concession by planters was, at the end of the New South era, a vicious tenure system for both blacks and legions of poor whites, who also had been drawn into it. Many of the problems and abuses that came to be associated with Southern tenant farming, especially with sharecropping, were due to its redefinition. Davis argued that the former slaves' preference for sharecropping in the 1860s and 1870s was not based on a lucid understanding of what the system might become.[53] But if the former slaves comprehended sharecropping as it existed in the South prior to the Civil War and in the years immediately following, then it is more apparent why they so readily accepted the tenure system.

During the antebellum period, sharecropping was not common, but it was practiced in the South.[54] Sharecroppers were tenant-partners of land-owners. One partner contributed land, work stock, and farm implements, and the other provided labor. Although the tenure arrangement between a landlord and a sharecropper was not legally defined during the Old South era, it had common-law standing in the courts.[55] After emancipation and the widespread emergence of sharecropping on plantations, the traditional definition of *sharecropper* was legally modified. Few planters wanted for-mer slaves as tenant-partners, and those who entered into partnership agreements with them often were condemned by fellow landlords. A Glynn County, Georgia, planter who farmed "successfully" with his former slaves as tenant-partners between 1866 and 1870 abandoned the practice "on account of prejudice in [his] neighborhood against having blacks as part-ners." In retrospect, he thought that he would have "done better to have continued on that system."[56]

The legal redefinition of sharecropper was accomplished through the crop-lien laws that were passed by Southern legislatures. Landowners and tenants had different standing under these laws, which specified who le-gally owned a crop and could give a lien on it. Georgia's first lien law was adopted by the state legislature in 1866. In an 1872 ruling on the law, the Georgia Supreme Court held that there was a distinction between a tenant and a sharecropper and that the share of the crop given to a cropper was merely a method of paying wages to a day laborer.[57] When the Georgia General Assembly passed a more comprehensive crop-lien law in 1873, the term "laborers" was used to describe sharecroppers.[58] In other Southern states the status of a sharecropper was also legally changed from that of a tenant to that of a laborer. Rulings of the state supreme courts in North Carolina in 1874 and Tennessee in 1884 held sharecroppers to be laborers paid with a share of the crop.[59] In Mississippi, sharecroppers occupied an ambiguous legal status, for they were considered laborers in criminal pros-ecutions and tenants in civil actions.[60]

Central to the resubjection of blacks in the plantation regions after Reconstruction was their loss of political power. In 1880, the majority of black adult males voted in every Southern state except Georgia and Mis-sissippi, where at least one-third voted. By 1910, fewer than 10 percent of black adult males in the South were registered.[61] At first whites tried to control and weaken the black vote through intimidation by the relatively short-lived Ku Klux Klan and by the passage of vote-dilution laws. Finally, blacks were removed from the political process by the enactment of dis-franchisement laws. A myth regarding disfranchisement is that although planters manipulated votes of blacks, they did not attempt to eliminate blacks from politics. White yeoman farmers resented blacks, and they initiated and led the campaigns to disfranchise them. Kousser, however, carefully demonstrated that most of the initiative to quell political participation by blacks came from the plantation regions. Planters, rather than yeoman whites, were the principal ones who undermined blacks' political participation. According to Kousser, "the vestiges of the antebellum ideology and social structure—the unqualified belief in the innate inferiority or even the inhumanness of the Negro, the contradictory impulses to violence and paternalism, the acceptance of the hegemony of a tiny white elite—retained their greatest strength after the War among whites" in the plantation regions. What affluent whites feared most was an alliance by which blacks and yeoman whites might take control of a county or a state. This would mean more black officials, including tax collectors, sheriffs, constables, and even black juries and judges. As long as blacks had any political power in plantation regions where they were a numerical majority, affluent whites had to be wary of their potential to seize government. Reinforcing this belief was the social reformation almost accomplished in parts of the plantation regions during Reconstruction.[62]

Nearly all of the proponents of disfranchisement were Democrats, and the principal leaders within the party were from the plantation regions. In Mississippi Senator J. Z. George was a leader.[63] In Alabama Judge Thomas W. Coleman of Greene County, a former slaveholder, Confederate officer, and a justice of the Alabama Supreme Court, was a principal leader. Greene County, which in 1900 was 84 percent black, voted 1,077 to 101 to ratify constitutional disfranchisement. Dallas and Wilcox Counties, which were 80 and 78 percent black, also supported disfranchisement by wide majorities. Not only were blacks kept from voting in the referendums, but black votes were fraudulently counted.[64]

Disfranchisement had to be accomplished without formal violation of the Fifteenth Amendment to the federal Constitution, which states that "the right of citizens of the United States to vote shall not be denied or abridged by the United States or by any state on account of race, color, or previous condition of servitude."[65] The only voting-age males specifically disfranchised were "idiots, insane persons, and Indians not taxed." However, an imposing set of obstacles was erected to eliminate black voters either by rewriting state constitutions or by adopting amendments to them. Among the methods were elaborate residence and registration requirements, literacy tests, and annual poll taxes. Mississippians took the initiative in the disfranchisement. Mississippi's 1890 constitution required a cumulative annual state poll tax of two dollars, "to be used in aid of the common schools." Counties were given the authority to levy an additional tax up to one dollar. Even though the tax was relatively low, it was a financial burden, not only for poor blacks, but also for poor whites. The new constitution also required a voter to register at least four months before an election and to have lived in the state for two years and in the district for one year. The lengthy registration and residency requirements discriminated against tenant farmers because they moved frequently. The new constitution stated, "Every elector shall . . . be able to read any section of the constitution of this state; or he shall be able to understand the same when read to him, or give a reasonable interpretation thereof." The oral interpretation clause was added as a loophole for illiterate whites. A registrar could read a section of the state constitution and ask an illiterate person for an explanation. Not only was a black intimidated by the test, but he passed or failed at the discretion of the white registrar. No matter what interpretation an illiterate white gave, he could be judged as having passed the test. South Carolina, Louisiana, North Carolina, Alabama, Virginia, and Georgia soon followed in the adoption of literacy tests and other methods designed to restrict the political participation of blacks.[66]

Blacks did not acquiesce to disfranchisement; they strongly opposed it. Black leaders, including Booker T. Washington, who is regarded as an advocate of accommodation, appealed to state legislatures. In six Southern states blacks organized campaigns against proposed disfranchisement and in ten filed court suits to overturn restrictions on voting. Washington, however, thought that more could have been done, for he complained that his appeal to the Georgia legislature in 1899 was not adequately supported by the state's black leaders.[67]

Disfranchisement and its consequences were severe and far reaching for blacks in the plantation South. For more than six decades, from the post-Reconstruction period until the civil rights movement in the 1960s, Southern blacks had no direct role in local, state, or national politics. "So few have been Negro voters in the South that to estimate their number seems futile," wrote V. O. Key Jr. in 1949.[68] In the first decade of the twentieth century, as the plantation approached its spatial and numerical zeniths, almost 90 percent of America's blacks were still concentrated in the plantation regions, where, as in slavery time, they were a defenseless people. Loss of political power meant not only that blacks were unable to win any new economic, social, and political concessions but that they had difficulty retaining those that had been attained.

One of the most significant effects of disfranchisement was its devastating impact on public education. The fervent desire for education among blacks which began with such enthusiasm and ambitious goals during Reconstruction lay crushed seventy years later. In the depths of the depression of the 1930s, Howard Odum wrote, "[Although] from almost no beginnings Negro education has grown into many multiples of its earlier status, in . . . the Southeast, many of whose indices reflect about half the achievement of the nation, the Negro's measures are no more than a fourth."[69]

The crusade for public elementary schools was waged in the North between 1830 and 1860. The South, however, lagged behind; the campaign had barely begun when it was interrupted by the Civil War. Following the war, former slaves took much of the initiative in the plantation regions to establish and fund public schools. The South's more affluent whites, including many planters, believed strongly in education for both their male and female children, but they generally favored exclusive private education. When planters began to reassume political control of counties and states in the 1870s, they sought to subvert the efforts of blacks to establish and fund public schools. Planters were opposed to public schools because of the taxes required to support them and the belief that education would have a negative impact on the supply of cheap, docile black labor. Emancipation did not change the idea that education ruined a Negro. In addition, children were integral parts of the new tenant labor system. Boys and girls as young as six were old enough to help weed and harvest cotton. Sending children to school, with the exception of the three winter months, interfered with agriculture, for the acreage of cotton assigned to a black tenant farmer was based on the size of his family.[70]

The post–Civil War public education system that developed in the plantation regions was a dual, racially segregated one of underfunded elementary schools. Schools for whites occupied substantial one- and two-room buildings in towns and rural communities. Occasionally, a public school occupied a building that originally had housed a private academy. In 1900, most schools for whites did not go beyond the eighth grade, and many went only to the fourth or sixth. Public schools for blacks frequently were housed in churches and poorly maintained one-room buildings, and they rarely went beyond the third or fourth grade.

In the 1890s, yeoman whites began to assume the initiative in demanding improvements in Southern public education. Political compromises between planters and yeoman whites in the plantation states resulted in greater financial support for public schools. However, segregation meant that states that could barely afford one school system had to support two.

Funds began to be diverted to the schools for whites at the expense of the ones for blacks, and disfranchisement meant that blacks had no power to stop the erosion of financial support.[71] In Alabama at the end of Reconstruction, annual per capita expenditures for black and white schools were almost equal. Per capita expenditures for teachers' salaries were $0.83 for whites and $0.82 for blacks. Between 1877 and the Populist revolt in the 1890s, a small inequity developed. In 1892, per capita expenditures for white and black teachers' salaries were $1.07 and $0.87. After disfranchisement, substantial inequity developed in funding dual school systems. By 1908 the per capita expenditure for black teachers in Alabama was $0.89, compared with $5.05 for white ones.[72]

As the New South plantation emerged, paternalism was redefined and continued to be employed to manage blacks and keep them subservient. Blacks retained certain privileges that they had under slavery. Traditional holidays, including Christmas and the few days without work after the crops had been "laid by" in early August, were preserved under the New South plantation system. On Sundays and frequently on Saturday afternoons, there was no work in the fields. Hunting and fishing privileges and the right to use plantation mules and farm implements to work gardens and to haul firewood were important aspects of paternalism. Medical care was provided as part of the annual furnish, with tenants repaying the planters from their shares of the crops. When blacks who lived for many years on a plantation grew elderly, the planter was expected to provide a dwelling and basic sustenance until death. The reforged paternalism was frequently mentioned by New South planters as the quintessence of the plantation system, but they also complained of the burdens that it imposed. "To live among a people whom, because of their needs, one must in common decency protect and defend is a sore burden in a world where one's own troubles are about all any life can shoulder," griped William Alexander Percy, one of the more literate New South planters.[73]

After legal rights were stripped from farm tenants, suffrage restricted, and an unequal dual school system created, the justifications for tenant farming and the reasons for lack of political participation and low educational attainment of blacks quickly became part of the mythical racial dogma of Southern whites. The 1890 Mississippi constitution was hardly a decade old when J. S. McNeily, a member of the state's disfranchisement convention, wrote: "The fact of disfranchisement was accepted by the masses of the sons of Ham without show of sorrow or sign of resentment. Suffrage had come to them unsolicited; it departed from them unregretted. As a demonstration of the Negro's incapacity and unworthiness for the equipment of political equality nothing more need be written."[74] Although the post–Civil War planters reluctantly accepted tenant farming, sixty years later, when the rental system came under attack for its social and economic injustices, their children and grandchildren rallied to defend it. A Washington County, Mississippi, planter declared in 1942 that sharecropping was "the fairest division between capital and labor that [had] yet been devised." His neighbor, William Alexander Percy, acknowledged the dependent and helpless situation of black tenants but argued that sharecropping was "one of the best systems ever devised to give security and a chance for profit to the simple and the unskilled."[75]

Discrimination in funding black schools helped to perpetuate and broaden the socioeconomic chasm between blacks and whites. Many of the whites in the plantation regions were relatively well educated, but into the second half of the twentieth century most blacks remained illiterate or semiliterate. Blacks' illiteracy became a vital part of the arguments whites used to explain their deplorable social and economic conditions and to justify segregation and discrimination. Whites asserted that spending large sums on education was a waste, for blacks were inherently incapable of retaining knowledge beyond an elementary level. Ulrich Phillips thought that blacks had little intellectual capacity. As a young ambitious academician, he argued that "a possible solution" to the "vital question" of "how is civilization to be promoted among the mass of Southern negroes who are beyond question in need of further and higher development" was the plantation—the institution for which blacks' rights were appropriated.[76]

CHAPTER TWO

From Old South to
New South Plantation

❦ ❦ ❦

Out of the North the train thundered. . . . The sun is setting,
but we can see the great cotton country as we enter it. . . .
This was indeed the Egypt of the Confederacy,—the rich
granary whence potatoes and corn and cotton poured. . . .
Sheltered and secure, it became the place of refuge for families,
wealth, and slaves. . . . Then came the revolution of war and
Emancipation, the bewilderment of Reconstruction,—and now,
what is the Egypt of the Confederacy, and what meaning
has it for the nation's weal or woe.

W. E. B. Du Bois, *The Souls of Black Folk*

The agriculture of a country is as dependent upon the knowledge
and power of the people as it is upon the qualities and characteristics
of the land. The boundaries of . . . agricultural regions . . . may not
all be permanent. This is no detriment, however, for nothing
is permanent in the world except change.

O. E. Baker, "Agricultural Regions of North America"

TENANCY AND THE PLANTATION

Because tenancy was a central component of the agricultural changes, the differences between tenant farmers and the degree of management land-lords exercised over each were critical to the evolution and viability of the plantation system between the Reconstruction period and the Second World War (table 2.1). Since 1880, the Bureau of the Census has employed various terms and definitions for tenant farmers. In addition, colloquial terms and definitions for tenants have varied from the ones employed by the Bureau of the Census. Three basic types of farm tenants existed in the Southern plantation regions during the century following the Civil War—*cropper*, *share*, and *cash*. A cropper, also called a *sharecropper*, owned no

TABLE 2.1. Types of Farm Tenants in the United States, 1880–1959

1959	1950	1940	1930	1920
759,973 All tenants	1,447,455 All tenants	2,364,923 All tenants	2,668,811 All tenants	2,458,554 All tenants
107,217 Cash	215,392 Cash	514,438 Cash	489,210 Cash	483,577 Cash (Includes standing renters for North and West)
132,524 Share-cash	193,141 Share-cash	278,605 Share-cash	*	127,822 Share-cash
287,475 Share	535,332 Share	815,799 Share	*	1,117,892 Share
201,046 Crop-share	420,049 Crop-share			
86,429 Livestock-share	115,283 Livestock-share			
121,037 Croppers (South only)	346,765 Croppers (South only)	541,291 Croppers (South only)	776,278 Croppers (South only)	561,091 Croppers (South only)
87,596 Other and unspecified	156,825 Other and unspecified	214,790 Other and unspecified	*	104,996 Standing renters (South only)
	48,333 Other			
	108,472 Unspecified			63,176 Unspecified

Source: U.S. Bureau of the Census 1962.

* = For 1930, tenants other than cash tenants and croppers were included in "other tenants."

1910	1900	1890	1880
2,357,784 All tenants	2,026,286 Tenants	1,294,913 Rented	1,024,601 Rented
715,188 Cash (Includes standing renters)	752,920 Cash (Includes standing renters and unspecified tenants)	454,659 Rented for fixed money value	322,357 Rented for fixed money value
128,466 Share-cash			
1,400,137 Share	1,273,355 Share (includes share-cash tenants)	840,254 Rented for share of products	702,244 Rented for share of products
(Standing renters included included with cash tenants)	(Included with cash tenants)	(Included in above tenant classes)	(Included in above tenant classes)
113,993 Unspecified			

farm implements or work stock. All that a sharecropper contributed to the production of a crop was labor, including that of his or her family. From a perspective that emphasized close supervision, sharecroppers were the preferred tenant farmers, for they worked under the immediate direction of a planter or a manager. Although called a *tenant* colloquially and defined and enumerated by the Bureau of the Census as a type of tenant, as explained in Chapter 1, legally sharecroppers were not tenants but laborers who were paid with a share of the crops. Under customary rental agreements the landlord and the sharecropper split the crop fifty-fifty. Each also paid for half of the fertilizer, insecticide, ginning, and other production costs. A comprehensive definition of *sharecropper* had evolved by the end of the New South era. In Mississippi a sharecropper was defined thus:

> A member of the plantation labor force who rents a parcel of land from the owner or operator under an agreement to farm the land in accordance with the supervision, direction, and instruction of the owner or landlord, and who pays as rental for the land an agreed share (usually ½ but frequently ⅓ and ¼) of the crops. . . . The plantation operator provides a house, farm power and equipment, planting seed, and supervision, and he usually provides credit for the cropper's current expenses up to harvest time, and . . . pays for one-half the fertilizer, insecticides, and ginning. The landlord retains a lien upon the sharecropper's share of the crops to secure such indebtedness as may accrue. . . . The sharecropper in reality occupies an anomalous legal status . . . in that croppers are regarded as "tenants" in civil actions and "laborers" in criminal prosecutions.[1]

A *share tenant* was a farmer who possessed farm implements and work stock. Because he provided more than a sharecropper to the production of a crop, a share tenant paid less rent, usually one-fourth or one-third of the crop. Production costs were likewise split between the share tenant and the landlord on a one-fourth/three-fourths or one-third/two-thirds basis. Ownership of implements and work stock and legal standing as a true tenant usually gave a share tenant greater flexibility and more freedom from supervision than a sharecropper. Possession of capital goods that could be used as collateral also made it easier for a share tenant to obtain an annual furnish from local merchants and even meant that he might be able to secure a small loan from a local bank.

A *cash tenant* was one who paid a pre-agreed amount in cash or agricultural commodities for use of land. Like the share tenant, a cash tenant owned farm implements and work stock, but because the amount he paid was set at the beginning of the crop year, it was not tied to his success as a farmer. The arrangement between a cash tenant and a landlord might have been for lease of an entire landholding or for a specific acreage. Locally, the lease of a cash tenant sometimes was stated in terms of his farming capacity, such as number of mules or plows. A "one-mule farmer" might pay 1.5 bales of cotton and have use of 10 to 20 acres. A farmer who "ran one plow" was a one-mule farmer, for the conventional type of plow used during the New South era was pulled by a mule down one furrow, or half of a row. A cash tenant who "ran two plows" owned two mules, farmed 20 to 40 acres, and paid 2 to 4 bales of cotton as rent. The control management exercised over a cash tenant usually was minimal.

Although the distinctions between the three principal types of tenants and the terms used to identify each type were well defined by the end of the New South era, this was hardly the case during the first half of the period. In the agricultural censuses from 1880 through 1910, no distinction is made between a *share tenant* and a *sharecropper* (or *cropper*) (table 2.1). Sharecroppers were enumerated only from 1920 through 1959. Moreover, in several of the early-twentieth-century censuses a *standing rent tenant*, who paid rent in agricultural commodities, is distinguished from a true *cash tenant*, who paid in currency.

Within the early-twentieth-century literature on the alterations in Southern agriculture are discussions of the relationship of the three types of tenancy to the survival and viability of the plantation system. The key to seemingly contradictory statements by Phillips and other scholars on the survival of the plantation lies in this relationship. In his 1914 study of the post–Civil War agrarian revolution in Georgia, Brooks distinguished between the "share system" and the "renting system" of farm tenure. What he termed the "share system" was sharecropping because "the landlord supplies everything necessary to make the crop, except the manual labor, and the owner and tenant are in a sense co-partners in the undertaking." In the "renting system" Brooks grouped both share and cash tenants, including those who paid standing rent. "Under the renting system the landlord furnishes only the land and house." The tenant paid for use of the land either in a pre-agreed amount of cotton or in currency.[2]

Because centralized management is a critical component of the plantation system, to both Phillips and Brooks deterioration of a landowner's close supervision of farm operations meant demise of the plantation. Estates that were worked by sharecroppers and/or wage employees required close supervision and were plantations. The turning of landholdings over to share and cash tenants, a form of agriculture reminiscent of the medieval feudal estate in which the landlord exercised little supervision of farming, meant that the renters had primary management responsibility. In the post–Civil War South the practice was regarded as evidence of poor farming, demise of the plantation, and agricultural decline. According to Phillips, the "concentration of labor under skilled management made the plantation system . . . practically the factory system applied to agriculture," but "the counter replacement of the plantation system by peasant farming" was "an industrial counter-revolution."[3] Writing of the changes in plantation agriculture in Georgia during the half century following the Civil War, Brooks stated, "Planters have been virtually unanimous during the entire post-bellum period in resisting the growth of renting [share tenancy and cash tenancy]. They feel that where the laboring class is of such a low order, complete control should remain in the hands of the capitalist class." From interviews with post–Civil War planters, Brooks concluded that "renting" developed because of "the negroes' desire for complete emancipation from control. By reason of the scarcity of labor they were able to realize their wishes. When the movement was once begun, it grew with great rapidity, for as soon as the negroes who were working as day laborers [wage workers] or as share hands [sharecroppers] saw the large degree of personal liberty enjoyed by those who had succeeded in attaining the position of renters, they speedily demanded like privileges, and, in a number of cases, the planters were not in a position to refuse."[4] A planter who witnessed the post–Civil War labor situation on Georgia and Alabama plantations told Brooks that "the negro would rather be his own master on scant rations than to have a white man supervise him on full rations. . . . In their effort to gain still 'more independent position,' [Negroes] began demanding rent privileges where they could enforce it."[5]

Because of labor problems, some landowners abandoned attempts to reorganize their landholdings as closely supervised plantations. A group of planters actually left their estates and moved to towns and cities. The absentee owners rented to share and cash tenants who worked their farms with little or no supervision.[6] In his first major plantation paper Phillips wrote: "The present system of renting, or cropping, can be but temporary. Under it the negro is superintended in but a half-hearted way. Whenever he fails to raise a good crop and to sell it at a good price, he involves his landlord and his creditor with himself in a common embarrassment. . . . The most promising solution for the problem is the re-establishment of the old plantation system, with some form of hired labor instead of slave labor."[7]

The growth of share and cash tenancy was facilitated by the widespread development of town and country merchants from whom renters could obtain their annual furnish. Some merchants became quasi managers of plantations, collecting the annual rent for absentee landlords as well as providing the annual furnish to tenants. The merchants sometimes employed "riders" who visited the tenants periodically and reported on the condition of the crops.[8] Some cash tenants even became managers. They paid a set annual rent to landowners and then subleased to share tenants or to sharecroppers. A former slave in Greene County, Georgia, rented the entire plantation on which he was born for ten bales of cotton. He purchased farm implements and six mules and as a tenant manager subleased the plantation to six sharecropper families.[9]

Gavin Wright emphasized that after the Civil War control of land more than control of labor was the critical measure of a planter. "Before they were laborlords, now they were only landlords."[10] But with the reorganization of plantations following the war, many planters were still laborlords, farming their closely managed estates with hordes of sharecroppers and wage hands. Only those who refused to govern their holdings carefully, those who leased to "renters," were just landlords.

IDENTIFICATION OF THE NEW SOUTH PLANTATION

Despite census statistics, which indicated the disintegration of the plantations following the Civil War, by 1900 a group of Southern scholars had begun to argue that the large farms survived. In 1901, Alfred E. Stone, who owned a plantation in the Yazoo Delta in Mississippi, wrote that "the census rule treating every tract of land on which agricultural industry [was] conducted as a farm . . . [was] misleading" and continued with a discussion of the renaissance of the plantation system.[11] The census office eventually acknowledged that post–Civil War alterations in Southern agriculture were obscured by the data collected. To remedy the situation, a special study was made of plantation farming as part of the 1910 Census of Agriculture. The investigation was focused on the regions where black tenants were concentrated and designed principally to address how tenant farming concealed the survival of plantation agriculture (map 2.1). It was not comprehensive in either geographical extent or definition of *plantation*. The survey was confined to 325 counties in the South where plantations were thought to exist in greatest number and was limited to *tenant plantation*, which was defined as "a continuous tract of land of considerable area under the general supervision or control of a single individual or firm, all or part of such tract being divided into at least five smaller tracts, which are leased to tenants."[12] Large estates composed of noncontiguous

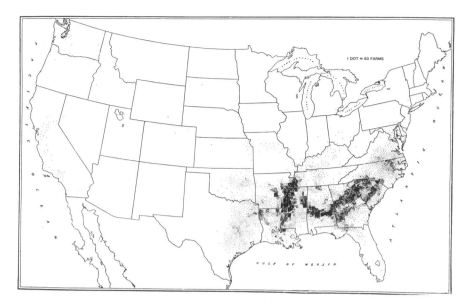

MAP 2.1. Distribution of "colored" tenants in 1910. Although the map reveals the extent of traditional plantation agriculture in the South, by 1910 several of the older plantation regions were in decline (U.S. Bureau of the Census 1913a).

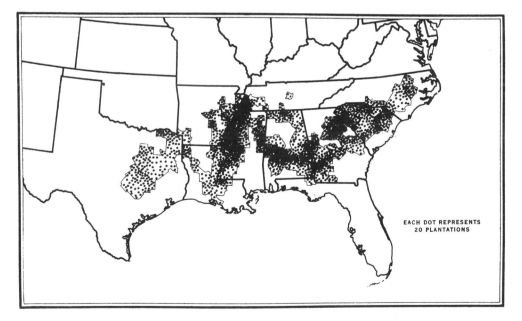

MAP 2.2. Plantations with five or more tenants in 325 Southern counties in the cotton regions in 1910 (U.S. Dept. of Agriculture 1918, 11)

tracts were excluded. Blacks constituted 50 percent or more of the population in most of the counties surveyed.

The special census was confined primarily to cotton regions of the South. The sugarcane region of Louisiana, the rice region of the South Carolina–Georgia coast, and most of the tobacco region of Virginia were not included. The Census Bureau's definition of *plantation* incorporated three critical elements that historically distinguished it from other types of farms: a large landholding, a large labor force, and centralized management. The 1910 census also attempted to enumerate the extent of the "wage-labor plantation system" for the entire country. Farms on which one thousand dollars or more was spent on "wage laborers" were considered "nearly comparable with tenant plantations operated by five or more tenants." Combined, the bureau's three sectional divisions that comprised the thirteen Southern states had the largest number of wage plantations, 18,043. But by sectional division, the three Pacific Coast states with 16,231, primarily California, together with the four western Midwest and four northern Great Plains states with 12,495, had the largest number of farms that spent one thousand dollars or more on wage labor. That the North and West had 58,708 of the nation's 76,751 farms that paid one thousand dollars or more in wages is evidence that the transformation of American agriculture from family farms to large-scale plantation-like operations was well advanced by 1909.[13]

The 39,073 tenant plantations enumerated in 1910 extended in a crescent across the South from southern Virginia into eastern Texas (map 2.2).[14] More than half, 68 percent, were relatively small with an average of 495 acres farmed by five to nine tenants. Only 4,102, 10.5 percent, were great plantations with an average of 3,535 acres and fifty or more tenants. Tenant plantations made up 31.5 percent of the land in farms and accounted for 32.8 percent of the value of land and buildings in the 325 counties surveyed. Variations existed among the Southern states in the relative importance of the plantation system. Mississippi, with 7,960 plantations, had the largest number; Alabama and Georgia followed with 7,280 and 6,627.[15] The special census confirmed the change in settlement pattern from a nucleated one of the Old South to a dispersed one of the New. Thirty-seven schedules from across the Georgia plantation regions survive in the Brooks Papers at the University of Georgia. On all of the thirty-seven, "grouped quarters" had been replaced by houses at "scattered sites."[16]

A number of technical studies of plantation agriculture conducted by state agricultural colleges, the U.S. Department of Agriculture, and other federal agencies followed the 1910 special census.[17] Several problems were dealt with in the definition of *plantation*, including whether wage laborers should be counted as part of the labor force and whether farms that met all criteria except that the land was in dispersed parcels were plantations. By the end of the Great Depression of the 1930s, Southern cotton plantations had been comprehensively analyzed. Two studies conducted by Thomas Jackson Woofter Jr. and his associates for the Works Progress Administration were the apogee of a genre of technical plantation studies. In *Landlord and Tenant on the Cotton Plantation* (1936), which built on the part of the definition in the 1910 special census, a minimum of five "tenant families" was modified to "five resident families of any type," including the landlord or manager.[18] This change recognized the significant role of wage labor on large commercial farms and also marked a distinct break with the limitations imposed by interpretation of the plantation in the context of post–Civil War adjustments from slave to tenant labor. The Woofter studies also dealt with the vagueness of "tract of land of considerable area." The minimum acreage of a cotton plantation was set at 260 acres. The 1936 study found that 39 percent of the cotton planters owned noncontiguous tracts and that there was a tendency to enlarge the size of plantations by leasing additional land, which in turn was rented to tenants. The definition of *plantation*, however, was not modified to include tracts that were not contiguous to the principal one.[19]

A more comprehensive attempt to enumerate plantations in the United States was conducted in 1940 as part of the sixteenth census. The definition of *plantation* was broadened from that for the 1910 census to include farms composed of dispersed tracts that were managed from a central headquarters. The more comprehensive definition was an improvement, but the limitations imposed by the controversy over the relation of tenant farms to reorganization of plantations following the Civil War were still quite evident. Any farm operation that had five or more families was considered a plantation if at least one was a tenant family. Not only was the definition of plantation modified, but the area for the study was considerably enlarged from that in 1910. Plantations were enumerated in 1,382 counties, including the 372 used for the 1910 census. In addition to counties in fifteen Southern states, several in Arizona, California, and New Mexico were included.[20] In 1940, 24,199 plantations were enumerated in the United States. Mississippi, with 7,485, had nearly one-third of the large farms. The closest rivals were Arkansas with 2,744 and Louisiana with 2,670.

At the local scale, no effort was made to accurately distinguish plantations from large farms that did not meet a precise definition. In some areas the term *plantation* was not even used or by the early twentieth century had faded from the provincial lexicon. Across the South the expression *place* with an attached family name came to be used as a synonym for *plantation*. That certain families, through their control of land and the agricultural infrastructure, were locally powerful and composed the upper or planter class is what was implied by plantation and place. William Faulkner, perhaps, best captured who this illusive, but locally overt, group was through Gavin Stevens' view of Yoknapatawpha County, Mississippi's plantation country: "the same fat black rich plantation earth still synonymous of the proud fading white plantation names whether we—I mean of course they— ever actually owned a plantation or not: . . . generals and governors and judges, soldiers (even if only Cuban lieutenants) and statesmen failed or not, and simple politicians and over-reachers and just simple failures."[21]

TRANSFORMATION OF THE
COTTON PLANTATION LANDSCAPE

Although some scholars have dismissed the agricultural technology and infrastructure of the plantation regions as essentially static between the Civil War and the depression,[22] significant changes occurred both in the spatial organization of farms and in the processing and marketing of cotton and other agricultural commodities, illustrated by the lower Georgia Piedmont (fig. 2.1). These modifications in turn greatly altered the landscape of the plantation regions. So sweeping were the changes that by 1910 the New South cotton plantation landscape had few features that survived from the Old South era. The landscape was reordered, not only by the transition of the plantation system from a slave to a tenant labor force, but also by the emergence of a railroad network, a central place hierarchy, and a new, more complex agricultural infrastructure. Even some plantation big houses that might appear "antebellum" are actually from the New South era.

Historians have paid significant attention to the crop-lien system and the rise of furnish merchants in the plantation regions following the Civil War.[23] Furnish stores appeared in towns and at strategic locations in the countryside, such as at crossroads, and also on a number of plantations (fig. 2.2). Essentially overlooked is the important revolution in the technology and spatial organization of cotton ginning which occurred following the war. As important as furnish merchants were to the agrarian economy of the New South, cotton gins were more fundamental than stores to the spatial reorganization of the landscape. Large, centrally located community cotton gins, known at the time as *ginneries*, replaced small plantation gin houses of the Old South era.[24] In almost every instance the new cotton

Old South Plantation Era	Reorganization of Plantation Agriculture
1800 → 1865	

Expanding plantation agriculture

Slaves

Gang system

Aggressive plantation management

Nucleated plantation settlement pattern

Plantation cotton gins

Increasing cotton acreage

Embryonic railroad and road network

Embryonic central place network

Embryonic agricultural infrastructure:

    Cotton factors

    Banks

    Guano companies

Decline of white yeoman farmers

Freedmen

Squad and gang systems

Cash wages

Tenant farmers

Management alterations

Disintegration of nucleated plantation settlement pattern

Expanding railroad network

Expanding central place network

Newly emerging agricultural infrastructure:

    Furnish merchants

    Banks

    Public cotton ginneries

    Fertilizer plants

    Cotton factors

FIG. 2.1. Changes in the agricultural infrastructure of the lower Piedmont cotton plantation region, 1800–1940

FIG. 2.2. The emerging New South landscape. The rural store owned by the Cornwell brothers in Jasper County, Georgia, in the 1890s is representative of ones that developed during the early part of the New South era. Simple Greek Revival architecture for such structures was common throughout the period. Like the Cornwell store, many rural stores were not painted. Courtesy of the Georgia Department of Archives and History

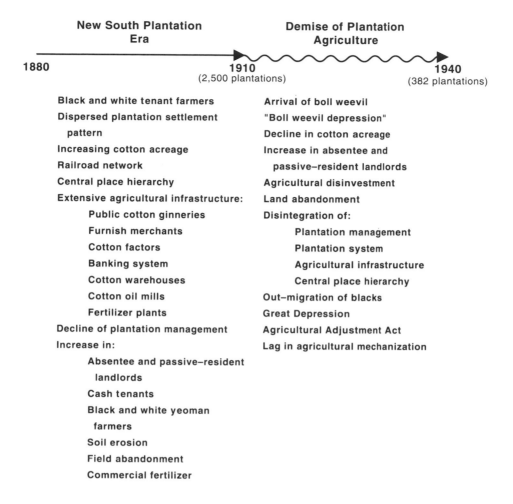

New South Plantation Era → **Demise of Plantation Agriculture**

1880   1910
(2,500 plantations)   1940
(382 plantations)

Black and white tenant farmers
Dispersed plantation settlement
   pattern
Increasing cotton acreage
Railroad network
Central place hierarchy
Extensive agricultural infrastructure:
   Public cotton ginneries
   Furnish merchants
   Cotton factors
   Banking system
   Cotton warehouses
   Cotton oil mills
   Fertilizer plants
Decline of plantation management
Increase in:
   Absentee and passive–resident
    landlords
   Cash tenants
   Black and white yeoman
    farmers
   Soil erosion
   Field abandonment
   Commercial fertilizer

Arrival of boll weevil
"Boll weevil depression"
Decline in cotton acreage
Increase in absentee and
   passive–resident landlords
Agricultural disinvestment
Land abandonment
Disintegration of:
   Plantation management
   Plantation system
   Agricultural infrastructure
   Central place hierarchy
Out–migration of blacks
Great Depression
Agricultural Adjustment Act
Lag in agricultural mechanization

ginneries were allied with merchandising and farming, and frequently they were associated with other businesses that included buying cotton, mixing and selling fertilizer, and processing cotton seed. Ownership of a store did not necessarily identify an individual or family as prominent in a plantation community, but ownership of a cotton gin meant command of a small domain.

The fall of slavery and the emergence of tenant farming caused serious problems in preparation of cotton for shipment to market. On an Old South plantation the entire cotton crop belonged to the planter. Before cotton was ginned, it was sorted by color and trash content. The development of tenant farming resulted in a decline in quality control, for each renter on a plantation picked his own cotton and had it ginned. Moreover, the tenant system did not provide labor for ginning. Shortly after the war, recommendations were made that large public gin plants replace the small plantation gins. In the 1866 report of the United States commissioner of agriculture, one correspondent wrote, "Another radical modification of the former system, which ought to be made immediately, . . . is the building of neighborhood gin-houses in well-chosen locations, so as to be central to large farming communities." Henry Grady, a major advocate of a "New South," thought that the gin was "the pivotal point around which the whole of manufacture of cotton revolve[d]." He estimated that improvements in ginning could add thirty million dollars to the annual value of the Southern crop. The increased value could be achieved by the removal of greater amounts of dust and trash and by the reduction of labor costs through the construction of larger, more automated plants.[25]

Significant innovations in the technology of cotton ginning were made by several persons during the two decades following the Civil War. Robert Munger of Mexia, Texas, was one of the most important. Between 1883 and 1885 Munger developed an automated steam-powered "ginning system" with a "battery" of two to six gin stands, which moved cotton by air, gravity, and belts from the time it left the wagon until the lint was baled (fig. 2.3).[26] Using fewer laborers, a Munger ginning system could produce twenty-four to forty 500-pound bales of lint cotton in the time that the Old South plantation cotton gin could process three or four 400-pound bales. By the turn of the century small, labor-intensive cotton gins were being rapidly replaced by larger, more automated "ginneries" (figs. 2.4 and 2.5). Between 1898 and 1903 the number of small cotton gins decreased more than 50 percent. In 1905, only 587 of the South's 28,757 cotton gins were still propelled by animals, and only 1,905 were water-powered. Approximately one-half were modern, steam-powered ginning systems.[27]

By the 1880s, the agricultural infrastructure of the plantation regions was rapidly evolving to a new, more extensive and complex stage (fig. 2.1). Steam engines, whose numbers multiplied rapidly after the Civil War, were central to the new infrastructure. In addition to the gin, a cotton ginner usually owned a plantation and a furnish store. Although on the larger landholdings the gin and store were primarily company-oriented, most were community enterprises that served surrounding tenant plantations and yeoman farms. In larger rural communities and in towns, additional enterprises were associated with cotton gins. Within small domains, ginners were the most powerful individuals. Frequently, their authority extended to banks on whose boards of directors they sat and to school, county, and town governing bodies to which they were elected members.

Prior to 1916, most farmers sold their cotton as soon as it was ginned, even if prices were not favorable. The crop-lien system was not conducive

FIG. 2.3. New South cotton ginnery. This type of two-story cotton gin plant was built from the late 1880s into the 1930s. Pneumatic suction lifted the seed cotton from the wagon to a cleaning device called a separator. A conveyor belt delivered the cotton to the feeders of a battery of two to six seventy- or eighty-saw gin stands. To permit continuous ginning, Robert Munger, inventor of the ginning system, devised a revolving double-box up-packing press to bale the lint. Additional trash was removed in the feeders and in the condenser, from which fine trash was blown through flues. A single-battery, steam-powered cotton ginnery required a minimum of seven adults to operate it, but in a twelve-hour period it could process enough cotton to make twenty-four to forty 500-pound bales of lint. A structure that was one story on the end with the gin stands and two story on the one with the up-packing press was a variation of the two-story building (Tompkins 1901).

to deferred sales, and baled cotton was accepted as security for loans only if it was properly stored and insured. High storage and insurance rates suppressed most advantages to be gained by holding baled cotton and awaiting improvements in market prices.[28] The difficulties in farmers' holding cotton and other agricultural commodities caused the National Farmers' Alliance and the Populist Party, both of which developed from agrarian discontent during the latter part of the nineteenth century, to advocate a "sub-treasury" plan. The plan called for the creation of government-sponsored warehouses and grain elevators in which farmers could store nonperishable crops such as cotton, tobacco, corn, and wheat and use the storage receipts to obtain loans of up to 80 percent of the market value. Although the Populist Party failed in its bid for national sanction, many of

FIG. 2.4. The vanishing Old South landscape. Ruins of the abandoned gin house and press on the Toombs Plantation, Wilkes County, Georgia, in 1902. By 1900, the principal components of the Old South plantation landscape were rapidly disappearing. Only a few hundred of these obsolete facilities were still in operable condition across the cotton plantation regions. Courtesy of the Georgia Department of Archives and History

FIG. 2.5. The mature New South landscape. Cotton ginnery of J. S. Clifton at Siloam in Greene County, Georgia, circa 1920. The suction pipe of this two-story ginnery was on the end rather than the side of the building, an arrangement used for ginning systems manufactured by the Gullett Gin Company. Courtesy of the Georgia Department of Archives and History

its goals, including the creation of a system of government-sponsored warehouses, eventually were enacted into law.[29] In recognition of the shortages of adequate storage facilities and the difficult problems of farmers in obtaining credit, the United States Warehouse Act was passed in 1916. The principal advantage of the warehouses was that the uniform receipts they issued could be converted at any time into liquid assets.[30]

In the cotton regions of the eastern South, including the Piedmont and the Black Belt, numerous small warehouses were constructed. Almost every town had at least one of the storage facilities, most of which were allied with a cotton ginnery. However, few small warehouses developed in the South's central and western cotton regions. In the Loess Plains and the alluvial Mississippi Valley, *compresses* developed. These were large warehouses that had special machinery for compressing cotton bales to a smaller size to facilitate storage and shipment. Most compresses were owned by large corporations, such as the Federal Compress Company with headquarters in Memphis, Tennessee.

Although in the central and western regions of the South manufactured fertilizer was not used in large quantities until the 1930s, farmers and planters in the eastern regions began spreading imported guano on their fields before the Civil War, and the use of manufactured fertilizer increased rapidly after the conflict.[31] Growth of the commercial fertilizer industry in the Piedmont, the Alabama Black Belt, and other agricultural regions of the eastern South was characterized by the development of numerous small plants. Local firms, such as the Blue Eye Oil and Fertilizer Company in Lincoln, Alabama, and the Chipley Home Mixture Guano Company in Chipley, Georgia, purchased phosphate, potash, and nitrite in bulk and mixed and sold their brands of *guano*, as all purchased fertilizer continued to be called in the eastern South.[32] At the time of the First World War, more than one hundred brands of fertilizer were sold in Georgia alone. Many of the small fertilizer mixing plants were components of cotton ginning firms, and some ginners were agents for large fertilizer corporations such as the Royster Guano Company.

The cotton seed oil industry is a significant by-product business associated with cotton production. This industry developed after the Civil War, stimulated by growth in markets for cotton and other vegetable oils. Its development was made economically feasible by the emergence of large ginneries and expansion of the railroad network, which facilitated the collection of cotton seed. In 1860, there were only 7 small cotton oil mills in the United States. By 1880, the number had increased to 45, and at the turn of the century there were 369. The number more than doubled between

1900 and 1909, a decade when 448 new cotton oil mills were built.[33] Nearly all cotton oil mills, except those in the larger cities, had ginneries at the plant, and most owned additional ones in the surrounding towns and countryside. Allied activities included the manufacture of ice in summer and the sale of coal in winter, which provided employment and income throughout the year. Other firms had gristmills and sawmills and sold horses, mules, and farm implements. The industry consisted of local oil mills, small regional companies that owned several mills, and large interregional corporations with many. The Macon County Oil Company in Tuskegee in the Alabama Black Belt was a locally owned firm that had six ginneries and a fertilizer mixing plant. The Hodson Oil and Refining Company in Athens, Georgia, had a refinery, three oil mills, and eight cotton ginneries on the lower Piedmont. Hodson manufactured soap stock, "compound lard" (vegetable shortening), cooking oil, fertilizer, and livestock feed.[34] The Southern Cotton Oil Company, which was organized in 1887, was one of the most important interregional corporations and eventually became nationally known by its Wesson Oil and Snowdrift Shortening. Although parts of a large corporation, Southern Cotton Oil mills in small towns across the cotton regions were operated as though they were community-owned enterprises. Each mill had a local manager, one or more cotton ginneries, and fertilizer and farm supply businesses.

By the first decade of the twentieth century, cotton ginning was less of the plantation-oriented activity than it was in 1860. Across the maturing New South landscape under the umbrella titles "ginner" and "planter" were a cluster of business activities in which the gin, while central, was not always monetarily dominant. Northern Morgan County on the lower Georgia Piedmont illustrates the New South landscape at the beginning of the twentieth century (map 2.3). The area was dominated by a few white families who owned a cotton ginnery or a furnish store, or both. The small agribusiness complexes at strategic locations in the countryside included those of the Nolans, the Ponders, and the Malcoms. The largest complex was at the emerging hamlet of Bostwick, much of which was owned by John Bostwick, an especially ambitious New South entrepreneur. By 1910, the hamlet had a small cotton oil mill, a ginnery, a hotel, and three stores. The railroad network in northern Morgan County had evolved to one that included the mainline of the Georgia Railroad and a short line built by John Bostwick from Monroe in adjoining Morgan County through Bostwick to Appalachie, a hamlet a few miles to the east. Most farmers in northern Morgan County were less than 2.5 miles from a cotton gin or a store or 4 miles from a railroad.

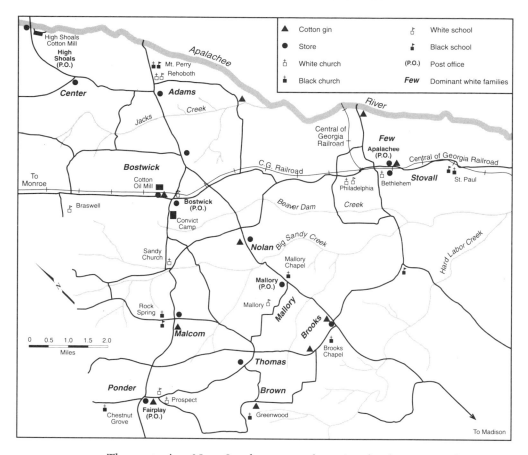

MAP 2.3. The maturing New South cotton plantation landscape, northern Morgan County, Georgia, about 1900 (compiled from Tufts 1897 and field investigations)

Ginning firms in towns were usually larger and more complex than those in the countryside. From companies that owned cotton warehouses and fertilizer mixing plants, the level of integration of ancillary enterprises grew to include cotton oil mills, ice plants, and farm implement businesses. The complexes were concentrated along the railroad and formed the industrial section of an agrarian town. Even small New South municipalities often had an impressive array of enterprises that drew hordes of wagons, farmers, and laborers; belched smoke, dust, and cotton linters; and noisily droned with steam engines and machinery from late summer into winter. Clustered about the railroad depot in Washington, Georgia, were two major agribusinesses, a branch of the Southern Cotton Oil Company, and the local Pope Manufacturing Company (map 2.4). The Southern complex consisted of a cotton oil mill, a double-battery ginnery (a building with two ginning systems powered by one engine), a fertilizer mixing house (guano house), and a coal yard. The Pope Manufacturing Company operated a cotton oil mill, a double-battery ginnery, a gristmill, an ice plant, and a fertilizer mixing house. Two cotton storage companies, the Wilkes-Lincoln Ware House and the Planters Ware House, a subsidiary of the Pope Manufacturing Company, were also major components of Washington's industrial district.

That cotton ginneries were more central to the economic spatial organization of the New South landscape than furnish-merchant stores is supported by the work of the early location theoreticians. In his innovative work on the economics of location, Lösch used the spatial distribution of cotton gins as an example of a belted market network in which the centers of service areas were close but did not coincide. According to Lösch, the market areas of cotton gins might "rub shoulders on every side with their neighbors," but the pattern was dominated, not by overlapping, but by

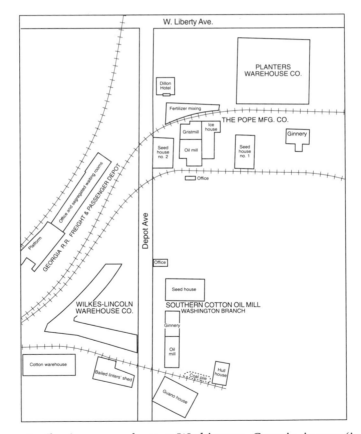

MAP 2.4. Agribusiness complexes at Washington, Georgia, in 1917 (insurance maps of Washington, Georgia; Sanborn Map Company, 1917)

dispersion and separation.[35] In another early work on location theory, Hoover recognized that cotton gins were components of more than one segment of the agricultural infrastructure. They were collection points for unprocessed cotton and units in the systems that supplied ginned cotton to the textile industry and cotton seed to oil mills. Cotton gins also were parts of agribusiness firms that supplied farmers in well-defined local markets.[36]

The actual sizes of cotton gin service areas varied. A 1920 study found that the average distances over which Piedmont farmers hauled cotton ranged from 2.5 to 5.1 miles.[37] In an ideal situation the sizes of cotton ginnery supply areas were adjusted to capital investment and operation expenditures in such a manner as to maximize profits. Whereas the Old South–era plantation cotton gins needed to process 100 to 250 or more bales annually to be economical, the larger public ginneries that emerged early in the New South era required 500 to 1,500 or more bales. However, across the cotton regions during the New South era greater ginning capacity developed than was needed.[38] The overbuilding of ginneries can be explained in part by the economic, political, and social power and prestige that were associated with ownership of the plants.

THE MATURE NEW SOUTH COTTON PLANTATION

The dwellings that initially composed the dispersed settlement pattern of the New South plantation were the first of four generations of housing which rural blacks occupied in the cotton plantation regions between the Reconstruction period and the end of the twentieth century. The first-generation housing consisted both of newly constructed dwellings and ones relocated from slave quarters. Although it might seem that former slaves would have sought to exert more self-expression in their dwellings, paintings and photographs of first-generation postwar houses reveal that they were primarily copies of houses in slave quarters (figs. 2.6 and 2.7). African architectural designs transferred to North America by slaves found expression in the buildings on Southern plantations during the seventeenth and eighteenth centuries. Most overt African influences, however, were suppressed and destroyed as blacks increasingly submitted to the ways of their masters, and they had practically disappeared by the last decades of slavery.[39] The use of the yard as an extension of the house for cooking, washing, and socializing was a primary way in which African influences survived.

Although house designs, construction methods, and room dimensions remained those of the pre–Civil War era, freedmen introduced certain

FIG. 2.6. The emerging New South landscape. An example of the first generation of housing occupied by blacks following the Civil War about 1890. The windowless, one-room log dwelling and a detached kitchen are similar to the types of homes of former slaves described by David Barrow Jr. on his family's Sylls Fork Plantation in Oglethorpe County, Georgia. The detached kitchen was an exception to the types of dwellings constructed. The photograph is from Union Point in Greene County, Georgia, less than 5 miles from the site of Sylls Fork Plantation. Courtesy of the Georgia Department of Archives and History

FIG. 2.7. The emerging New South landscape. William Aiken Walker's composite painting from the 1880s is representative of tenant farms of former slaves within plantation landholdings during the three decades following the Civil War. Private collection

modifications to tenant dwellings. On the Barrow family's Sylls Fork Plantation, the houses that the freedmen built on their tenant farms were single-pen log cabins with chimneys made of sticks and clay. A "separate kitchen," which in "architectural design" was "a miniature of the house," was built next to each (fig. 2.6). The small kitchen was apparently detached from the dwelling because kitchens of plantation big houses were commonly separate. David Barrow Jr. thought that the kitchen, which was about the size of a "chicken coop," was "really ridiculous in its pretentiousness."[40]

The first generation of post–Civil War housing of blacks was transitional, and by the 1890s it had begun to give way to a second generation of dwellings (fig. 2.8). Whereas former slaves appear to have had considerable authority in the selection of sites and construction of first-generation houses, by the last decade of the nineteenth century planters who operated closely managed landholdings had begun to exercise more authority in the spatial arrangement. The second generation of housing was part of the maturing New South plantation infrastructure. Log construction was prominent in the first generation of black housing, but second-generation dwellings were primarily frame. Technological advances in sawing technology lowered the cost and increased the number of sawmills, which often were powered by the same steam engine or used the same boiler as the ginnery. A 1913–14 study of blacks in rural Clarke County, Georgia, under the direction of David Barrow Jr., who had risen to chancellor of the University of Georgia, found that "the old style of Negro cabin, built of hewn logs . . . was passing out of existence." The typical house was a "boarded up" one. The 1910 plantation census schedules included questions on the type, condition, and features of houses. Almost no log houses remained on the thirty-seven Georgia plantations for which schedules survive. Recent construction of tenant dwellings is indicated by most being in "new" rather than "medium" or "old" condition. Like indoor plumbing a half century later, in 1910 glass windows were regarded an index of improved housing for farm laborers. Nearly all of the "new" tenant houses and most of the ones in "medium" condition on the thirty-seven plantations had glass windows rather than wooden shutters.[41] From the Georgia schedules, Brooks selected an example of a newly reorganized and carefully managed plantation on the lower Piedmont which had "been cultivated for many years":

The present owners are two young men, both of whom live on the place, giving their entire time to the work of supervision. The total acreage is 3,750, of which 2,500 acres are improved. The two residences of the owners, set in a fine oak grove, are modern structures,

FIG. 2.8. The mature New South plantation landscape. The dwelling, in Clarke County, Georgia, 1925, is an example of the second generation of housing occupied by blacks following the Civil War. By 1900 the first generation, built mostly of logs, had begun to be replaced by the second generation of dwellings, constructed primarily of rough-sawn planks. Although the house was a single-family dwelling, its design, with a front door for each of the two principal rooms and a single flue to serve fireplaces in each, is a replica of a two-compartment slave dwelling. Houses built on roads usually had little space between the dwelling and the thoroughfare. Hargrett Family Papers, Hargrett Rare Book and Manuscript Library, Courtesy University of Georgia Libraries

with every convenience, such as screens, waterworks, and acetylene lights. . . .

Scattered over the plantation at convenient places are the tenant houses, eighty-six in number. All are comparatively new frame structures, costing on the average $300. No log houses were seen.[42]

The 1936 study directed by Woofter summarized the spatial characteristics of the well-managed, mature New South cotton plantation with a map (map 2.5). With a commissary, which was found on 25.9 percent of the plantations, and a ginnery, the plantation of the 1930s had evolved well beyond the embryonic stage depicted by the Barrow family's Sylls Fork Plantation in 1881, but the elemental geography was the same. Tenant house, church, and school had come to occupy their ordered places on the landscape. As on Sylls Fork half a century earlier, tenant families were

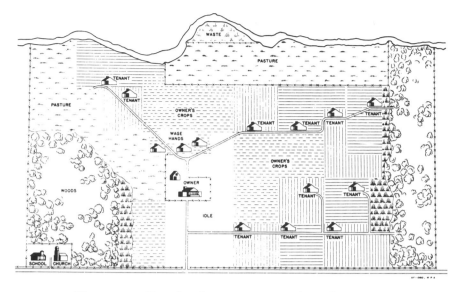

MAP 2.5. The mature New South tenant cotton plantation. A cotton ginnery and commissary, which were found on larger plantations, are not included (Woofter et al. 1936).

FIG. 2.9. The mature New South plantation landscape. Dispersed houses on tenant tracts, St. Francis Basin, Arkansas, 1938. Dorothea Lange, FSA Collection, Library of Congress

MAP 2.6. The locality of the Percy family's Trail Lake plantation in the Yazoo Delta in 1935 is representative of the dispersed settlement pattern of the mature New South cotton plantation landscape (War Department, Corps of Engineers, Trailake Quadrangle, 1:62,500, 1939).

dispersed across the plantation in small dwellings (map 2.6; fig. 2.9). Churches and schools, which had begun to emerge on plantations by 1881, were present in greater numbers fifty years later. In 1936, as in 1881, schools for blacks rarely went beyond the fourth grade. Wage hands, who were paid by the day, lived in houses that were grouped near the store or the cotton ginnery or at the edge of the fields. Management's control was exercised in three ways, each symbolized by a prominent landscape feature. Financial control was through the company store, which secured the annual "furnish" with liens on the crop (fig. 2.10). Day-to-day supervision of work in the fields was through allocation of work stock and farm implements. The central mule barn and the equipment sheds were near the owner's or manager's residence. Each morning sharecroppers had to come to the barn to obtain mules and implements for the day's work and had to return them in the evening. Command of marketing was exercised through the ginnery, where cotton was prepared for sale. Harvested cotton was temporarily stored in cotton houses, small sheds in the fields. To haul cotton from the field to the ginnery, a sharecropper had to obtain mules and a wagon from the central barn. Once the cotton was sucked from the wagon into the gin, the commodity passed into the control of the planter. Sharecroppers and, ofter, share tenants did not see their cotton again. The tenants were paid their percentages after the planter had sold the cotton and deducted their debts. For blacks, footpaths, more than roads, were the thoroughfares of their small insular worlds. The paths connected tenant house, church, and school, landscape symbols of their freedom, to big house, commissary, mule barn, and ginnery, visual reminders of their servitude.

UPPER COUNTRY TO ALLUVIAL EMPIRE

If there ever was a Southern "cotton belt," it existed during the first two decades of the twentieth century. The land of cotton, 30 million acres of the staple, stretched for more than 1,000 miles across the South from southern

FIG. 2.10. The mature New South plantation landscape. Commissary for the Sterlingwell-Plantations, Coahoma County, Mississippi, 1936. Dorothea Lange, FSA collection, Library of Congress

Virginia to the Rio Grande (map 2.7). Not only was the crop the foundation of the economy in the plantation regions, but legions of white yeomen farmers grew cotton in areas that bordered them. A cotton civilization to which life and social institutions, including churches and schools, were geared was firmly established. But significant variations existed among, as well as within, the regimented cotton plantation regions. The agrarian New South often is presented as a homogeneous place without significant economic, cultural, or geographical differences. However, the cotton plantation regions differed considerably in their origin, evolution, and spatial characteristics. Although the basic landscape components were similar, subtle variations existed in topography, soil, settlement patterns, associated crops, and the viability of the plantation system. The cotton plantation regions of the central South—Georgia, Alabama, and Mississippi—illustrate the diversity (map 1.1).

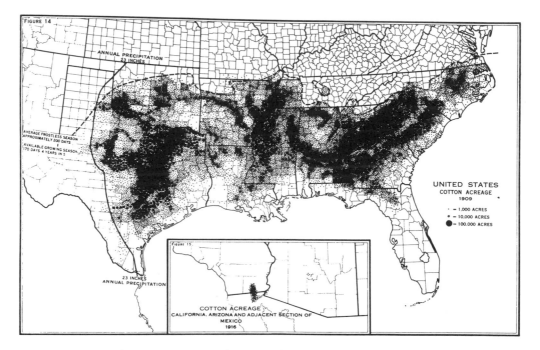

MAP 2.7. Cotton acreage in the United States in 1909. At the end of the first decade of the twentieth century, cotton production was near its apex in the South. Not only did major regions such as the Southern Piedmont produce the staple, but cotton was an important crop in the areas on the periphery of them. Cotton was in only the initial stages of a major shift from the Southeast to the Southwest. Into the second decade of the twentieth century, the 23-inch precipitation line was thought to mark the westward limit of where the staple could be grown profitably (U.S. Dept. of Agriculture 1918, 9).

The lower Georgia Piedmont was a part of one of America's earliest cotton regions. The problem of mechanically extracting seeds from the variety of cotton which could be grown on the Southern Piedmont, or Upper Country, which in elevation lay above the Coastal Plain, was the impetus for Eli Whitney's invention of his gin.[43] The first American cotton region, the Sea Island region, developed in the infant United States at the end of the eighteenth century. The commencement of the Industrial Revolution greatly increased the demand for cotton in Great Britain. During the 1780s, long-staple black seed cotton from the Caribbean began to be grown commercially along the South Carolina–Georgia coast. The cotton was processed using what was known as a *gin*, or *roller gin*, to remove the seed. This device originated in the Indian subcontinent in antiquity and diffused into the Mediterranean Basin, from where it was taken into the Caribbean and the subtropical fringe of the North American continent. Piedmont farmers, however, found that black seed cotton, a tropical variety, would not mature on the Piedmont. A cotton variety known as *green seed* could be successfully cultivated, but the seeds clung so tightly to the lint that roller gins could not remove them. In 1792 Eli Whitney, a Connecticut schoolmaster who had come south to be a tutor for a South Carolina planter's children, invented another method of ginning cotton, which was based in part on modification of the roller principle. Although the Whitney gin, what became known as the *saw gin*, was never accepted in the Sea Island region and the western long-staple cotton regions because it shortened the fibers, it opened the Piedmont, the Upper Country, and other interior regions of the South to cotton production. The term *upland cotton* superseded green seed as the name for the heartier varieties of the staple.[44]

By 1860, the Southern Piedmont was one of the major agricultural regions of the South and the nation. The lower Piedmont had evolved into a cotton plantation region of the Lowland South, while the upper Piedmont had developed as part of the Upland South, which was dominated by white yeoman farmers. Wilkes County on the lower Piedmont in 1860 had 7,953 slaves and ninety-eight farms with 500 or more acres. In contrast, nearby Hall County on the upper Piedmont had only 1,216 slaves and two farms with 500 or more acres. Although after the Civil War cotton became a major crop on the upper Piedmont, the cultural differences between the plantation and yeoman farmer sections persisted. The differences took various forms in their landscape expressions. In 1922, Roland Harper reported that white women could be seen "working in the fields almost any day in the upper Piedmont, but rarely if ever in the lower Piedmont."[45]

By 1860, the political, economic, and cultural heart of Georgia had shifted from Savannah and the Atlantic coast inland to the lower Piedmont. The state capital was at Milledgeville and the state university in Athens. The lower Piedmont also was a culture hearth of major splinter church denominations, including the Methodist Episcopal Church, South, and the Southern Baptist Convention, which originated in sectional controversy over slavery. Well before the Civil War, capital created by plantation agriculture began to be invested in an infant manufacturing complex that consisted principally of textile mills at scattered sites on rivers of the lower Piedmont. To facilitate the region's cotton economy and infant textile industry, in the 1830s planters and industrialists chartered one of the world's first railroads, which began laying track along a carefully surveyed route from Augusta on the Savannah River westward across the lower Georgia Piedmont. Although the Georgia Railroad originally was intended to run from Athens to Augusta, one of the world's first railroad junctions was created at Union Point in Greene County. From Union Point, the Georgia Railroad was extended westward to Madison and then farther westward to meet the state-owned Western and Atlantic, which was built southward from Chattanooga to a place on a Piedmont interfluve called Terminus.[46] By the outbreak of the Civil War, the importance of the lower Georgia Piedmont extended well beyond the state, for it was one of the significant regions in the creation and leadership of the Southern Confederacy. Terminus, under its new promotional name Atlanta, had become a strategic railroad hub.

The rapid early growth of Atlanta and its strategic importance during the Civil War forecast a significant future. The quick rebuilding of the city following its burning by Federal troops in 1864 and its rapid postwar growth signified that Atlanta had superseded the rural plantation Piedmont as Georgia's economic and political heart. The state capital was relocated from Milledgeville to Atlanta in 1868, and in 1916 the Presbyterians reopened in the city Oglethorpe College, which had closed in Milledgeville in 1872. Subsidized by a large endowment from the Candlers, the first family to make a fortune in Coca-Cola, a beverage that an Atlanta pharmacist originally concocted as a tonic, the Methodists relocated Emory College from Oxford in Newton County to the city's suburbs in 1924 and upgraded it to university status. Although the legislature left the University of Georgia in Athens, the university's engineering college, the School of Technology, which opened in 1888 and was a critical part of the state's industrial New South strategy, was built in Atlanta because of the city's promise of $2,500 annually for twenty-five years.[47]

At the time the lower Piedmont achieved its apex, early in the twentieth century, it was an old cotton kingdom by American precepts of "old" (map 2.7). Four hundred miles to the west was a new kingdom, the alluvial Mississippi Valley, the Delta, a rapidly emerging plantation region. The Old South plantation did not develop in most of the Delta; the region was a cotton kingdom of the New South.[48] In 1910, only 52 percent of the alluvial valley in Mississippi, the Yazoo Delta, was in farms, and only 38 percent was improved for agriculture. With its fertile soils, the alluvial plain of the Mississippi River and its tributaries was potentially "the cream jug of the continent," but poor drainage, devastating floods, and debilitating malaria made the region "*terra incognita.*"[49] Despite the agricultural potential, the environmental hazards, especially the "miasma," made the region uninviting to white yeomen. Much of the Delta was developed by persons who acquired large tracts on which they cut the hardwood timber, dug drainage canals, and created plantations. During the Old South era, the settled areas were along the Mississippi River on the west and the bluffs of the Loess Plains on the east. The interior of the basin remained a wilderness, one of North America's great cypress-tupelo forests, which was destroyed as the plantation advanced. Federally funded drainage projects and great levees along the Mississippi River and the Yazoo, Tallahatchie, and other tributaries, together with the construction of railroads, encouraged agricultural development. The railroads shifted the regional foci from the region's margins to its interior.[50]

By 1940, more of the Mississippi alluvial valley's land area was in plantations than that of any other part of the South, and the grandiose "Alluvial Empire" had been coined to convey the scale and perceived importance of the region. William Faulkner, one of the most perceptive and eloquent of the Yazoo Delta's critics, described the mature New South landscape at the end of its era:

> Now the land lay open from the cradling hills on the East to the rampart of levee on the West, standing horseman-tall with cotton for the world's looms—the rich black land, imponderable and vast, fecund up to the very doorsteps of the negroes who worked it and of the white men who owned it; . . . the land in which neon flashed past them from the little countless towns and countless shining this-year's automobiles sped past them on the broad plumb-ruled highways, yet in which the only permanent mark of man's occupation seemed to be the tremendous gins, . . . the land across which there came now no scream of panther but instead the long hooting of locomotives: trains

of incredible length and drawn by a single engine, since there was no gradient anywhere.[51]

Rupert Vance characterized the Alluvial Empire as "Cotton obsessed, Negro obsessed, . . . the deepest South, the heart of Dixie, America's super-plantation belt."[52] Over large expanses of the Yazoo Delta, blacks constituted 90 percent or more of the population. Although Alfred Stone, who owned Dunleith Plantation near Greenville, held traditional planter racial beliefs, he did comprehend the important role of blacks to the region:

> The railroad rights of way through its forests have been cut out by the Negro, and every mile of track laid by his hands. These forest lands have been converted by him into fertile fields, and their subsequent cultivation has called for his constant service. The levees upon which the Delta depends for protection from floods have been erected mainly by the Negro, and the daily labour in field and town, in planting and building, in operating gins and compresses and oil mills, in moving trains, in handling the great staple of the country— all, in fact, that makes the life behind these earthen ramparts—is but the Negro's daily toil.[53]

A well-developed central place hierarchy of plantations, hamlets, towns, and small cities had evolved in the Delta by 1940. However, differences existed among the settlement patterns of the alluvial Mississippi Valley and older plantation regions to the east, illustrated by the Yazoo Delta and the lower Georgia Piedmont. The New South agribusiness structure of the lower Piedmont evolved, in part, to serve relatively large numbers of yeoman farmers and share and cash tenants. Because Delta plantations were larger, New South crossroad agribusiness complexes were less important than on the lower Piedmont. On the Piedmont, however, New South buildings were more substantially constructed than in the Yazoo Delta. As a consequence, more New South relic and fossil agrarian structures remained on the Piedmont than in the Delta at the close of the twentieth century.

To a greater extent than the lower Georgia Piedmont, the Yazoo Delta was a region whose landowners actively managed agriculture from without. Whether landlords were individuals who commuted to their plantations from Memphis or were corporations such as Fine Spinners, Ltd., the British company that owned the 37,000-acre Delta and Pine Land Company at Scott, from initial settlement the importance of alien proprietorship made the Yazoo Delta more like the typical world plantation region

FIG. 2.11. The mature New South plantation landscape. "Main Street," Wood-ville, Georgia, June 1941. Although Woodville is a municipality, it has never been more than a large hamlet. The population was 458 in 1940 and 415 in 1990. The brick building in the foreground housed the Bank of Woodville, which closed in the 1920s during the boll weevil depression. A two-story cotton ginnery is in the background. The smoke is billowing from the smokestack of the boiler for the steam engine that powered a planing mill, usually a winter business, which was still operating in June as the nation geared up for the Second World War. The Great Depression had ended; lumber and cotton were essential war materials. This scene was repeated without much variation at strategic crossroads through-out the cotton plantation landscape. Jack Delano, FSA Collection, Library of Congress

than the lower Piedmont. In addition, the average size of a plantation in the Yazoo Delta was larger than on the lower Georgia Piedmont.[54] This does not mean that the Piedmont did not have large plantations. The 20,000-acre Smithsonia Plantation in Oglethorpe County, Georgia, just north of the Barrow family's Sylls Fork Plantation, was farmed in 1910 by more than one thousand wage and leased-convict laborers and had its own cotton warehouses and short-line railroad. The R. F. Strickland Company in Con-cord, Georgia, owned 10,000 acres. Its agribusiness complex, which in-cluded a cotton ginnery and a "cradle-to-the-grave" store built in 1908 at the apex of the New South era and operated until 1992, dominated the town of Concord, located a few miles south of the Atlanta's Hartsfield International Airport.[55]

Whereas an important diversified socioeconomic structure that included the state capital and university, factories, and regional cities developed within the lower Georgia Piedmont, beyond its agriculture and a few unique cultural attributes, the Yazoo Delta was never a prominent region. The Delta was the heart of Dixie in a locational sense but not in cultural or economic ones. The region had no major cities, universities, or corporate headquarters. In the heyday of the nation's and the region's rail net, no mainline crossed the Delta, and the Delta's two largest cities, Memphis and New Orleans, lay beyond its borders. David Cohn declared that the "Mississippi Delta begins in the lobby of Hotel Peabody in Memphis,"[56] but the region was more a part of the city than the city a part of the region. The Delta was an important source of a few prominent aspects of black culture. For the blacks entrapped in the backbreaking toil of Delta plantations, blues music became an expressive release from their hardships. Robert Johnson, Muddy Waters (McKinley Morganfield), B. B. King, and other legendary artists took the blues from the Delta to urban America, where the music had profound repercussions in the nation and eventually the world.[57]

Over the years the geographical characteristics and relative importance of the South's cotton plantation regions changed. At the time of the depression, the cotton plantation regions had many similarities, but significant differences also existed (fig. 2.11). In addition to the emphasis given to one dominant crop, in cotton plantation counties blacks made up a high proportion of the population, agriculture was the dominant occupation, the rural nonfarm population was relatively small, and farm tenancy and poverty rates were exorbitant.[58] There also were a number of contrasts between the cotton plantation regions. Most significant were differences in the viability of agriculture and the plantation system. By the 1930s, the Georgia Piedmont plantation region was in chronic demise. In contrast, the Yazoo Delta was a viable, innovative plantation region that was still expanding agriculturally. Other Southern cotton plantation regions depicted various stages of agricultural vigor which ranged between that of the Piedmont and that of the Delta.

CHAPTER THREE

The Demise
of the Plantation

❧ ❧ ❧

*For the last thirty-five years the most progressive men in the country
districts of the South have been moving to nearby towns or to the
Northern cities. This is disastrous to agriculture, and a reverse
tendency should be set to work. . . . A new plantation system must
offer profitable and attractive careers to well-equipped men, or the
pine thicket and the sedge field will continue to be conspicuous
features of the landscape in the cotton belt.*

Ulrich Bonnell Phillips, "The Economics of the Plantation" (1903)

*I was "bawn an' bred" in the cotton belt, and thought it
a goodly land until riper knowledge taught me that my red hills
were niggard for all pecuniary purposes.*

Ulrich Bonnell Phillips, *Life and Labor in the Old South* (1929)

REGIONAL VARIATION IN THE
VIABILITY OF THE PLANTATION SYSTEM

During the twentieth century the plantation system in the American South
took two divergent directions. One route led to mechanization and modernization, the other to decline and the demise of plantation agriculture.
There was significant regional variation in what happened to the plantation. In some regions, the plantation became extinct, but in others it
survives in new structural and spatial forms. Generally, scholars have failed
to recognize the great geographical variations in the viability of the plantation system between 1900 and 1940. By focusing almost exclusively on the
impact of the Agricultural Adjustment Act and mechanization, they miss
one of the most critical changes in Southern plantation agriculture during
the first half of the twentieth century, its decline. In the 1930s, the agrarian
South was far from homogeneous. There were regional differences not
only in crops and types of farms but also in the viability of agriculture.

According to Mandle, "the period from World War I until 1940 was one in which an incremental chipping away at the structure of the plantation economy occurred. The structure remained intact with the planters continuing to control large numbers of dependent workers in the production of the cotton staple." Mandle also credits blacks with instigation of the "chipping away." Blacks began to escape from the plantation system by migration to cities and to resist planter tyranny by formation of collective bargaining organizations such as the Southern Tenant Farmers Union. The great migration that commenced during the First World War is assumed to have been in response to the pull of jobs in Northern cities which were opened to blacks by industrialists because of the loss of foreign immigrant labor.[1] That the great migration could also have been instigated by changes in the plantation system is not considered.

O. E. Baker thought that the viability of agriculture was not necessarily durable and that study of the "dissolution of an agricultural region" was "likely to be one of the most fruitful in . . . geography."[2] Carl Sauer conjectured that in an agrarian society, "there may arise loss of productive energy . . . between primary producers and those who are carried as the leisure class. There may be a shift of comparative advantage to another people and area." Between the end of the Civil War and the onset of the First World War, the plantation system in the lower Georgia Piedmont, Alabama Black Belt, and Natchez district reached what Sauer would have termed a "cultural climax," a peak followed by "stabilization, and . . . cultural decline."[3]

The contrast in the regional viability of the plantation system is portrayed in Arthur Raper and Ira Reid's *Sharecroppers All*, which appeared at the close of the Great Depression and was the culmination and synthesis of a genre of studies on the social and economic problems of the rural South published by the University of North Carolina Press. Raper and Reid compare the situation of an elderly sharecropper, Seab Johnson, and his wife, Kate, who live on a decrepit estate on the lower Georgia Piedmont in the vicinity of the Barrow family's Sylls Fork Plantation, with that of one of their children, who is a sharecropper on a plantation in the Yazoo Delta in Mississippi.[4] Seab and Kate subsist in two rooms of a crumbling, antebellum big house (fig. 3.1). A column of the old mansion leans, the upper half of a chimney has collapsed, and the driveway is a gully. Five unoccupied rooms of the house are filled with fertilizer sacks, mildewed cotton, pea hulls, and other rubbish. Of the many outbuildings that once surrounded the big house, only the smokehouse remains. On what was once a plantation that averaged a hundred bales of cotton annually, only two bales are harvested

FIG. 3.1. The disintegration of the lower Georgia Piedmont plantation land-scape. Decaying pre–Civil War plantation house in the vicinity of the Barrow family's Sylls Fork Plantation, 1937. Dorothea Lange, FSA Collection, Library of Congress

from eight of the best acres. One goes to the absentee landlord to pay cash rent. The surrounding landscape is one of neglect and decay. Abandoned fields are covered with broom sedge and pine trees. Networks of red gullies and remains of chimneys that mark the sites of former tenant houses are scattered throughout the fields. The scene is repeated across the coun-tryside: "red gullies shine under the sun where prized green fields once were; the big houses have been swallowed up by the pines. With few exceptions, the scattered families still on the land are renters and spend as much time getting berries and rabbits for their tables as in growing cotton to sell." In contrast to "their place on the map," Seab thinks of the Yazoo Delta, where his sharecropper son lives. The land is so rich that it can produce a bale of cotton per acre without fertilizer, and it is so level that the rows are a mile long. In addition, mechanization of cotton production is well under way in the Delta with the introduction of tractors.[5]

During the latter part of the depression, Dorothea Lange traveled widely across America with Paul Taylor, her husband. She was a photographer for the Farm Security Administration and its predecessor, the Resettlement Administration, and he was a researcher for the Social Security Board. In *An American Exodus*, a book based on their travels, a distinction is made

FIG. 3.2. The disintegration of the lower Georgia Piedmont plantation landscape. In 1937, a chimney was all that remained of a Greene County, Georgia, plantation big house. That the site of the former yard and gardens was plowed to plant a small patch of cotton indicates the continual search for fresh land in the wake of erosion and increased need for commercial fertilizer. Although such solitary chimneys were sometimes called Sherman's sentinels, those that stood in the 1930s were artifacts of the demise of agriculture near the end of the New South era, not relics of the Old South left in the wake of the Union army's pillage across the lower Georgia Piedmont more than seventy years earlier. Dorothea Lange, FSA Collection, Library of Congress

between the "Old South," where plantations were in "decay," and the alluvial Mississippi Valley (the "Delta"), where "large plantations . . . survived more vigorously . . . than anywhere in the Cotton Belt."[6]

Raper and Reid and Lange and Taylor accurately portray the variations in the viability of the plantation which existed across the South at the time of the depression. In some regions the plantation was virtually extinct; in others it was in rapid decline; and in others it remained viable. So great and rapid was the decline in agriculture across the lower Piedmont from North Carolina into eastern Alabama after the First World War that the United States Department of Agriculture conducted a special study of the phenomenon. Between 1919 and 1924 harvested cropland in the fifty-nine counties studied decreased 29 percent, from 5,632,000 to 3,980,000 acres. The relative decline in cotton was even greater; it dropped 40 percent, from 3,083,000

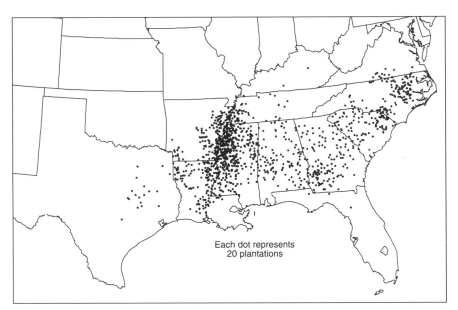

MAP 3.1. Plantations in 1940. Despite the inclusion of plantations operated with wage labor together with those with five or more tenants, the number declined significantly in most plantation areas of the eastern South between 1910 and 1940 (source of data: U.S. Bureau of the Census [1948]).

TABLE 3.1 Number of Plantations in 325 Southern Counties, 1910 and 1940

	Number of Plantations 1910	Number of Plantations 1940	Difference 1910–1940	Percentage Difference 1910–1940
Alabama	7,287	1,801	−5,486	−75.3
Arkansas	2,674	2,499	−175	−6.5
Florida	84	10	−74	−88.1
Georgia	6,627	1,840	−4,787	−72.2
Louisiana	2,480	2,292	−188	−7.6
Mississippi	7,960	6,668	−1,292	−16.2
North Carolina	1,775	1,513	−262	−14.8
South Carolina	5,105	1,737	−3,368	−66.0
Tennessee	1,413	647	−766	−54.2
Texas	3,468	359	−3,109	−89.7
Virginia	200	132	−68	−34.0
Total	39,073	19,498	−19,575	−50.1

Source: U.S. Bureau of the Census 1916; [1948].

acres to 1,472,000. The landscape depicted the degeneration. Johnson and Turner, the study's authors, described the decay: "Most of the larger houses that might once have been seen on larger land holdings have burned or been ruined through neglect. Some landowners have been unable to look after their property, others have deliberately neglected their property. In either case much of the land is idle without occupants" (fig. 3.2).[7]

In the 325 counties included in the 1910 plantation census, the number of plantations declined by 50 percent from 39,073 to 19,498 between 1910 and 1940 (table 3.1). The decrease occurred despite the more inclusive definition of *plantation* for the 1940 census. The largest declines were in the eastern cotton belt. In Alabama, Georgia, and South Carolina the number of plantations decreased 75, 72, and 66 percent, while in Arkansas, Louisiana, and Mississippi the declines were 7, 8, and 16 percent. Changes in number of plantations were not uniform among regions. A comparison of the distribution of plantations in 1910 and 1940 reveals significant declines in the Piedmont, the Alabama Black Belt, and the Natchez district (maps 2.2 and 3.1). Across the Coastal Plain of South Carolina and Georgia and the north Mississippi Loess Plains, the decreases were not nearly as great, and in the Yazoo Delta the number of plantations actually increased.

THE DECLINE OF PLANTATION MANAGEMENT

Reasons usually cited for the decline of Southern agriculture, including farm tenancy, soil erosion, and the cotton boll weevil, were actually more symptoms than causes of disintegration. Management failure was the underlying basis of the decay. Landowners increasingly ceased to exercise the attentive supervision that was critical to plantation agriculture. A major conclusion of the U.S. Department of Agriculture's 1930 study of the lower Piedmont was "a great scarcity of real ability to manage large holdings of farm land" under adverse conditions. "Any success" on plantations was "certain to mean that the owners live[d] on their land and carefully direct[ed] their tenants and hired hands."[8] Gibson, who studied the Alabama Black Belt in the 1930s, thought that the decaying plantation big houses, abandoned fields, and eroded hillsides made it obvious that the region had "never offered many opportunities for people with professional training and ambition; hence, the bulk of the more promising ones . . . sought opportunities in the cities of Alabama and of the North and East."[9] During the depression Arthur Raper carefully analyzed the historical process of plantation occupance in two older cotton regions of the South. "When the plantation is flourishing, one finds a few large, well-built dwellings where the owners live, and a great number of small, twisted, unpainted cabins which house the landless agricultural workers—usually wage hands and croppers. As the plantation deteriorates, the big houses go without paint, the roofs leak, the porches tumble down, one field and then another is abandoned to brambles and gullies. . . . As the plantation crumbles, most of the erstwhile owners and some of the more alert tenants abandon the scene."[10]

The demise of prudent plantation management on the lower Georgia Piedmont was an evolutionary process that originated after the Civil War with the difficulties planters encountered in the transition from a slave to a free labor force. Because centralized management is critical to the plantation system, deterioration of landowners' ability or desire to exercise close supervision of farm operations resulted in decline of the plantation system. An increase in the number of absentee landlords and the growth of cash tenancy were signs of an impending leadership crisis. Ample evidence indicates that following the Civil War many planter families increasingly turned their attention from the land to other pursuits, including leadership roles in industrialization and urban development. Billings concluded that "the supposedly 'new men' of politics and industry in late nineteenth-century" North Carolina were actually members of "the landed upper classes" and "were the principal agents of industrial development." In his study of post–Civil War Alabama, Wiener found that Black Belt planters played a major role in the growth of Birmingham, the only large industrial city of the New South. A coalition of industrialists and planters took "the Prussian Road" to a restrictive, closely controlled type of factory system.[11] A local journalist bragged in 1931 that Greene County in the Alabama Black Belt "gave to Birmingham Webb Crawford, the banker; William McQueen, the industrialist; Sam Murphy, the judge; and John McQueen, the lawyer; all four of whom made their mark in the Magic City and helped it to go forward."[12] Russell found that most of the economic leaders of Atlanta between its creation in 1847 and 1890 were born in the South, primarily in Georgia, and had family ties to the Southern elite. According to one resident, Atlanta was "a city fashioned and controlled by 'native' men . . . born and reared within a hundred miles of its streets." The Washington, Georgia, weekly boasted in 1926 that the Danburg community in Wilkes County gave Atlanta "a number of distinguished sons," including a prominent pediatrician and the superintendent of schools.[13] Henry Grady, who as the editor of the *Atlanta Constitution* was one of the most vehement champions of an industrial "New South," was reared in a majestic Greek-revival house in Athens.

Even urban planter families of old coastal cities lost their ambitious children. Charleston, with its "urban focused plantation society" of the Sea Island region,[14] after the Civil War became a stagnant decaying city as new interior cities, such as Atlanta, Memphis, and Birmingham, grew. In 1940, Robert Goodwin Rhett, a member of Charleston's elite, bemoaned that the "flower of the youth of Charleston" moved to the South Carolina Piedmont and to other states seeking better opportunities. "Their departure drained the city of much of its vitality."[15]

Although agriculture initially was the principal underpinning of Southern wealth, with growth of cities and economic modernization, planter families increasingly paid less and less attention to landholdings, especially ones in the older plantation regions. The concern for agriculture eroded with each successive generation. Planters' children who left the land may have retained a significant interest in agriculture, but knowledge of and concern for farming declined with each succeeding urban-born generation, even though its members may have inherited estates. Contrary to the popular notion that planter families had nostalgic ties that kept them on the land, creation, preservation, and increase of wealth were more primary considerations than the family landholdings.[16] Reasons given by planters for abandoning careful management of their plantations after the Civil War varied. However, a common theme was that children were not available to assume management because, educated for occupations other than agriculture, they had left the land. In 1881, the David C. Barrow family lived in Athens rather than on Sylls Fork or one of the family's other three plantations. David C. Barrow Jr. was a faculty member and, eventually, chancellor of the University of Georgia. The experiences of three Hancock County planters are representative of those who ceased to manage their plantations. After the death of her husband, Edgeworth, in 1867, Sallie Bird rented Granite Farm and moved to Athens and then Baltimore, where she remarried and then died in 1910. Frank White successfully made the transition from a slave to a sharecropper labor force. White, who did not have any children to take over his plantation, in 1887 leased the land to "renters" because his "health became such as to interfere with close supervision." In 1896, Moses Harris "moved family to town to educate children and practically abandoned plantation to renting."[17]

Brooks identified absentee ownership as a major problem across the lower Georgia Piedmont in 1911, with large numbers of "renters" and lack of adequate supervision common to plantations that landlords had left. "The share system [sharecropping] was used where the landlords still lived on their plantations and exercised supervision over the operations of the tenants. . . . As time went on, more and more planters continued to move to the towns and the renting system [cash rent] tended to supplant the share system." Describing Meriwether, Troup, and Coweta Counties in the western part of the region, Brooks wrote that the big plantations seemed to be "on the wane in this district, great readiness being reported to sell lands at fair prices to small farmers."[18] Raper found that absentee ownership was greater in Greene County on the Piedmont, where agriculture was in decline, than in Macon County on the Coastal Plain, where it remained viable. In 1934, the 115 landholdings with 500 or more acres in Greene County contained 117,000 acres, approximately half of the county. Thirty-six of the 115 were owned by absentee landlords.[19]

In itself, the high percentage of land held by absentee landlords was not the cause of the demise of plantation agriculture. What was critical was failure to insure careful management of their properties. Raper concluded, "Absentee landlordism is most harmful to a community when it persists irrespective of the revenue received from the land; that is, when large tracts are held for sentimental rather than economic reasons." He also found that deterioration of management was common even among resident owners. These passive or "quasi-absentee" landlords, like absentee owners, paid "very little personal attention" to "cultivation" of their holdings. The failure of resident owners to manage their properties effectively was most prevalent in the "red land area," the principal plantation section of Greene County.[20]

In 1935, the Georgia Agricultural Experiment Station published a lengthy, detailed study of the lower Piedmont plantation region. Among the findings were the concentration of a high percentage of the land in ownership units of 500 or more acres, ownership of a large percentage of the land by a few persons, a high percentage of absentee owners, lack of careful management of properties by both resident and absentee owners, and decline of both agriculture and the plantation system. Forty-one percent, 1,778,000 acres, of the land in twenty-four counties of the lower Georgia Piedmont in 1932 were in ownership units of 500 or more acres. Absentee ownership was even higher than Raper found in Greene County. Less than two-fifths of the land in the twenty-four counties was owned by resident operators. Administrators and executors of estates controlled 10.5 percent and banks and mortgage companies 7.7 percent of the land. "A material proportion of the land" was "left pretty much to operation by tenants and croppers with only occasional attention by owner or manager. Consequently, the farm operations on many of these holdings" was "very slip-shod." Some resident owners permitted destructive agricultural practices in an effort to drain "every dollar possible in order to maintain standards and customs of living based on an outworn order of the old plantation period." The cultivated acreage declined year by year. On many landholdings "only the especially favorable fields" were still farmed. "The big owner-operated plantation in the Lower Piedmont" was "the exception rather than the rule." The striking conclusion was that "the old planters and their families and heirs" were "through with farming on the old estates."[21]

The relative importance of cash tenants is among the evidence that supports the premise that by the early twentieth century many plantations on the lower Georgia Piedmont had inept leaders. Richard Ely, Charles Galpin, and W. J. Spillman conceived the controversial "agricultural ladder" as a method of conceptualizing the relationships among farmers with respect to land ownership.[22] The Midwestern family farm model that they developed held large-scale agriculture and farm tenancy to be economic and social evils. The paragon American farmer was a *family farmer* who owned no more land than he and his wife and children could till. Tenancy, which the mythical ladder was conceived largely to explain and justify, was tolerated but only because it was the sole route by which some young men could become farmers (fig. 3.3). After entering agriculture as a "hired man," a young man could accumulate enough capital to purchase implements and work stock and climb to a higher rung and become a share tenant. As the young farmer matured, he was able to negotiate better terms and become the acme of renters, a cash tenant. Climbing up through part owner, he finally became a full owner, the apex of the agricultural ladder. Growth of cash tenancy, which gave a renter more freedom, and decline of share tenants indicated improvement in agriculture.

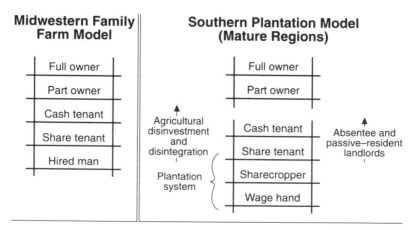

FIG. 3.3. The agricultural ladder, 1880–1950

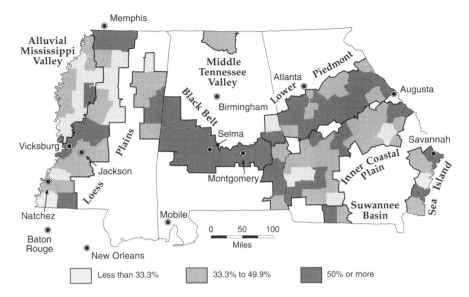

MAP 3.2. Percentage of tenants who paid cash rent in 1910. Tenants who paid standing rent are included (source of data: U.S. Bureau of the Census 1913a).

On the lower Georgia Piedmont in 1910, a large percentage of the tenant farmers paid cash rent (map 3.2).[23] In Meriwether County most cash tenants paid a "bale-per-plow" annual rent. Daniel interpreted this method of renting land in Meriwether as one that "encouraged initiative."[24] Therefore, he supported the Midwestern model. In his studies of Greene and Macon Counties, Georgia, Raper also found evidence that appeared to sustain the model. He thought that the role of a cash tenant approximated that of a landowner because he did not move often, owned capital goods, and had greater freedom. He also found that a "considerable number" of black and white cash tenants had been able to reach the apex of the agricultural ladder and buy farms.[25] Plantation agriculture, however, is a different context for interpretation of the role of cash tenants (fig. 3.3). Like the traditional factory system, efficient, centralized management that closely supervised inexpensive, docile labor was vital to the plantation. The plantation system does not foster small family farms, and the New South–era plantation did not promote the progression of wage hands and sharecroppers up tenancy rungs toward land ownership. Unlike the Midwest, in mature Southern plantation regions large numbers of cash tenants were actually evidence of the decline of management and the consequent disintegration of agriculture. The context of Meriwether, Greene, and Macon Counties was the Southern, not the Midwestern, model. Raper concluded that "cash renters reflect[ed] the disintegration of plantation farming," for "fixed renting naturally emerge[d] on the acreage of absentee owners and of resident owners who care[d] little about their land."[26]

What was especially unfortunate about the growth of cash tenancy in the plantation South was that the centralized decision making, together with the planter philosophy that the labor force must be kept uneducated and submissive, did not prepare sharecroppers, share tenants, and laborers to function effectively as independent farmers. They lived in small, isolated worlds that rarely extended beyond the horizon. Although share and cash tenants were capable of successfully farming from year to year, their limited education, isolation, and lack of capital prevented most from responding rapidly and effectively to a major crisis. In addition, the local economic infrastructure of furnish stores, cotton ginneries, warehouses, and banks was not transformed to one of yeoman agriculture but remained that of the plantation. Although poor blacks and whites initially had more freedom as cash tenants, ultimately those on the lower Piedmont were in worse circumstances, for the deteriorating agricultural system eventually came crashing down around them.

By the second decade of the twentieth century the plantation system on the lower Piedmont had evolved to a stage in which management varied significantly from landholding to landholding. Some landholdings were carefully and efficiently overseen and were profitable; they were the embodiment of the ideal New South plantation. Such plantations were worked by sharecroppers and wage laborers. The largest had commissaries and cotton gins. Management practiced aggressive erosion control and maintained soil fertility through the annual application of commercial fertilizer. Fields were circled by numerous bench terraces, and, in addition to heavy applications of "guano," crops were rotated to the extent that rotation was possible in an agricultural system that emphasized one commercial crop. A step down in management were plantations worked by a combination of sharecroppers and share tenants whose resident owners paid little attention to the operation. Mules and farm implements were sold to tenants who wished to buy them; the plantation was in transition from one worked by sharecroppers to one worked by share and cash tenants. Usually such plantations were owned by elderly persons who had no children or whose children had chosen occupations other than farming. In the absence of a manager, one of the tenants might be given primary responsibility for supervision of the plantation. In 1920, the agricultural infrastructure of the lower Piedmont was an intricate one that focused on small agribusiness complexes in towns, hamlets, and crossroads more than on

FIG. 3.4. The disintegration of the lower Georgia Piedmont plantation land-scape. Abandoned big house on Pharr plantation near Social Circle on the lower Georgia Piedmont in 1937. The house was constructed by slaves in 1840. Cotton is planted up to the house and covers the former sites of gardens, orchard, and auxiliary buildings. According to Dorothea Lange, who took the photograph, the plantation, which formerly had 150 slaves, was abandoned by the one remaining family member and was rented to small farmers. FSA Collection, Library of Congress

plantations. Such an infrastructure had evolved in response to the growing numbers of cash and share tenants. Although many landlords no longer provided the annual "furnish," tenant farmers easily obtained it from one of the local merchants. The final stage in the transition was a landholding on which management had ceased to function (fig. 3.4). The land was leased to cash tenants who, in turn, might sublease to sharecroppers. Absentee ownership of such properties was high. The annual rent, which was paid after the crops were harvested in the fall, was collected by the landlord on one of his infrequent visits or by a local ginner or merchant who acted as the landlord's agent.

Inertia causes organizations, including farms and agricultural regions, to continue to function long after leadership begins to decline. Buchanan argued that one of the characteristics of an agricultural region is the tendency to sustain itself as a result of accumulation of capital investments, special skills, and ancillary institutions geared to a particular type of agriculture.[27] Precipitous demise of an agricultural region, though, can be produced by a major crisis with which an apathetic and inept leadership is not prepared to deal. During the twentieth century the Piedmont cotton region confronted two major crises. One was the arrival of an insect, the boll weevil; the other was the development of competitive disadvantage for cotton, which necessitated modernization of production by the adoption of new technology.

THE COLLAPSE OF AGRICULTURE
ON THE LOWER GEORGIA PIEDMONT

The lower Piedmont plantation region survived civil war, emancipation of its slave labor force, and Federal occupation only to succumb a little more than half a century later to another invasion, that of a tiny insect, the boll weevil. The insect entered Texas from Mexico about 1892 and year by year, flying from field to field, spread across the South. By 1915 it had reached the Georgia Piedmont and by 1923 had spread to the northern limit of the cotton belt in southern Virginia.[28] The migration of the boll weevil was carefully monitored and mapped by the U.S. Department of Agriculture. Farmers were warned of the weevil's impending arrival several years before the insect reached a cotton region.

The Department of Agriculture developed methods whereby farmers could control and economically cope with the insect. Weevil control, however, required cotton growers to make considerable changes in traditional farming methods and to invest more capital in a crop. As soon as cotton was harvested, stalks were to be cut and plowed under to destroy both weevils and winter hibernation places. To diminish the weevil population further, fields and areas bordering them were to be burned. During the growing season, weevil infestation was to be monitored. If infestation increased, insecticides such as calcium arsenate had to be dusted or mixed with molasses and mopped on cotton plants (fig. 3.5). In addition, new varieties of cotton which could be planted earlier than traditional ones were introduced. Because weevil populations became progressively larger during the growing season, earlier planting resulted in cotton growing beyond its most susceptible stage for weevil damage when populations were smallest. Heavier applications of fertilizer caused cotton to grow faster and to produce more fruit, which helped compensate for losses to weevils.[29]

That the boll weevil was expected on the lower Georgia Piedmont and educated farmers knew aggressive measures had to be taken if cotton was to remain economically viable is revealed in what a student at the state college of agriculture in Athens wrote in 1912:

FIG. 3.5. Poisoning for boll weevils in Clarke County, Georgia, with hand-powered dusters circa 1930. The insecticide was calcium arsenic or DDT. Applications were made with no effort to prevent humans or animals from breathing the poison. Another three decades passed before the effects of agricultural chemicals on humans and the environment began to be questioned seriously. Unidentified photographer, FSA Collection, Library of Congress

That the Mexican cotton boll weevil will have entered Georgia within the next three years is as certain as it is that cotton will continue to be produced in this state. . . . If we shall meet this enemy in a manner to result in a minimum of loss, it is essential that we make preparation fast in the little time that may intervene. . . . It would be well for us to profit by the experience of Texas, Louisiana, Mississippi, and Alabama, for which they have paid so large a price. Active operation of the fundamental improvements in agricultural practice which has been worked out in the last few years in the path of the weevil is our strongest weapon against this steadily marching enemy.[30]

Dissemination of weevil control techniques was primarily through newspapers, agricultural journals, and Department of Agriculture and agricultural college bulletins. Although educated planters and yeomen were aware of the threat the boll weevil posed, most farmers, including cash tenants and new landowners who had risen from the tenant class, were not. In Greene and Macon Counties, Georgia, only 10 percent of black families subscribed to agricultural papers.[31]

The spread of the boll weevil across the South was a test of the health of agriculture and the viability of leadership at local and regional scales. So devastating was the insect on the lower Georgia Piedmont that its invasion was likened to that of Sherman's march across the region a little more than half a century earlier. In regions with competent management, including the Yazoo Delta and the northern portion of the Mississippi Loess Plains, cotton production survived the onslaught of the boll weevil (map 3.3). Cotton yields decreased for a few years after the arrival of the insect, but as control methods were employed, production recovered. At the close of the twentieth century, the boll weevil was not eradicated from much of the South, and cotton production still required that the insect be aggressively controlled.

Cotton regions that lost most of their competent and aggressive agricultural leaders prior to the arrival of the boll weevil underwent rapid decline from which they never recovered. The decline of cotton production and acreage between 1909 and 1929 was more pronounced in counties with the largest percentages of cash tenants. On the lower Georgia Piedmont, the insect initiated what became known as the "boll weevil depression." A severe decline in cotton production followed the insect's arrival. Hancock County produced 25,077 five-hundred-pound bales of lint cotton in 1914 but only 710 in 1922. In Oglethorpe and Wilkes Counties, production

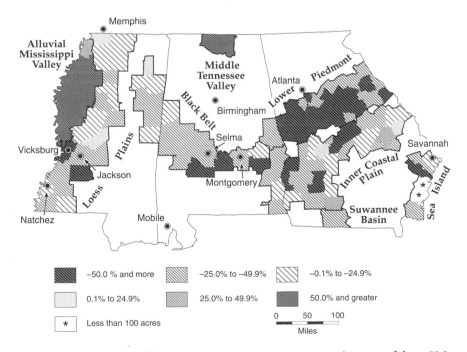

MAP 3.3. Percentage of change in cotton acreage, 1909–29 (source of data: U.S. Bureau of the Census 1913a, 1932a)

dropped from 26,403 to 2,828 and 33,760 to 2,492 bales. The suddenness and severity of the decline was even worse in Greene County, where production decreased from 21,479 bales in 1919 to 14,025 in 1920 and then plummeted to 1,420 bales in 1921 and 326 in 1922.[32]

A number of scholars have incorrectly assumed that the boll weevil is what destroyed much of the cotton economy as it spread across the South. The impact of the insect, however, was more a spectacular symptom of agricultural demise than a direct cause of it. When the insect reached the western edge of the lower Georgia Piedmont in 1915, the region was ripe for disaster, for the seeds that precipitated it were sown years before. Raper concluded that "for years the load had been accumulating" and the weevil was merely the "straw . . . which broke . . . cotton['s] back." The scenario of demise began with management failure, which led to soil erosion and the decline of soil fertility. "With the land becoming poorer and poorer and the surpluses from farming smaller and smaller and the dependence upon cash crops greater and greater, the population movement of rural dwellers was already decreed long before the coming of the weevil."[33] A similar conclusion was reached by Hartman and Wooten. Decrease of agricultural land on the lower Georgia Piedmont actually began in the 1880s. "Although that decrease was more or less precipitous after the boll weevil invasion, it started as a natural consequence of exploitative land use practices prevailing under the cotton plantation system of farming." That the decline of agriculture "was not the result of . . . the boll weevil invasion is indicated by the fact that the agricultural plant began to break down in seventeen of the thirty-five counties [of the lower Piedmont] before 1910." In "all but 5 of the 35 counties" there were "decreases in area of agricultural land before 1920."[34]

That much of the agricultural leadership and management capability disappeared from the lower Piedmont is illustrated by what the Wilkes County Agricultural Extension agent chided in 1927. Six years into the boll weevil disaster, Wilkes had "very little dusting machinery," and "not more than one farmer out of sixty" had "carried on the boll weevil fight in the systematic way" that was "successful."[35] Because many landowners failed to fight the insect aggressively, much of the initiative originated with the owners of businesses that depended on cotton (fig. 3.6). Although owners of ginneries, furnish stores, banks, and other businesses that made up the agricultural infrastructure attempted to advise semiliterate farmers on how to control the boll weevil, they had no direct supervisory role except on their own properties.

The heyday of New South agriculture was relatively brief on the lower Georgia Piedmont, lasting from approximately 1890 to 1920. Two factors especially accelerated agricultural decline with the advent of the boll weevil: agricultural disinvestment and concentration of land ownership. The insect initiated an economic domino effect that destroyed a substantial part of the financial infrastructure. The most serious result of several consecutive years of paltry cotton harvests was destruction of the credit of many tenant farmers and in turn that of the merchants and planters who furnished them and banks from which the merchants and planters borrowed. The failure of small-town banks rippled up to regional banks, some of which also crashed or were badly shaken. "We do not know who is going to finance white and negro tenants on many gully-washed plantations," lamented an editorial in Morgan County's weekly just before Christmas, 1922. "The landlord is likely land poor and cannot help, while even if he waives the rent, no cotton factor or banker will lend a hand for the simple reason that the tenant can offer no collateral for monies advanced except an uncertain crop of cotton." In Greene County, both banks in Greensboro and the ones in Woodville and White Plains had closed by 1928.[36]

Capital investment in the agricultural infrastructure began to decline significantly after the arrival of the boll weevil. Few new tenant houses, mule barns, ginneries, warehouses, cotton oil mills, and fertilizer plants

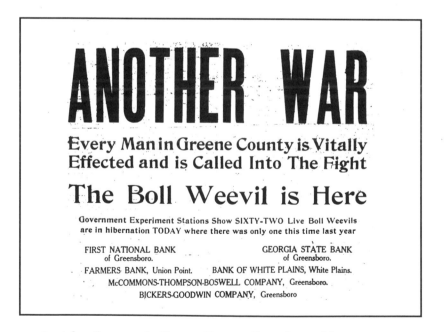

FIG. 3.6. Advertisement in Greene County, Georgia, weekly newspaper sponsored by agricultural infrastructure businesses encouraging farmers to control the boll weevil more aggressively. By 1930, all of the four banks had failed. *Herald-Journal*, May 23, 1925

were built after 1920. Even with the partial recovery of cotton, by 1930 the lower Piedmont grew only enough to support one-third of the existing infrastructure, which was essentially used up with little new investment after cotton renewed its decline.[37] When planters abandoned their land-holdings, they sold work stock and farm implements. In 1926, 41 percent of the owners of 500 or more acres in Morgan, 47 percent in Hancock, and 53 percent in Putnam Counties had no work stock. In Morgan County, which with respect to topography and soil was among the best agricultural areas of the lower Piedmont, landowners with 500 or more acres who did not own work stock increased from 37 percent in 1920, just prior to the boll weevil disaster, to 60 percent in 1932.[38]

Share and cash tenants lost the most. Unable to pay annual bills for planting seed, fertilizer, and food and clothing, they had to sell the mules and farm implements, which they had worked so long and hard to buy. Loss of work stock and equipment meant that the tenants either had to become sharecroppers or had to migrate. Because few planters had work stock and equipment with which to increase the number of sharecroppers and because many had no desire to operate sharecropper plantations, loss of work stock and farm implements meant that many share and cash tenants were finished in agriculture on the lower Piedmont. Johnson and Turner's study confirmed "that the intensity of the decline in the . . . acreage . . . harvested . . . varied, in general, directly with the percentage of tenants . . . who were standing renters [cash tenants]." The conclusion was that an area that lost its renters could not easily replace them. "[The] renters furnish capital goods of a type that landlords worry most about. . . . There were about as many croppers in the 59 counties in 1925 as in 1920, a decline of only 6 per cent. . . . Other tenants, the more important class because of their ownership of work animals, declined much more—36 per cent. The effective strength of these other tenants to operate farm lands, as shown by the numbers of their work animals, declined even more."[39]

The decline of the plantation system on the lower Piedmont initiated a large exodus of blacks. Most who left for Northern and Southern cities faded into the swelling black urban masses, finding neither fortune nor fame. The lower Georgia Piedmont's black population peaked just prior to World War I and the arrival of the boll weevil. During the 1920s, Greene County's black population dropped from 11,000 to 6,628, a 43 percent decrease. The decline was 67 percent in the five minor civil divisions where most of the plantations were located. By 1933, 20 percent of the rural dwellings in Jasper, Morgan, and Putnam Counties in 1918 had vanished, and fewer than half of the remaining 80 percent were still occupied.[40]

The exodus of large numbers of blacks from the lower Georgia Pied-mont during the 1920s meant that the labor-intensive cotton system could not recover to pre–boll weevil levels. The editor of the *Madisonian* wrote in 1922, "Every negro who leaves Morgan County cripples our only industry just that much more and lessens our chance of making a crop. It is esti-mated that two to three thousand have already left the county."[41] The editor of the *Augusta Chronicle* also failed to perceive the desperate situation of poor blacks in the agricultural collapse and lambasted Northern indus-trialists who had begun to recruit cheap black labor. "From one end of this country to the other and to be iterated and reiterated by every newspaper and magazine in this land should be this tragedy, and a wholly unnecessary one, that is going on as regards the migration of common labor from the farm, thus taking out of agriculture its prime means of sustenance and the taking over of this common labor, which is equivalent to robbing agricul-ture by the industrials [*sic*] of this country and let the public press not mince matters as regards causation."[42]

Blaming blacks for the agricultural demise was an attempt to make them the scapegoats. Accustomed to an abundant supply of cheap labor, whites panicked when blacks began to leave in droves. But no serious effort was made to hold black labor by increases in wages, improvements to housing and schools, or amelioration of segregation. The editor of Greene County's weekly newspaper placed the blame for the boll weevil disaster on the large white landowners. "The trouble is more or less with our people. It isn't the fault of our possibilities, our climate, . . . our magnificent low-lands. . . . The negroes are leaving by the thousands. Maybe, after all, this will force the white man to go to work. He has been depending upon 'old cuff,' to make him a living and abusing the black man when he didn't do it."[43]

Declines in cotton yields caused by the boll weevil meant that landlords had to decrease the rent of cash tenants. Most cash tenants in Meriwether County traditionally paid one and a half bales per plow, but after the arrival of the weevil, landlords had to lower the rent to one bale. In the early 1930s, "numerous renters" on properties of absentee landlords in Greene County had paid little or no rent since 1921. Some landlords decided that under conditions in which they received little or no rent it was to their advantage to terminate farming rather than to have the worry of a few tenants trying to eke out a living. Many simply "let their land lie idle, rather than guaran-tee a living to farmers . . . who, since 1920, [had] hardly in any year produced goods equal in value to what a landowner . . . had to expend."[44]

The high percentage of land in large holdings significantly affected the stability of agriculture on the lower Piedmont. Decisions by a few planters in a county to reduce crop acreage or to abandon farming removed a considerable acreage. During the 1920s, the decline in agriculture was much less severe on the upper Piedmont, with its numerous small yeoman farms. Johnson and Turner speculated that had the lower Piedmont plantation region "been farmed mainly by those working the family size farms, . . . most of the land would [have been] occupied."[45] In 1940, the editor of the Greene County weekly described the effect of concentration of land ownership on agricultural demise more candidly: "The trouble with Greene County is that we have had too many large plantations. Too many white men sat around town on their 'fannies' while the nigahs did all the work. The nigahs didn't know how to look after things, so they let the land wash away. When the boll weevil came, there was nothing left. We can't go back to the old cotton system. We've got to make the people live at home on the farms, and you can't do that with the plantation system."[46]

Planters who continued to farm often reduced cotton acreage and attempted to diversify. The 6,000-acre Orr Plantation was one of the largest in Wilkes County and produced more than one thousand bales of cotton annually. In 1920, Orr employed about 150 tenants and planted more than 1,500 acres of cotton. Two years after the weevil arrived, the harvest dropped to eight bales. By 1926, the number of tenants had declined to 50 or 60 and cotton acreage to 500. Although cotton yields had improved, only two hundred bales were produced. To compensate for the loss in cotton, Orr grew more corn and wheat and raised pigs, which he sold for five dollars each. Loss of two-thirds of the tenants, however, meant that abandoned fields covered with broom sedge stretched across the great plantation. Six years after the weevil's arrival, Orr thought that his "small kingdom" had become a "gigantic burden."[47]

In the face of the onslaught of the boll weevil were well-managed farms and plantations that survived the agricultural collapse around them. A Wilkes County farmer who purchased a Feny Mule Back Duster and followed the Department of Agriculture's advice on its use made twelve bales of cotton on 17 acres. Some of the neighboring planters who did not apply calcium arsenate properly and at the critical times made less than twelve bales on 100 acres.[48] By the mid-1930s only ten large landholdings in Greene County remained genuine plantations. The ten "were without exception, operated by tenants under the close supervision of resident owners." The ten were located in various parts of Greene County, and several were surrounded by vast abandoned landholdings that had once been prominent plantations.[49] Across the lower Piedmont, Johnson and Turner encountered successful farmers who continued profitable cotton production:

There are farmers who seem able to grow cotton regardless of their negligent neighbors. By their efforts these men market cotton every year and do not miss in the years when prices are high because damage has been unusually severe to the crop as a whole.

Possibilities of controlling weevil damages are illustrated by the yields obtained by one Piedmont cotton farmer whose production on his weevil-controlled 335 acre farm has been so much more uniform than for his county of Wilkes that he has paid off the debt he had in 1920 with profits obtained in years when most cotton farmers have been getting deeper and deeper in debt.[50]

THE PINE THICKET AND THE SEDGE FIELD

When the New South plantation reached its apex on the lower Georgia Piedmont about 1910, the area was one of the oldest and most important cotton regions of the world. At least 2,500 of the 6,627 plantations the Bureau of the Census enumerated in seventy Georgia counties in 1910 were on the lower Piedmont, and tenant families toiled in red clay fields that stretched from horizon to horizon. In 1940, only 382 plantations survived (table 3.1; map 3.1). Nowhere is the theme of the decline of plantation agriculture as the result of lack of effective management more vividly depicted than in fictional literature. So striking was the decline of plantation agriculture on the lower Georgia Piedmont that writers of fiction captured it with vivid descriptions in enduring stories. *Gone with the Wind*, published in 1936, was written by Margaret Mitchell, in part, to explain contemporary reality, including the abandonment of plantations for Atlanta. The decadent early-twentieth-century landscape of the lower Georgia Piedmont was the genesis of the novel. In 1936, Mitchell wrote Henry Steele Commager concerning *Gone with the Wind*'s origin:

> How happy I was that you were impressed by Rhett's [Butler's] remark about the upside-down world.... For in this paragraph lies the genesis of my book and that genesis lies [29] years back [in 1907] when I was six years old and those words ... were said to me ... by ... my mother. I didnt want to go to school ... and I saw no value at all in education. And my mother took me out on the hottest day I ever saw and ... [took] me [from Atlanta] down the road toward Jonesboro— "the Road to Tara" and showed me the old ruins of houses where fine and wealthy people had once lived....
>
> And she talked about the world those people had lived in, such a

secure world, and how it had exploded beneath them. And she told me that my own world was going to explode under me some day, and God help me if I didnt have the weapon [of education] to meet the new world.[51]

Mitchell was prone to dramatization and probably enacted much of this childhood confrontation with her mother. She wrote *Gone with the Wind* during the 1920s and essentially completed the novel by 1929, just prior to her thirtieth birthday. The plantation landscape from which she drew was the contemporary one that revealed the impact of the boll weevil. Mitchell incorrectly believed that the decadent New South Piedmont plantation landscape that she knew was that of the Old South near the end of a disintegration that began during the Civil War. Actually, the New South plantation landscape had few features that, even in 1907, survived from the Old South era. The landscape that Mitchell knew was the one reordered after the Civil War by the alteration of the plantation system to a tenant labor force and by the emergence of a railroad network, a central place hierarchy, and a new complex agricultural infrastructure (figs. 2.1 and 3.7).

Mitchell had worked as an obituary writer for the *Atlanta Journal* and wrote *Gone with the Wind* like an obituary. She knew the ending; all she had to do was explain it. The ultimate conclusion, however, was not in the 1870s, when she terminated the novel, but in the twentieth century, the time she knew personally. Not only did Mitchell have the corpse of the Georgia Piedmont plantation landscape from which to draw, but she had a host of persons on which to base her fictional characters. Scarlett O'Hara was taken from Annie Fitzgerald Stephens, Mitchell's maternal grand-mother, who died in 1934. Annie, who was about the same age as the fictional Scarlett, had moved from a plantation in Clayton County on the lower Piedmont to nearby Atlanta as a young woman, and she eventually lived in a great Victorian house. She was hardly a loving matriarch. Family members, including Mitchell, whom she disinherited, remembered her "as grasping, possessive, and materialistic; . . . devoid of sympathy as a mother, friend, or landlord."[52]

Mitchell accurately depicts an exodus of planter families—the O'Haras, the Wilkes, and the Hamiltons—from the rural Georgia Piedmont to At-lanta after the Civil War. Scarlett marries her second husband, Frank Ken-nedy, and becomes a shrewd Atlanta businesswoman with a store, lum-beryard, and sawmill. Although still possessing a love for Tara, she is an absentee owner who abandoned the family plantation and the past for Atlanta and the future. After Scarlett marries Rhett Butler, she becomes even more of an urbanite with her new Atlanta mansion, which is elite compared with the crudely build, decaying big house on Tara which she left. The migration to Atlanta is important in the geographical symbolism of *Gone with the Wind*. Mitchell confirmed that Scarlett is the personification of Atlanta, a vibrant tumultuous city with a future.[53] The disillusioned Ashley Wilkes, whom Scarlett thinks she loves but really does not, personifies the plantation, the past. Tara Plantation, which Scarlett regards as home, is really an illusory sanctuary. Toward the conclusion of the novel, Rhett Butler, whose role is that of interpreter of reality, tells Scarlett what the plantation actually is, a "white elephant in Clayton County."[54]

Erskine Caldwell, who briefly worked with Margaret Mitchell at the *Atlanta Journal* in the 1920s, was fascinated with other aspects of the decaying Piedmont plantation system, especially the severe poverty and desperate plight of people who were discarded. *Tobacco Road*, published in 1932, is set at the edge of the Piedmont near Augusta. The eroded fields are abandoned to broom sedge, pine, and blackjack. The boll weevil arrived, and a significant exodus occurred from the land. Jeeter and Ada Lester, who bear striking resemblance to Seab and Kate Johnson, whom Raper and Reid introduced a decade later, are destitute white tenants. Captain John, Jeeter's

FIG. 3.7. The "new" big house on the former Nolan plantation, Morgan County, Georgia, 1968. Although the Greek Revival structure may seem stereotypical antebellum, the cheaply constructed dwelling was actually New South, built about 1900 to replace the original big house, a Carolina-I across the highway. Charles S. Aiken

FIG. 3.8. The disintegration of the lower Georgia Piedmont plantation landscape. Abandoned fields on the Jackson plantation, Greene County, Georgia, in 1941. The topsoil was removed by sheet and rill erosion. The fields in front of the big house evidence no bench terraces or other methods to prevent and lessen erosion and were destroyed before 1900. Jack Delano, FSA Collection, Library of Congress

landlord, sold "all the mules" and went "up to Augusta to live." He told the Lesters and the others who remained on the old plantation that "they could stay in the houses until the buildings rotted to the ground and that he would never ask for a penny of rent." The Lesters are on the verge of starvation, but "nobody is taking on share-croppers." A bitter Jeeter gripes: "My children all blame me because God sees fit to make me poverty-ridden. . . . I ain't had nothing to do with it. It ain't my fault that Captain John shut down on giving us rations and stuff. It's his fault. . . . I worked all my life for Captain John. . . . Then the first thing I knowed, he came down here one morning and says he can't be letting me be getting no more rations and stuff at the store." At the conclusion of *Tobacco Road*, the Lesters die in a house inferno, which caught from the broom sedge fire that Jeeter set to prepare the land for planting an illusory cotton crop. "There were several hundred acres of land to burn; the fields that had not been cultivated, some of them for ten or fifteen years. . . . He went to sleep . . . with a new determination to stir the earth and cultivate plants of cotton."[55]

Because few plantations on the lower Georgia Piedmont made the tran-

FIG. 3.9. The vanishing New South landscape of the lower Georgia Piedmont. Abandoned two-story cotton ginnery and warehouse complex at Crawford in Oglethorpe County, Georgia, in 1965. The complex was constructed about 1910 and remained in business until the 1950s. The structures were razed in the 1970s. Charles S. Aiken

sition from the New to the Modern South, the cultural landscape long confirmed the decline of the old. Well into the second half of the twentieth century, the cultural relics and fossils of the New South plantation littered the landscape among the pine thickets and sedge fields, gradually disappearing as destruction and decay of big houses, tenant shacks, furnish merchant stores, and, especially, cotton ginneries, cotton oil mills, and other types of agricultural industrial buildings took their toll (figs. 3.8–3.11). Ironically, the landholding that serves as the classic spatial example of the transition from the Old South to the New South plantation also is a paradigm of the agricultural collapse. The Barrow family's Sylls Fork Plantation failed to make the conversion to Modern South agriculture. Sylls Fork was never mechanized, and farming on the landholding had ceased by the end of the Second World War. So great was the agricultural demise in Oglethorpe County that by 1940 only twelve landholdings met the definition of *plantation*.[56]

Sylls Fork passed from David Barrow to three ensuing generations of the family. The plantation's demise was slow but persistent. In 1932, the original 2,365-acre landholding was divided into 1,684- and 681-acre tracts, and by 1942 both were covered by forest and abandoned fields (map 3.4).[57] The landholdings, which sloped toward the Little River, exhibited severe sheet, rill, and gully erosion, despite the numerous bench terraces that circled some of its slopes. Although the landholdings were no longer organized as

FIG. 3.10. The vanishing New South landscape of the lower Georgia Piedmont. Abandoned furnish merchant store buildings at Maxeys in Oglethorpe County, Georgia, in 1966. During the boll weevil depression, the population of Maxeys dropped from 356 in 1920 to 209 in 1930. Charles S. Aiken

FIG. 3.11. The vanishing New South landscape of the lower Georgia Piedmont. Burdette Mill and Gin Company, Washington, Georgia, in 1966. The company remained in business until 1970. The agribusiness complex of which the ginnery building was once a part belonged to the Southern Cotton Oil Company at the time of its heyday in 1917 (see map 2.4.) The three-story structure was built about 1890 and was originally designed to house a machinery arrangement that was transitional between the Old South gin house and press and the New South ginning system. The concept was that of the early factory system, with the process of cleaning, ginning, and baling cotton beginning on the top floor and proceeding downward. The far end of the building housed the cotton oil mill. The building in the background is the guano house shown on map 2.4. Charles S. Aiken

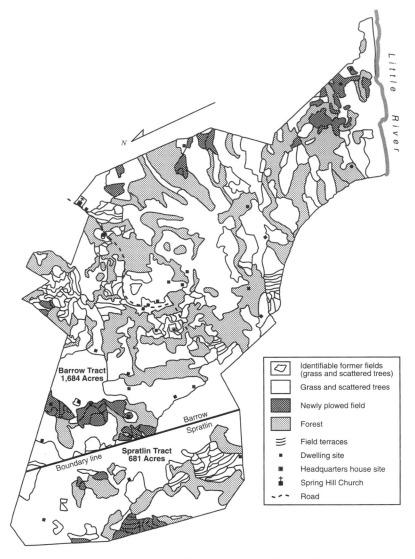

MAP 3.4. Land use on Sylls Fork Plantation, Oglethorpe County, Georgia, in 1942 (compiled using U.S. Department of Agriculture aerial photographs taken April 12, 1942, and field investigation)

a plantation in 1942, a few small choice plots were still planted to cotton and corn by several blacks who farmed with mules and paid cash rent.

The agricultural collapse on the lower Georgia Piedmont in the 1920s greatly deflated land prices, and the economic depression of the 1930s decreased them further. A golden opportunity was created for the timber products industry, whose leaders foresaw the end of the nation's virgin forest frontier before the end of the twentieth century. Land on which to plant rapidly growing pine trees could be purchased and leased exceptionally cheaply in the declining agricultural regions of the South. In 1944, the Barrow family sold the larger tract to a local company that planted the landholding in pine. That the old plantation brought only five dollars an acre, a low price even for the time, illustrates the disastrous impact of agricultural collapse on land prices. The Barrow tract changed ownership thrice over the next decade. In 1954, it was one of sixty-three tracts in Oglethorpe County, with a combined acreage of 18,797, purchased from Georgia Forest Farms by Champion Paper and Fiber Company (now Champion International Corporation). Champion and other timber products companies were attracted to the lower Georgia Piedmont and other declining plantation regions, not only by cheap land prices, but also by the large sizes of tracts, low property taxes, and probusiness county and state governments. Only five of the sixty-three tracts that Champion purchased in Oglethorpe County in 1954 were less than 100 acres, and the average tract contained 298 acres, substantially larger than the typical yeoman farm of the upper Piedmont or the 160-acre paragon Midwestern family farm.[58]

The landscape on Sylls Fork Plantation and across the lower Georgia Piedmont evolved from forest to fields and back to forest. When geographer J. Russell Smith visited the University of Georgia in the 1920s and gazed out across the Piedmont, he was struck by the magnitude of land use changes. "From the top of the State University buildings at Athens in the hilly country of North Georgia, one sees varied landscape of field and woods spread out in all directions. But professors at the University tell me that every bit of the woods covers land that has at some time been a cotton field."[59] The cultural process of landscape change, however, dramatically altered the physical character and the grandeur of the forest. In the late 1700s, just prior to its settlement, William Bartram described the great hardwood wilderness near the north branch of the Little River in the vicinity of the site of Sylls Fork Plantation as "the most magnificent forest [he] had ever seen. . . . Black oaks measured eight, nine, ten, and eleven feet in diameter five feet above the ground." And "the tulip tree . . . and beech, were equally stately.[60] Two centuries later the area is covered by a monotonous planted pine forest in various stages of growth, beneath which are the scars of two centuries of erosion.

THE DEMISE OF OTHER
COTTON PLANTATION REGIONS

Like the lower Georgia Piedmont, other mature cotton plantation regions where planter management was in decline by the early twentieth century are identified by high percentages of cash tenants and by catastrophic effects of the boll weevil (maps 3.2, 3.3).[61] In 1910, a large percentage of the tenants in the Natchez district, the Black Belt, and the northeastern fringe of the Mississippi Loess Plains paid cash rent. The weevil had a disastrous impact in the Natchez district and the Alabama Black Belt, but cotton production in the Yazoo Delta and large areas of the Loess Plains survived the insect. The Natchez district and the Yazoo Delta were among the first areas in the central South to experience what Richard Wright, a child refugee from ravaged Adams County, Mississippi, remembered as "the eveel boll weevil."[62] The insect reached the Natchez district in 1907 and one year later spread into the Yazoo Delta. The effects of the boll weevil on the two regions varied significantly. In Adams County, the location of Natchez, cotton production dropped from 20,455 bales in 1907 to 14,124 in 1908 and 1,592 in 1909. Production remained below 2,500 bales until 1917, and the county never again consistently produced more than 8,000 bales annually. The decline of agriculture in the Natchez district was so severe that by the Second World War it was considered a historic plantation area.[63] In the Yazoo Delta and other parts of the Mississippi Loess Plains the repercussions from the boll weevil were less severe. The shock of dramatic crop losses caused planters to adopt quickly the measures required to control the insect. The Delta, which was still developing as an agricultural region, substantially increased cotton acreage and production between the arrival of the weevil and the imposition of acreage restrictions under the Agricultural Adjustment Act in 1933 (map 3.3). The destruction of several consecutive cotton crops by the insect created a scenario in the Natchez district similar to that which occurred later on the lower Georgia Piedmont. Some planters and merchants, who could not collect rent or furnish debts, became bankrupt. The agribusiness infrastructure began to collapse, and a great exodus of blacks commenced. In Mississippi's six Natchez district counties, the black population, which peaked at 93,327 in 1900, had declined by more than a third by 1930. Cotton decreased from 255,000 acres in 1899 to 86,000 in 1929 (map 3.3).[64]

Deep South, a classic, detailed study of the plantation society and economy of Natchez and Adams County, was conducted during the 1930s by social anthropologists, who used the pseudonyms "Old City" and "Old County."[65] The agricultural situation was similar to that of the lower Georgia Piedmont. Land ownership was concentrated, absentee owners held a significant proportion of the land, indifference prevailed in management of plantations, and agriculture was in demise. Thirty-six families owned or controlled nearly half of the total farmland and almost a third of the cultivatable land in Adams County. Seven of the thirty-six controlled a fourth of the farmland; one owned more than 30,000 acres. Some plantations were estates and pooled holdings of relatives managed by one member. The 1940 plantation census confirmed the concentration of land ownership. The forty-eight plantations enumerated in the county contained 46 percent of the land in farms and 41 percent of the harvested cropland.[66]

The deterioration of plantation management in Adams County was well advanced at the time of the depression. Half of the planters lived in Natchez and paid little attention to their properties. Even most resident owners were passive managers who "maintained little or no supervision of the actual work." The large number of absentee and passive resident landlords encouraged cash tenancy. More than half (56.6%) of the tenants in Adams County in 1935 paid cash rent, ranging from six to ten dollars per acre of cotton. Rent was paid either in dollars or in an amount of cotton which approximated the monetary value. Tenants were given free use of pasture and woodland, additional evidence of the economic collapse of plantation agriculture. The few remaining well-managed plantations employed sharecroppers who "worked by the bell" and whose fields were examined daily by the owner or manager.[67]

Like the lower Georgia Piedmont and the Natchez district, by the early twentieth century the Black Belt was ripe for disaster. The plantation system had begun to disintegrate; more than half of the tenants in 1910 paid cash rent (map 3.2). The boll weevil reached the western portion of the Alabama Black Belt in 1911 and four years later had spread across the region. In 1916, the state agricultural college warned, "All cotton planters . . . in Alabama should plan to take up the fight against the boll weevil immediately, even if they have not yet been forced to do so." Losses were "sure to follow if cotton culture" was "continued in the usual way."[68] Despite the warning, farmers did not promptly adopt or effectively use methods by which they could continue profitable cotton production. The scenario was similar to that in the Natchez district a few years earlier and the lower Georgia Piedmont a few years later. Several successive years of disastrous cotton crops destroyed much of the agribusiness infrastructure, beginning with tenant farmers who could not pay landlords and merchants and rippling up to local banks that became insolvent. Large landholders who survived the boll weevil depression aggressively managed their cotton plantations, shifted to less capital-intensive beef cattle production, or abandoned farming.

Between 1910 and 1930 the Black Belt's Negro population decreased from 279,900 to 232,300.[69] As in other plantation regions, loss of cheap black workers contributed to the further demise of agriculture. The state agricultural college made no recommendations as to economic and social improvements that might help planters retain black labor. It merely warned that retention of labor was "a matter of the utmost importance as land without labor to work it becomes nonproductive and unprofitable." "But if the labor once moves out of a community, the fields are allowed to become brushy, the unoccupied cabins decay rapidly, roads are neglected, the value of land goes down. . . . The sections which have suffered most heavily from the weevil invasion lost far more because they let their labor go than from any direct injury done by the weevils."[70]

By the depression, large areas of the Black Belt, which a few years earlier were "fertile, gently rolling prairie with negro cabins standing in the open fields," had become the landscape of which the state agricultural college forewarned. A Birmingham journalist wrote of Greene County in 1930, "Many colonial homes in the rural sections evidence the prosperity of former days, but their white owners, for the most part, long ago moved to the town or elsewhere in the state, and they now are occupied largely by negroes." The fields had been "permitted to grow up in native grass pastures and considerable numbers of beef cattle" were "fattened on these for the market."[71] Another journalist described his trip across the Black Belt in 1940:

> As you drive back through the prairies, rolling up and away, green as a pasture in the *National Geographic*, you begin to notice how many abandoned cabins there are. You stop for gas at a crossroads trading center and ask, . . . "How's Business?" For an answer the clerk points across the road to three padlocked buildings. Each used to house a thriving supply merchant.
>
> "People all moving out," he says. "It's hit us too. We used to have a couple of hundred croppers on our account that worked around here. Now about all we get is the gas trade from the highway."[72]

Two studies of Alabama Black Belt counties reveal the disintegration of the plantation system. *Shadow of the Plantation*, Charles Johnson's classic analysis of Macon County, was published during the depression. In 1930, 3,114 black farmers lived in Macon County; in 1910 there had been 3,842. Cotton production plummeted from 36,768 bales in 1914 to 17,425 in 1934.[73] Johnson depicted Macon County's rural blacks as an isolated and backward people trapped in a dying plantation system that white landowners had largely abandoned:

> The white of the mansion house has gone, symbolizing, it would almost seem, the inevitable abandonment by the lords of the manor to the Negroes who still hang on, trying to nurse a living out of the earth. Foxtail and broomsedge, harbingers of senility, and the ubiquitous boll weevil, a new pestilence, keep this black labor alive, vainly fighting against the approaching final desolation. Everywhere there are the sad evidences of an artless and exhausting culture of cotton.
>
> The hard white highways of Alabama have drawn a ring as distinct as the color line around these decaying plantations—each with its little settlement of black peasantry. Here they live almost within sight of the passing world, dully alive, in an intricate alliance with a tradition which has survived the plantation itself. The plantation of olden times has gone, leaving them a twin partner of the earth, and upon these two—black man and the earth—the proprietor himself, growing ever poorer, depends for the mutual preservation of all.[74]

Morton Rubin's *Plantation County* is a study of Wilcox County immediately after the Second World War. The agricultural history of Wilcox was similar to that of Macon. Wilcox's black population decreased from 27,602 in 1910 to 18,558 in 1950. The county produced 30,849 bales of cotton in 1914 but only 7,656 in 1947.[75] Rubin, a "Yankee, born and reared in Boston, of Jewish religion," brought a refreshing detachment to his study, but he did not fully grasp the geographical situation of Wilcox County and the Black Belt, especially the scope of agricultural decline. Rubin conducted his study in the midst of the initiation of a set of economic, social, and political forces that would change forever the plantation and blacks' relationship to it. Both blacks and whites were "on the threshold of a new technological and social era." However, their "security system which depended on the full power of white supremacy and the paternalistic plantation owner" was "being shattered." In its place, Rubin thought that "the government" was "taking over financial support."[76]

The severe impact of the boll weevil and other problems that plagued the Southern plantation regions caused innovative planters and farmers and agricultural scholars and bureaucrats to reevaluate the production system. By the mid-1920s, most believed that if cotton was to remain a viable crop in the South, a renaissance would have to come through the adoption of new labor-efficient technology. The need to mechanize production was a second major crisis in Southern cotton plantation agriculture during the twentieth century. Regional variations in agricultural leadership and viability of the plantation system affected adoption of new technology just as they affected response to the boll weevil. Both the acceptance of new technology and failure to embrace it caused the number of plantations and persons who were intimately tied to the agricultural system to continue to decline.

CHAPTER FOUR

Mechanization
of the Plantation

⚜ ⚜ ⚜

There is impending a violent revolution in cotton production
as a result of the development of the mechanical cotton picker.
Cotton has awaited this event with the eagerness that it awaited
the development of the gin. When it comes it will automatically
release hundreds of thousands of cotton workers particularly in
the Southeast, creating a new range of social problems.

Charles Johnson, Edwin Embree, and Will Alexander,
The Collapse of Cotton Tenancy

LORD IN HEAVEN! Good God Almighty! Great Day in the
Morning! It's here! Our time has come! We are leaving! We are angry
no more; we are leaving! We are bitter no more; we are leaving!
We are leaving our homes, pulling up stakes to move on.
We look up at the high southern sky and remember all the sunshine
and the rain and we feel a sense of loss, but we are leaving. . . .
We scan the kind black faces we have looked upon since we first saw
the light of day, and, though pain is in our hearts, we are leaving.
We take one last furtive look over our shoulders to the Big House—high
upon a hill beyond the railroad tracks—where the Lord of the Land
lives, and we feel glad, for we are leaving.

Richard Wright and Edwin Rosskam, *Twelve Million Black Voices*

THE PROCESS OF MECHANIZATION

Cotton production in the South was among the last of the American agricultural systems to be modernized with labor-saving machinery and techniques. In 1930, across the cotton regions more than 150 hours of labor went into every acre of cotton. Workers on cotton plantations were integrated into three distinct phases that required significant amounts of labor

FIG. 4.1. A sharecropper family plowing a cotton field with mules on the Wilborn plantation in Tate County, Mississippi, in 1961. Children not only provided labor, but they were also apprentices who were taught the obsolete art of cotton farming employing animal power. At the end of a row, the father yelled "haw" to his mule, the command to turn to the left. Charles S. Aiken

FIG. 4.2. Five tenant families weeding cotton on one of the viable plantations that remained in Greene County, Georgia, in 1941. Cooperative work among families, especially extended families, was common. Jack Delano, FSA Collection, Library of Congress

—land breaking, planting, and cultivating; thinning and weeding; and harvesting (table 4.1).

Mules and horses, especially mules, which were preferred over horses by cotton planters, were the primary sources of power. Breaking plows and cultivators that were drawn by two or more animals were used on some plantations, but the common method of plowing was one mule directed by one person pulling one plow down one furrow (fig. 4.1). Spring plowing, planting, and several cultivations performed from late spring until the "laying by" of the crop in midsummer required approximately thirty-four hours of labor per acre. To insure a sufficient number of cotton plants, seeds were sown in continuous drills. Thinning excess plants and weeding between young cotton stalks in places where cultivator sweeps did not reach were performed effectively only by laborers wielding hoes (fig. 4.2). Thinning excess plants was executed once, but cotton had to be weeded several times from May into July. Approximately 20 percent of the labor input was for plowing, planting, and cultivating and another 20 percent for thinning and weeding. The remainder, more than half, was for harvesting. Cotton picking began in late summer and continued into late autumn (fig.

FIG. 4.3. Picking cotton on the Marcella plantation in the Yazoo Delta in 1939. Cotton was placed into sacks dragged by the pickers, weighed, and emptied into wagons, which were designed to hold the 1,200–1,400 pounds of seed cotton required to gin one 500-pound bale of lint. Marion Post Wolcott, FSA Collection, Library of Congress

TABLE 4.1 Number of Man-hours in Cotton Production, Yazoo Delta, 1939

Activity	Hours	Percent
Breaking, planting and cultivating using mules	33.6	20.9
Thinning and weeding	33.0	20.5
Harvesting	94.5	58.6
Total	161.1	100.0

Source: Langsford and Thibodeaux 1939.

4.3). Cotton was placed in baskets and sacks as it was picked from the stalks and was temporarily stored in wagons or in small cotton houses that dotted the fields (fig. 4.4). In the Mississippi Loess Plains and the Yazoo Delta each tenant unit usually had a cotton house.

The intensive character of cotton culture, together with the relatively simple nature of part of the tasks, meant that all members of a black tenant household, including children from the time they were old enough to wield hoes and pull cotton from open bolls, were part of the production system. Men and older boys normally performed the plowing, but thinning, weeding, and harvesting involved women and younger children. The cotton acreage allotted to a tenant by a planter was largely determined by the size of his family. Because cotton culture was a family production system, the labor force was substantially greater than that indicated by the number of farmers. In 1920, several million persons participated in cotton production across the Southern plantation regions.

Modernization of cotton production required reduction of the labor in each of the three major phases (fig. 4.5). Tractors mechanized plowing, planting, and cultivating, and mechanical cotton pickers automated the harvest. Application of herbicides before and after planting became the principal method used to eliminate hand weeding. Critical to comprehension of the impact of the new technology in the plantation regions is the realization that the methods to reduce labor in each of the three phases were not developed simultaneously. The tractor was introduced first and herbicides last. In addition, not all cotton regions adopted the new technology simultaneously. The alluvial Mississippi Valley, including the Yazoo Delta, led in adoption, and the other plantation regions lagged at varying rates. As regions, the lower Piedmont, the Black Belt, and the Natchez district did not complete the transition to modern mechanized cotton

FIG. 4.4. The mature New South Landscape. The wife of a sharecropper transferring cotton from a cotton house to wagon number 7 on the Marcella plantation in the Yazoo Delta in 1939. Houses the size of this one could hold two bales of unginned seed cotton. Marion Post Wolcott, FSA Collection, Library of Congress

production. With each phase of mechanization, more and more farmers ceased growing the crop. Only in remnants of these once great agricultural regions, small "islands" where cotton culture remained, were the three phases of mechanization concluded.[1] The mechanization era extended from the 1930s through the 1960s, with the most significant impact occurring after the Second World War.

The history of the development of agricultural tractors, which began in the nineteenth century, was marked by important technological breakthroughs and milestones in design. Until the mid-1920s, the tractors available to American farmers were intended as general sources of mechanical power. From their power takeoffs, auxiliary machinery such as hay bailers, threshing machines, and saws could be operated. In fields they were limited, suited for land breaking and general plowing, but not well designed for row-crop cultivation. Although a boon on family farms of the Middle West, such tractors were thought to be of little value on labor-intensive cotton plantations of the South. In 1925, Illinois had 43,325 tractors, but Mississippi had only 1,871.[2] With abundant cheap labor, Southern planters had little incentive to purchase labor-saving machinery. Moreover, because a large labor force was needed for weeding and harvesting and because mules would still have to be kept for cultivating, planters had little reason to purchase tractors for land breaking and general plowing.

A significant event in tractor technology occurred in 1924 with International Harvester's introduction of the McCormick-Deering Farmall, the first all-purpose tractor. The Farmall had sufficient power for land breaking, and its tricycle design with high rear-axle clearance and braking of either rear wheel for quick, sharp turns at the ends of rows made it suitable for cultivation. The International Harvester Farmall Works at Rock Island, Illinois, was opened in 1924 to mass-produce the new tractor. By 1930, other farm implement companies had begun to introduce their versions of an all-purpose tractor, using such names as Row Crop, General Purpose, All Crop, and Universal. Antiknock gasolines, pneumatic tires, and further design improvements made all-purpose tractors versatile and dependable agricultural machines by 1935.[3]

The alluvial Mississippi Valley was among the leading cotton regions in the adoption of tractors (figs. 4.5, 4.6). All-purpose tractors began to be purchased in the Yazoo Delta shortly after they became available. Adoption accelerated at a rapid rate between 1935 and 1946. Even during the Second World War, tractors, which were considered strategic because of the drain of agricultural labor by the military and factories, continued to be manufactured in large numbers. Between 1940 and 1945 the number of tractors in the Yazoo Delta increased from 5,277 to 8,717 (65.2%).[4] By the end of the war, the demand for tractors had outstripped the production. Plantations with innovative management converted completely from mules to tractors.

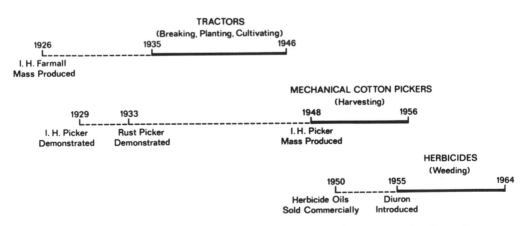

FIG. 4.5. The three critical stages of modernization of cotton production. The dates are for acceptance by the innovators and the early adopters in the alluvial Mississippi Valley (Aiken 1978). Courtesy of School of Geoscience, Louisiana State University

FIG. 4.6. Tractors in the fields of the Hopson Planting Company, Coahoma County, Mississippi, in 1940. The owners of Hopson were among the innovators in the use of tractors and mechanical cotton pickers. Marion Post Wolcott, FSA Collection, Library of Congress

The 4,000-acre Hopson Planting Company near Clarksdale sold all but five of its 140 mules and replaced them with twenty-two four-row tractors. The plantation's mule-drawn equipment was disposed of in scrap drives during the war.[5]

Although more than half of the farms and plantations in the Yazoo Delta did not have tractors in 1946, between 1935 and the end of the Second World War, knowledge of how to use the machines in cotton fields was perfected and disseminated. By 1946, tractors, even if not ubiquitous, were accepted in cotton culture. In the five-state area of the lower Mississippi Valley, cropland broken with tractors increased from 16 percent in 1939 to 42 percent in 1946. More important was the advancement made in cultivation with tractors, which increased from 6 percent to 18 percent of the cotton acreage.[6] That tractors rapidly replaced mules and horses in the Yazoo Delta once they were accepted is illustrated by a 1949 report by the region's farm implement dealers, which stated that since 1945 they had quickly sold every tractor they could obtain.[7] Even some advocates of small family farms understood the significance of the new agricultural technology. Although the Tallahatchie Leasing Cooperative Association, a Farm Security Administration project for black farmers near Glendora, Mississippi, purchased thirty-six mules when it was established in 1940, it also bought six all-purpose tractors.[8]

Mechanization of the cotton harvest was not as simple as mechanization of land breaking, planting, and cultivating. Not only did the mechanical harvester have to be a device that significantly reduced labor required to pick cotton, but it also had to operate in such a way that fibers were not broken and large amounts of trash were not collected. Several mechanical-harvest principles were patented, but only two, "spindle picking" and "stripping," were perfected. Mechanical pickers remove seed cotton from the open bolls with revolving spindles, whereas strippers take entire bolls by combing plants with extended teeth, stationary slots, or revolving rollers. Each type of harvester was accepted in definite regions, with strippers employed almost exclusively on the Red Rolling Plains, the High Plains, and the Black Prairies in Oklahoma and Texas and spindle pickers in the other American cotton regions.[9]

The first patent for a spindle picker was issued to Jedediah Prescott of Memphis, Tennessee, in 1850, but almost a century passed before spindle pickers were perfected and manufactured in large numbers (fig. 4.5). Although the spindle picker is among the most complex of agricultural machines, the lack of market among cotton planters was just as significant as the intricacy of the design in retarding development. As with the tractor, no market existed for a machine that would reduce labor in but one phase of cotton production. International Harvester played a major role in the perfection of the spindle cotton picker. In 1924, the company purchased the patents of Angus Campbell, who began work on a cotton harvester in 1885, and Theodore H. Price, who joined Campbell in 1910. By 1929, a harvester had been refined sufficiently that trial machines designed to be pulled by Farmall tractors were ready to be introduced. Two brothers of a yeoman-farm family, John and Mack Rust, began work on a spindle picker in 1924. With meager funds, by 1931 they had developed a machine that was capable of harvesting a bale of cotton a day, and in 1933 they demonstrated an improved model that could pick five bales a day.[10]

Despite the progress that had been made on mechanical cotton harvesters by the early 1930s, fifteen years passed before they were mass-produced. Three major factors delayed their introduction. Although early trial machines were successful, they were not dependable because they still had several design flaws. A major breakthrough was made by International Harvester engineers when the one-row picker was mounted backwards on a Farmall tractor and the machine driven in reverse down a cotton row (fig. 4.7). The attitude against labor-saving machinery which prevailed during

FIG. 4.7. International Harvester McCormick mechanical cotton harvesters on a plantation in the Yazoo Delta in 1949. The one-row pickers were mounted on Model M tricycle tractors which were driven in reverse. Early pickers such as these could harvest approximately 4,000 pounds of seed cotton a day. Courtesy of the National Cotton Council

the Great Depression was a second factor that prevented adoption of mechanical harvesters. Although International Harvester was ready to begin field trials for spindle pickers in 1929, the autumn that the depression began, company officials believed it was unwise to advertise them because of increasing unemployment.[11] Ironically, the Rust brothers, who were motivated to develop a picker from a deep desire to relieve one of the drudges in cotton culture, thought that "thrown on the market in the manner of past inventions," their machine "would mean, in the share-cropped country, that 75% of the labor population would be thrown out of employment."[12] The Rust brothers' efforts to control the use of their picker resulted in the machine not being manufactured in large numbers until after the Second World War.

The third factor that retarded manufacture of spindle pickers was problems encountered in ginning mechanically harvested cotton. Adoption of mechanical cotton harvesters had to be accompanied by development of new machinery and major investments in cotton gin plants. Just as a revolution in cotton ginning occurred after the Civil War in the process of adjustment from a slave to a free labor force, a second revolution in ginning happened after the Second World War as planters mechanized the harvest.[13] Cotton harvested with a spindle picker has a higher moisture content and contains more trash than cotton harvested by hand. To process machine-harvested cotton, gin plants had to be equipped with devices for drying seed cotton before it entered the gin stands and with machinery to remove additional trash. In 1945, 40 percent of the cotton ginneries in the lower Mississippi Valley had seed cotton driers, but most were not equipped with elaborate devices to remove trash, and the lint cleaner, an essential machine that extracts small bits of trash from ginned cotton, had not been developed.[14]

Mass production of spindle-type mechanical cotton harvesters began in 1948 with the opening of an International Harvester factory in Memphis. Other firms, including Ben Pearson, a Pine Bluff, Arkansas, sports equipment firm, and Allis-Chalmers, both with leased rights to the Rust patents, and Deere and Company, which in 1943 had purchased the spindle picker patents of Hiram and Charles Berry of Greenville, Mississippi, soon followed in commercial production of cotton harvesters. The alluvial Mississippi Valley, including the Yazoo Delta, led other Southern cotton regions in the adoption of spindle cotton harvesters (map 4.1). Although two decades passed before virtually all of the region's cotton was picked mechanically, by 1956 the mechanical cotton harvester had become an important part of cotton production. More than one-fourth of the crop in the Yazoo Delta was picked by machines in 1956, and in the area surrounding the Stoneville Agricultural Experiment Station near Greenville, more than

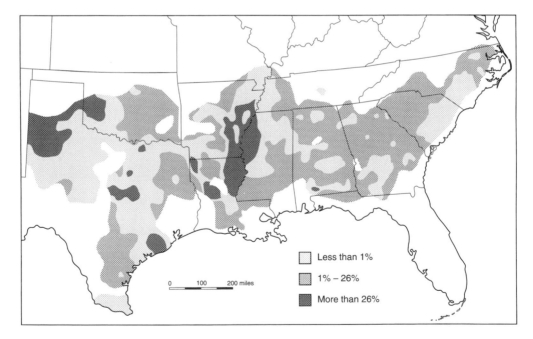

Less than 1%

1% – 26%

More than 26%

MAP 4.1. Percentage of cotton harvested mechanically in 1956 (source of data: Cotton Division, Consumer and Marketing Service, U.S. Department of Agriculture)

FIG. 4.8. The emerging Modern South plantation landscape. Large, high-capacity cotton gin with machinery designed to process mechanically harvested cotton rapidly in the Yazoo Delta in 1965. The new section of the building, indicated by the brighter corrugated steel, was added to accommodate new improved lint-cleaning machinery, which was developed shortly after the gin plant was built. Charles S. Aiken

half was harvested by them. Significant progress was made in cotton gin technology during the decade following the Second World War. By 1956, 57 percent of the gin plants in the Yazoo Delta were equipped to process machine-picked cotton (fig. 4.8).[15]

Tractors and spindle-type cotton harvesters are complex agricultural machines, but their development did not approach the problems encountered in the elimination of the simple hoe as the instrument for grass and weed control. The second of the three peak labor phases of traditional cotton production, thinning and weeding, was the last to be modernized (fig. 4.5). Elimination of labor for thinning the excess cotton plants was accomplished rather easily, but significant reduction in the labor needed to weed the crop was not achieved until the late 1950s.

Hand thinning was replaced by "hill dropping," setting the mechanical planters to space the seed rather than to sow solid drills. Because the planting technique requires seeds with high germination rates, elimination of labor for thinning was accompanied by significant improvements in the quality of seed. Beginning in the 1920s, a largely unheralded revolution in plant genetics initiated by the boll weevil resulted in high-yield cotton varieties with high seed-germination rates. Although labor for thinning was almost eliminated, the problem of finding a method to control grass and weeds that grew on the rows between the cotton plants persisted. Among the approaches to solve the problem was that which sought to perfect a hoeing machine. The Tillervator, the Kosik Crab Weeder, and the Dixie Cotton Chopper were among the various devices marketed. "Rotary hoes" became parts of the equipment inventories of some farmers, but chopping cotton with a machine was never perfected.

The complexities of finding a method of grass and weed control and, perhaps, the desperate nature of the problem are revealed by "flame cultivation" and "weeder geese." Invented by an Alabama farmer in the 1930s, flame cultivation is a method whereby fire or fire and water are shot around cotton stalks from small jets on a device mounted behind a tractor. The principle of flame cultivation is not to burn grass and weeds but merely to expose them to fire so that cells expand and the plants die. Interest in flame cultivation was principally in the alluvial Mississippi Valley, and it grew and waned several times between 1945 and 1970. In the mid-sixties when excitement was at a peak, approximately 26 percent of the cotton in Mississippi, almost all of which was in the Yazoo Delta, was flamed.[16] A limitation of flame cultivation is that it cannot be used on young cotton plants. Reluctance of even innovative farmers to employ a technique whereby fire is taken into a field also was a limiting factor. Flame cultivation was left to the bold and more adventuresome.

"Weeder geese" was another method used to attack the problem of weed

FIG. 4.9. Tractor planting cotton and applying preemergence herbicides to a field in the Yazoo Delta in 1965. Rust that formed in steel tanks proved to be a problem in clogging the spray nozzles. By the 1970s, plastic tanks had begun to replace steel ones. Courtesy of the National Cotton Council

control. Flocks of geese were placed in cotton fields enclosed by fences. A brace of geese can weed approximately one acre. Use of geese was handicapped by several problems, including trampling of young cotton plants and poisoning of flocks by insecticides.[17] Weed control with geese was never widely practiced in the Southern cotton regions, but in a few areas in the alluvial Mississippi Valley, such as the one centered on Weona, Arkansas, some planters employed them in the 1950s and 1960s. Not least among the drawbacks to the method were the hee-haws neighbors often gave planters who put geese in cotton fields.

The solution to the elimination of hand labor for weed control came with the introduction of herbicides in the 1950s (figs. 4.5, 4.9). Preemergence herbicides are sprayed as cotton is planted and, under ideal weather conditions, inhibit grass and weeds for up to four weeks. Postemergence herbicides are applied after preemergence chemicals lose their effectiveness. Herbicide oils were first sold in 1950, but it was not until diuron was introduced by DuPont in 1955 that the chemicals were effective enough that they began to be adopted rapidly. In 1955, preemergence herbicides were applied to less than 5 percent of the cotton acreage in the alluvial Mississippi Valley. Ten years later the chemicals were used on more than 80 percent of the cotton acreage in the Southeast.[18]

REORGANIZATION OF PLANTATIONS

A complete set of machines and chemicals that effectively reduced labor in all three phases of cotton production was not available to cotton planters until 1955. A planter who in 1935 wished to reduce labor could purchase tractors, but they mechanized only one phase of cotton production. Spindle-type cotton harvesters became available in 1948, but cotton still had to be weeded by hand. Because large labor forces were still required for weeding or weeding and harvesting, from the 1930s through 1960s some planters saw little need to purchase tractors or mechanical harvesters even though the machines were readily available. In addition, the new labor-saving machinery and techniques initially were not as satisfactory as traditional cotton production methods. With the increase in the percentage of cotton harvested mechanically came a decrease in the quality of the fiber. Cotton picked by hand and ginned in antique plants remained superior in quality to that produced with sophisticated labor-saving methods and was aggressively sought by certain textile companies. Careless application of herbicides killed cotton plants, and extremes in rainfall rendered the chemicals ineffective. Many cotton planters resisted or cautiously experimented with the new technology that involved such revolutionary practices as taking fire and plant-killing chemicals into fields and harvesting cotton with machines that lowered quality and price.

Because all machinery and methods to modernize cotton production were not developed simultaneously and because many planters, even in regions with large numbers of early adopters such as the Yazoo Delta, accepted them cautiously, elimination of tenant farmers and reorganization of plantations were evolutionary processes. Just as the geographical form of the Southern plantation was altered after the Civil War in the change from slave to tenant labor, so it was altered again after the Second World War in the transition from tenants to machines, chemicals, and wage labor. And just as there are scholars who fail to comprehend that labor and spatial alterations did not destroy the plantation following the Civil War, there are those who fail to understand that labor, technological, and geographical changes did not annihilate it after the Second World War. Mandle, for example, in discussing mechanization, states that "the plantation economy did not collapse at once," arguing that "the movement of black labor in response to wartime demand" is what "doomed the plantation economy."[19] Mandle is correct to the extent that the New South plantation came to an end, but he fails to recognize that a new, modernized plantation economy emerged.

The modern spatial form of the Southern plantation is what in 1955 Prunty identified and termed the *neoplantation* (map 4.2).[20] In the alluvial Mississippi Valley the transition from the fragmented tenant plantation of the New South era to the neoplantation of the Modern South began during the 1930s and accelerated after the Second World War. The small tenant farms within plantations vanished with the shift to wage labor. Because houses scattered across fields were obstacles to efficient use of the new machinery, those that were abandoned were razed, and the remaining ones were moved to form a line of dwellings near the plantation headquarters. The relocated dwellings in combination with newly constructed ones were the third generation of houses for blacks after the Civil War (fig. 4.10). High, open-sided buildings for storage of tractors and mechanical pickers replaced mule and horse barns. The furnish system disappeared with the demise of tenancy. Plantation commissaries and furnish merchant stores were closed, and the buildings were razed or converted to other uses (fig. 4.11). Even cotton ginneries vanished from many plantations as large, expensive high-capacity plants capable of rapidly processing machine-harvested cotton replaced small obsolete gin plants. With the consolidation of black schools after the Second World War, small one- and two-room plantation schools disappeared. The only New South–era landscape objects that remained on most neoplantations were big houses, churches, and cemeteries. With the reemergence of the village, the spatial form of the neoplantation of the Modern South resembles that of the Old South slave plantation more closely than that of the fragmented tenant plantation of the New South.

Both individual plantations and the plantation landscape were reorganized in stages. Because tractors did not reduce labor requirements for cotton production substantially, their introduction on a plantation usually was accompanied by minor reductions in the number of tenants and mules. Tractors were used for breaking and planting on a "through-and-through basis." Fields were plowed and planted as though tenant farm boundaries did not exist. Only after the young plants "showed green in the row" did sharecroppers assume control of their farms, cultivating their

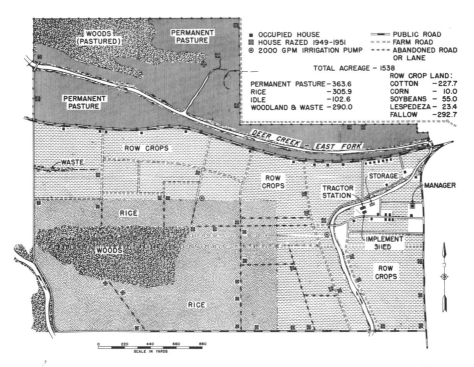

MAP 4.2. The emerging Modern South cotton plantation landscape. Fallback Plantation in Bolivar County, Mississippi, in 1954. A unit of the 37,000-acre Delta and Pine Land Company, Fallback was among the first to exhibit the removal from the fields of the tenant houses that were characteristic of the New South plantation. With the relocation of houses to a nucleated pattern, the village, an attribute of the Old South plantation and the world plantation, reemerged (Prunty 1955). Courtesy of the American Geographical Society

FIG. 4.10. The emerging Modern South plantation landscape. Row of tenant houses moved from the fields to the county road through the Roseborough plantation in Tate County, Mississippi, in 1961. Charles S. Aiken

FIG. 4.11. The vanishing New South landscape of the Yazoo Delta. Abandoned stores at Dundee, Tunica County, Mississippi, in 1978. A sundry store occupied what had been a large furnish merchant store on the bottom floor of the brick structure. The stores closed with the demise of tenant farming and the great exodus of blacks after the Second World War. The hotel that occupied the second story of the brick structure closed in the 1930s when automobiles and improved roads rendered train travel obsolete for the numerous salesmen from Memphis wholesale companies who roamed the region until the 1960s. Charles S. Aiken

crops with mules and thinning, weeding, and harvesting by hand. Tractors allowed a portion of the cotton to be grown by planters without tenants. The "company crop," or "day crop" as it was commonly called, was cultivated with tractors. Workers for weeding and harvesting were drawn from resident tenant and day-labor families and paid by the day for chopping and by the hundredweight of seed cotton for harvesting.[21] The introduction of mechanical harvesters on plantations was accompanied by a further decrease in the percentage of cotton grown by tenants but usually did not result in their complete elimination.[22] To permit time for day labor in the company crop, acreages assigned sharecropper families were smaller than under the traditional system. Even after the perfection of herbicides, a modified sharecrop system persisted on some plantations that had adopted methods for reducing labor in all phases of cotton production. To retain skilled workers and supplement wages, planters assigned machinery operators small, sharecrop cotton acreages that were planted and cultivated with tractors but weeded and picked by wives and children.[23] Because of the blend of the old and the new, persons traveling across the cotton plantation regions from the 1930s through the 1960s were confronted with the paradox of seeing in adjoining fields modern machines and antique mule-drawn implements.

LEAD-LAG IN ADOPTION OF THE NEW TECHNOLOGY

INTRAREGIONAL LEAD-LAG

A lead-lag pattern existed among the plantation regions and among plantations within regions in the elimination of tenant farmers and the adoption of new technology. Owners and managers of particular plantations, such as the Hopson Planting Company, were innovators who rapidly adopted new machinery and techniques and modified spatial morphology, while at the opposite extreme were laggards who chose to continue traditional methods as long as possible. Until his death in 1942, William Alexander Percy resisted mechanization of his Trail Lake plantation, which was only 50 miles south of Hobson. Tractors were introduced to break the land in the spring, but mule-drawn equipment continued to be used for planting and cultivating.[24]

In 1961, the plantations of Tate County in the north Mississippi Loess Plains depicted the transition from the New South tenant plantation to the Modern South neoplantation. The date is significant because the entire technology to modernize cotton production had been available for more than five years. Moreover, 1961 was the last year before the civil rights movement began to affect significantly traditional black and white relationships in the plantation regions. The location of Tate County is important. Although the north Mississippi Loess Plains lagged behind the Yazoo Delta in the adoption of new technology, agriculture and the plantation system were viable. The number of plantations had declined from the 168 enumerated in 1940, but Tate still had 51 farm operations that met the definition of *plantation*.[25] Because the Loess Plains lagged in adoption of new technology, Tate County depicted various phases of the transition from the classic New South tenant plantation to the Modern South neo-plantation.

Seven of the forty plantations studied had completed the transition from a tenant labor force and were neoplantations. The Double J Ranch,

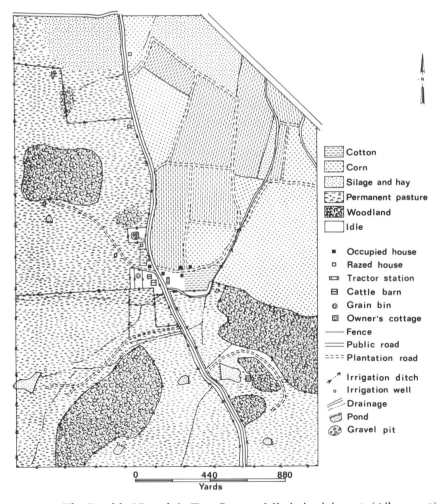

MAP 4.3. The Double J Ranch in Tate County, Mississippi, in 1961 (Aiken 1978). Courtesy of School of Geoscience, Louisiana State University

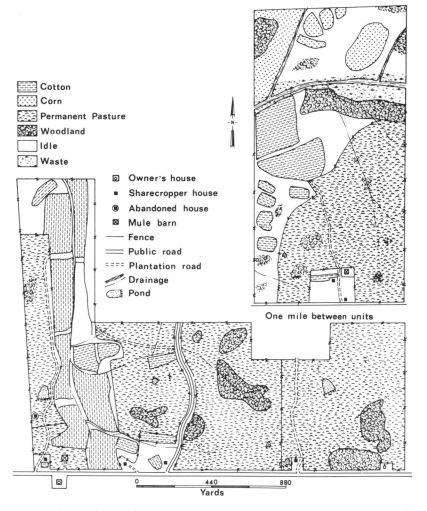

MAP 4.4. The Wilborn plantation in Tate County, Mississippi, in 1961 (Aiken 1978). Courtesy of School of Geoscience, Louisiana State University

named "ranch" to emphasize its registered cattle operation, was one of the seven (map 4.3). With its large fields, nucleated settlement pattern, and mechanized cotton production system, the Double J Ranch exhibited all the principal traits of the neoplantation. Approximately 130 acres were planted in cotton, 145 in corn, and 400 were improved pasture. The equipment inventory included five tractors, a mechanical cotton picker, a flame cultivator, and machinery for application of pre- and postemergence herbicides. Five resident black employees, who were paid weekly salaries, made up the resident labor force. In the late 1940s the plantation was devoted exclusively to cotton and had nineteen black sharecropper families and

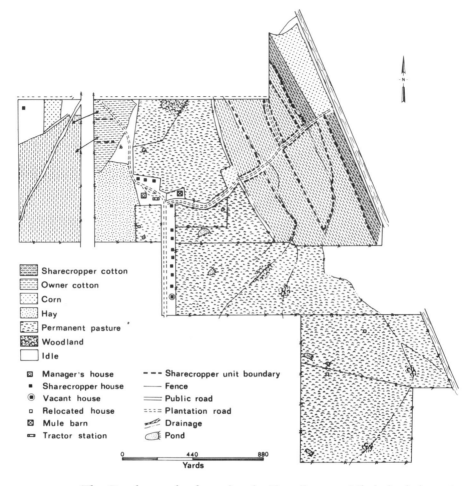

Sharecropper cotton
Owner cotton
Corn
Hay
Permanent pasture
Woodland
Idle

▣ Manager's house   - - - Sharecropper unit boundary
■ Sharecropper house   ——— Fence
◉ Vacant house   ═══ Public road
□ Relocated house   ==== Plantation road
⊠ Mule barn   ⤢ Drainage
⬭ Tractor station   ⬭ Pond

0   440   880
Yards

MAP 4.5. The Roseborough plantation in Tate County, Mississippi, in 1961 (Aiken 1978). Courtesy of School of Geoscience, Louisiana State University

thirty mules. The transition to a neoplantation began in 1948 and was completed in 1955 when the last sharecropper families left and the last mules were sold.

Three of the plantations in Tate County were still operated by traditional methods and contrasted sharply with the neoplantations. Morphologically and functionally the three differed little from the tenant plantation of 1910. They were cultural relics that persisted in the face of forces that would change them. Seven black sharecropper families worked the entire crop acreage on the Wilborn plantation (fig. 4.1, map 4.4). There were no tractors; fifteen mules supplied the power. Fifteen acres were planted in corn and 110 in cotton. In contrast to the large fields and nucleated settlement pattern of the Double J Ranch neoplantation, the dispersed house and field pattern on the Wilborn landholding was similar to the Barrow family's Sylls Fork Plantation in 1881. The plantation was divided into subunits, and each sharecropper family worked its own farm from the spring plowing through the autumn harvest. Because mules were adaptable to cultivation of choice patches, a few fields contained less than 2 acres.

Twenty-seven of the forty plantations were neither classic tenant nor neoplantations. They combined elements that were common to both and represented various stages in the transition from tenants to mechanized farming. On the basis of the technological stage and the role of sharecroppers, two major types of transitional plantations, *initial* and *advanced*, were identified. Tractors had been introduced and fields consolidated on plantations in the initial phase of the transition. The advanced stage was characterized by the use of mechanical cotton harvesters and herbicides and by a further reduction in the percentage of the cotton crop worked by sharecroppers. The Roseborough plantation in 1961 was in the initial phase (map 4.5). In certain respects it was similar to the Wilborn plantation. Twenty-six acres were planted in corn and 180 in cotton. Fields were divided into ten small sharecropper units that were cultivated with twenty-six mules. But there were also significant differences between the two. On the Roseborough, the nucleated settlement pattern characteristic of the neoplantation had emerged through the relocation of dispersed houses, and two tractors were used to plow and plant fields (fig. 4.10). The landholding depicted a stage in the transition reached on a number of plantations in the Yazoo Delta by the mid-1940s. Use of tractors resulted in consolidation of small fields and elimination of idle strips. Stakes at the end of particular rows were the only visible evidence that large cotton fields were actually subdivided into small sharecropper farms. Fifteen acres were "company cotton," worked for the owner using tractors for cultivation and day laborers from the sharecropper families for weeding and harvesting. The tractors were driven by the plantation's manager and by a resident employee who was paid a weekly salary.

INTERREGIONAL LEAD-LAG IN MECHANIZATION

Whereas planters in the alluvial Mississippi Valley took the initiative in the mechanization of cotton production and reorganization of plantations and those of the north Mississippi Loess Plains were not far behind, Southern Piedmont farmers significantly lagged in the adoption of new technology. Only a few landholdings were reorganized as neoplantations. Because of the loss of its agricultural leaders, by 1935 the Piedmont had few innovative farmers. In addition to the leadership problem, the hilly terrain of the

Piedmont posed a greater challenge to mechanization than did the flood-plains of the Mississippi River and its tributaries or the river and creek bottoms and gently undulating interfluves of the Loess Plains. In addition, by the mid-1930s, a number of the agricultural college faculty and agricultural extension service personnel of the University of Georgia and Auburn University believed that cotton production on the Piedmont should be discouraged and alternative types of agriculture and land use promoted, including planting of eroded cropland to pine trees.

The rolling terrain of the Piedmont presented greater challenges to mechanization than that of the alluvial Mississippi Valley or the Loess Plains, but use of new machinery was hardly impossible. The best supporting evidence is that by 1970 most of the cotton remaining was produced employing four- and six-row tractors, two-row mechanical harvesters, and herbicides.[26] The way in which the terrain of the Piedmont was perceived was more significant than the actual land surface. At the time of the introduction of tractors in the lower Mississippi Valley, the Johnson and Turner study assessing cotton production on the Piedmont stated, "There is a scarcity of large level fields and it is difficult to combine fields because of the irregular nature of the land, therefore tractor operated machinery to reduce labor costs scarcely seems practicable."[27] Twenty years later, tractors had been accepted on the Piedmont with fields redesigned for them. Planters in the alluvial Mississippi Valley were rapidly adopting mechanical cotton harvesters, but Piedmont farmers were advised: "Mechanical equipment that is currently used in other cotton-producing areas is not satisfactory for extensive use in the Piedmont. . . . This is particularly true of those machines required for chopping and hoeing, and for harvesting the crop. . . . In the face of increasing competition from cotton-producing areas that can adopt low-cost production practices, many cotton farmers in the area may find it advisable to consider alternative enterprises in which they have better competitive opportunities."[28]

Adoption of the new technology on the Piedmont required not only purchasing expensive machinery but also learning to use it. Bench terraces of the horse-mule era were replaced with channel terraces over which tractors and lumbering mechanical harvesters could travel, and the pattern of cotton rows on hillsides had to be redesigned to permit effective use of mechanical harvesters while preventing soil erosion. However, within an atmosphere in which many farmers believed that cotton would not survive, there was little desire to invest in expensive new machinery and to learn new skills. Mechanization was even more difficult in a milieu in which agricultural specialists continually told planters and farmers that the new technology could not be used on the hilly Piedmont. The additional burden of adapting terrain to machines and machines to terrain contributed to discouragement of all but the most enterprising farmers. With each of the three stages in the mechanization process, additional farmers quit growing cotton.

By 1970, all that remained of the once great lower Piedmont plantation region were a few agricultural islands in the midst of a sea of pine forests and abandoned fields.[29] Among the remaining cotton growers were small farmers, including part-time ones, who planted small cotton acreages to supplement income, and elderly black tenants on former plantations, who used mules to grow a few acres of cotton amid the ruin. Most of the cotton, however, was grown by the few farmers who planted 100 or more acres in the two larger agricultural islands of the lower Georgia Piedmont, one centered on Pike County south of Atlanta and the other on Morgan, Oconee, and Walton Counties near Athens. Small agribusiness firms to which cotton gins were central were the foci of the islands. Among them were the remains of John Bostwick's small empire in northern Morgan County, which was purchased by a local family from the cotton oil company that owned the complex. In part, cotton survived because of promotional efforts by owners of agribusinesses. To sustain their companies, which consisted of various combinations of gins, warehouses, farm supply, and certified planting-seed businesses, owners had to keep local farmers producing cotton. There were three types of large cotton growers: agribusiness owners, traditional planters, and *multitenants*. Multitenants were a new type of renter who leased cotton allotments from several landlords, creating a large dispersed farm operation. The largest of the dispersed farms met the definition of *plantation* and were *fragmented neoplantations*.[30] The multitenant system achieved economies of scale to support large, expensive machinery inventories.

MECHANIZATION AND THE DISPLACEMENT OF TENANTS

Public concern over displacement of tenant farmers in Southern agricultural regions by mechanization occurred during two brief periods separated by more than two decades. The first period was the latter part of the depression; it lasted from 1935 to 1942. The other closely coincided with the second and third phases of the civil rights movement. It began about 1963 and had essentially ended by 1974.[31] The periods differed, however, in their foci. During the thirties, almost all of the interest was in the plight of poverty-stricken white tenant farmers. A quarter century later, concern was over migration of blacks, who poured from the rural South into the nation's metropolises.

In his award-winning book on Southern sharecroppers and the New Deal, David Conrad termed the tenants "the forgotten farmers."[32] Southern sharecroppers may have been neglected by federal programs, but they hardly were a forgotten people. The depression was the decade of the Southern tenant farmer. Never before or since have tenant farmers received the attention they were given during the 1930s. The image of the Southern tenant farmer was continually before the American public throughout the depression decade. A play adapted from Erskine Caldwell's *Tobacco Road* opened on Broadway in December 1933 and ran continuously for the next seven and a half years.[33] John Steinbeck's *The Grapes of Wrath* was published in 1939, at the decade's close. A widely read controversial novel, *The Grapes of Wrath* is the story of the Joads, a white Oklahoma tenant family evicted from the land who migrate to California and become itinerant farm laborers. So fashionable were Southern tenant farmers by 1938 that James Agee confided in his journal that "the cheap use of the word" had made him "unable to hear, say or think, 'sharecropper' without a certain amount of nausea."[34]

During the 1930s, the documentary, a new type of media that combined narrative and photographs, came into its own as an art form, and Southern tenant farming was a major topic. Erskine Caldwell and Margaret Bourke-White's *You Have Seen Their Faces* (1937) employed photographs, captions, and text to depict the plight of the Southern rural poor. The impetus for *You Have Seen Their Faces* was Caldwell's belief that "the best way to prove that there were men, women and children existing in the backlands in the miserable squalor as described in" *Tobacco Road* was to travel the South's roads with a sensitive photographer and "produce a book which would portray real life of the rural South during the depression."[35]

Caldwell and Bourke-White's book was followed by Herman Clarence Nixon's *Forty Acres and Steel Mules* (1938), Archibald MacLeish's *Land of the Free* (1938), Dorothea Lange and Paul Taylor's *An American Exodus* (1939), and Richard Wright and Edwin Rosskam's *Twelve Million Black Voices* (1941). All used Farm Security Administration photographs and employed Southern tenant farmers as a theme or a major topic. James Agee and Walker Evans' *Let Us Now Praise Famous Men* (1941), the book often regarded as the classic of Great Depression documentary genre, was published at the conclusion of the first period of interest in Southern tenant farmers. Fewer than six hundred copies were sold because of loss of public concern with the Southern rural poor at the commencement of the Second

World War.[36] *Let Us Now Praise Famous Men* is the detailed study of three white tenant families in Hale County, Alabama. Seventy-two percent of Hale County's population and the majority of the tenants were blacks. But editors of *Fortune*, who sent Agee and Evans south for the original purpose of producing an article on sharecroppers, instructed them to focus on poor whites. Blacks had always been impoverished, and their situation was of little interest to the magazine's readers.[37] Agee was an Upland Southerner from Knoxville, Tennessee, who did not know the Lowland South. Concerning the *Fortune* assignment Agee confided that he "knew little or nothing about the cotton country, beyond a rough idea of the look of it and an even sketchier idea of just what the situation was there, beyond what [he] had got out of *Tobacco Road*, some passages in Faulkner, and a few meetings of the Committee for the Defense of Southern Workers."[38] Agee and Evans did not actually get beyond the Upland South; the boundary between the two culture regions crosses northern Hale County. The duo happened upon the three white tenant families whom they made infamous on Mills Hill near Moundville, a few miles north of the Black Belt.

The depression Southern tenant farmer literature contains several major themes, including the economic and social plight of tenants, the effects of New Deal programs on tenancy, the impending economic collapse of the plantation system, and the eviction of tenants from the land. The depression and the decline in the price of cotton caused some planters to reduce the number of tenants. The first widespread publicity given to the eviction of tenants concerned ones expelled as a result of the Agricultural Adjustment Act. When Franklin Roosevelt became president in 1933, a surplus of 13 million bales of cotton existed, and the fiber was selling for only five cents a pound. The hastily written Agricultural Adjustment Act was an attempt to reduce agricultural surpluses, increase prices, and guarantee farmers minimum prices for crops. To decrease the glut of cotton, farmers plowed up 10 million acres, approximately one-fourth of the 1933 crop. In 1934, 15 million acres were not planted.[39]

Acreage reductions under the Agricultural Adjustment Act were initiated without serious consideration as to effects on tenant farmers. Although more attention was paid to tenants in 1934 than in 1933, the Cotton Section of the Agricultural Adjustment Administration was run by persons whose perspective was that of the planter. Thoroughly familiar with the Southern plantation system, officials knew that legally sharecroppers were not tenants but laborers who were paid with a part of the crops. They also accepted the planter position that many share tenants were not actually independent renters. Any landowner, cash tenant, or "managing" share tenant who controlled a cotton farm could sign a 1934–35 AAA contract, but nonmanaging share tenants and sharecroppers could not. Paternalism of planters, rather than legal constraints, were relied upon to protect the interests of sharecroppers and nonmanaging share tenants. Determination of who were managing and nonmanaging share tenants was left to the discretion of county committees, which were dominated by planters.[40]

Under the Agricultural Adjustment Act, two types of payments were made to cotton farmers, rental and parity. The rental payment was for acreage taken out of production, or "rented" to the government. The parity payment was the difference between the price of lint per pound which the government guaranteed the farmer and the market price that the farmer received. The nature of the AAA cotton program motivated planters to reduce the number of tenant farmers. Cotton growers who had sharecroppers and nonmanaging share tenants were to give them their fair proportions of the AAA payments.[41] A decrease in cotton acreage also meant that

FIG. 4.12. Cotton choppers on a plantation in the Yazoo Delta in 1936. Comparison with figure 4.2 reveals subtle differences that illustrate the beginning of the transition from the New South tenant plantation to the Modern South neoplantation. With the exception of two men, the choppers in this photograph are middle-aged women, and unlike those in figure 4.2, they are under the direct supervision of a white foreman on horseback. Complete to a driver, the day-labor system was essentially a return to squad and gang methods that planters tried unsuccessfully to introduce immediately after the Civil War. The shift from tenants to day laborers and tractors was not fully comprehended at the time this photograph was taken by Carl Mydans. FSA Collection, Library of Congress

planters had to reduce the size of tenant farms. However, by keeping the farms the same size and decreasing the number of tenants, a planter could lower overhead and management costs. By eliminating tenants and working a larger share of the cotton acreage with day labor, a planter could keep a larger part of AAA payments.

With the commencement of the 1933 cotton plow up, letters from tenant farmers began arriving at the White House, the Capitol, and the Department of Agriculture, and they continued throughout the depression. Although failure to understand the complexities of the AAA program was the basis of many letters, there were genuine problems, including planters not sharing federal payments with tenants and the elimination of tenants in response to cotton acreage reductions. Elimination of tenants occurred in two ways. One was the outright eviction from a plantation; the other was reduction in status from tenant farmer to day laborer. Because few non-agricultural job opportunities existed in the plantation regions, tenants who were evicted and moved to local towns and cities also often became farm laborers paid by the day. Resident day laborers were permitted to live in a plantation house, usually without payment of rent and with the right to a garden plot and free firewood. Day laborers worked for a planter as needed, primarily for weeding and harvesting cotton. Elderly tenants and black tenant families headed by females were especially vulnerable to eviction or reduction in status. Increased use of day labor was essentially a return to the squad and gang systems that planters tried to introduce without much success immediately after the Civil War. Day laborers usually were worked in gangs or extended-family squads, but unlike the former slaves, who often were paid with a share of the crop, they were paid in cash daily or weekly (fig. 4.12). In the Yazoo Delta, the total black population peaked in 1940 at the commencement of the Second World War. The region's black farm population, however, climaxed in the 1930s.

The idea that farm tenants were evicted from the land by new agricultural machinery developed after 1935 and received much of its impetus from two sources.[42] One was the effort to explain the migration of down-and-out farmers from Oklahoma and Texas to the agricultural valleys of California. The other was publicity given to events surrounding the Southern Tenant Farmers' Union. Dorothea Lange and Paul Taylor worked throughout the latter part of the depression studying and publicizing the plight of the "Okies" who became itinerant agricultural laborers in California. During the spring and summer of 1937 the couple traveled across the Southwest and Southeast to study and photograph farm tenants and dust bowl refugees and to discover the factors that caused the migration. In June, Lange wrote Roy

Stryker, chief of the Resettlement Administrations' Historical Section, that she and Paul had discovered in Hall County, Texas, "a very interesting situation on tenancy—people put off farms, tenants, often of years standing and established—with tractors coming in purchased by the landowners with Soil Conservation money." She continued, "Not just a few cases. It's the story of the county. . . . How far across Texas the same story holds we do not yet know. . . . This may be an important part of our California story. In part I know it is now."[43] As Lange and Taylor continued their journey, they looked for additional evidence of eviction of tenant farmers by tractors. They were among the first to document the beginning of black migration from plantations of the alluvial Mississippi Valley to local towns and cities and to distant metropolises as a consequence of agricultural mechanization.[44] Archibald MacLeish used a number of Lange's pictures in *Land of the Free*. When MacLeish's book appeared, the expression "tractored off" was added to the depression lexicon.[45]

The idea of tenant farmers "tractored off" is more thoroughly developed in Lange and Taylor's *An American Exodus*. The couple wrote that "the record of power farming in cutting cotton workers from the land [was] already impressive" and that "this problem, originating in the South, [was] national in its repercussions." A former black tenant farmer supposedly said: "Tractors are against the black man. Every time you kill a mule you kill a black man. You've heard about the machine picker? That's against the black man too." The most striking testimonies of eviction by machine are not the prose but Lange's pictures. Among the most memorable photographs is one of an abandoned house in the middle of a Texas cotton field with freshly plowed furrows up to the door step (fig. 4.13).[46]

The most powerful and enduring expression of the idea of tractors pushing tenant farmers from the land is *The Grapes of Wrath*. Although Steinbeck did not meet Dorothea Lange until after publication of the novel, he was influenced by her photographs and Taylor's articles.[47] The novel synthesized several depression themes: poverty, tenant farming, drought, migration, and agricultural mechanization. Steinbeck vividly depicted two forces pushing tenants from the land, drought and the tractor.

> And at last the owner men came to the point. The tenant system won't work any more. One man on a tractor can take the place of twelve or fourteen families. Pay him a wage and take all the crop. We have to do it. . . .
>
> The tenant men looked up alarmed. But what'll happen to us? How'll we eat?

You'll have to get off the land. The plows'll go through the door-yard. . . .

The tractors came over the roads and into the fields, great crawlers moving like insects. . . . Snub-nosed monsters, raising the dust . . . across the country, through fences, through dooryards, in and out of gullies in straight lines. . . .

The houses were left vacant on the land, and the land was vacant because of this. Only the tractor sheds of corrugated iron, silver and gleaming were alive; and they were alive with metal and gasoline and oil.[48]

The persistent role of *The Grapes of Wrath* in the portrayal of eviction of tenant farmers by agricultural mechanization is significant. Among depression novels, *The Grapes of Wrath* was surpassed in sales only by Margaret Mitchell's *Gone with the Wind*. Steinbeck's most famous novel is still widely read and still controversial. Both the novel and John Ford's award-winning motion picture version of *The Grapes of Wrath*, which was filmed

FIG. 4.13. Dorothea Lange's photograph of an abandoned house in Childress County, Texas, in 1938, one of the most enduring symbols of eviction of farm tenants by machine during the Great Depression decade, contains no machinery. "Tractors Replace Not Only Mules, But People. They Cultivate To the Very Door of the Houses of Those Whom They Replace" (Lange and Taylor 1939, 73). Although cultivation also extends up to the big house shown in figure 3.4, its size and architecture convey a different symbolism than that of Lange's photograph of the less assertive tenant dwelling. FSA Collection, Library of Congress

in such a way as to give it documentary character and which is continually shown on television and at classic film festivals, present the idea of eviction of tenants by tractors to younger generations.

The alluvial Mississippi Valley in northeastern Arkansas and southeastern Missouri was among the areas where evictions and reduction in status of tenants as a consequence of the Agricultural Adjustment Act were most severe. Here, however, tenants did not accept the situation complacently. The alluvial valley in northeastern Arkansas and southeastern Missouri was a recently settled plantation area, and the tenant farmers included large numbers of whites as well as blacks.[49] The area also was one of the few rural places of the nation where the Socialist Party had a relatively large membership. When Norman Thomas, leader of the Socialist Party, visited eastern Arkansas in 1934 to give an address, he encouraged two young members, H. L. Mitchell and Clay East, to organize a union to fight for the rights of tenant farmers who were evicted from plantations or denied their shares of AAA payments. The Southern Tenant Farmers' Union, which began in Arkansas in 1934, by 1939 had spread to Missouri, Tennessee, Oklahoma, Mississippi, and Alabama and had thirty-five thousand members.[50] By convincing influential persons such as Eleanor Roosevelt and Fiorello LaGuardia of the plight of sharecroppers and by effective use of the media in publicizing their situation, the union had a greater role than its modest membership suggests. Publicity of tenancy problems eventually led Franklin Roosevelt to create the Committee on Farm Tenancy. The committee's report contributed to congressional passage of the 1937 Bankhead-Jones Farm Tenant Act and to the creation within the Department of Agriculture of the Farm Security Administration from the Resettlement Administration.[51]

From its inception, the Southern Tenant Farmers' Union waged a successful publicity campaign. Books such as Thomas's *The Plight of the Sharecropper* (1934) and Howard Kester's *Revolt among the Sharecroppers* (1936), a "March of Time" film *The Land of Cotton* (1936), and a National Sharecropper's Week launched annually from New York City were highly effective. However, in retrospect, Mitchell considered a highway demonstration in southeastern Missouri during January 1939 the "most spectacular event to occur in the sharecropper movement."[52] To protest eviction from plantations, 251 families composed of 1,161 persons camped at dispersed intervals along U.S. Highway 61. An investigation by the Federal Bureau of Investigation concluded that "the principal reason" for the evictions was "to enable the landowner to retain all of the AAA benefit money." But Lange and Taylor put a different spin on the roadside demonstration in *An*

American Exodus, interpreting it as evidence that "plantations of the Delta" were "coming under the machine" and "the sharecropper system . . . [was] collapsing at its advance."[53]

The roadside demonstrations caused Secretary of Agriculture Henry Wallace and other high-ranking officials in the Department of Agriculture to reassess the effects of New Deal farm programs on tenants. One official stated in a letter to Wallace that although "some of the displacement of tenants" going on in the South at the time traced directly back to that part of the Agricultural Adjustment Act program which paid a landlord "to make day laborers out of his tenants," he also believed that mechanization had become a factor in the reduction of tenants.[54] In 1939 and 1940, the Farm Security Administration sent Marion Post Wolcott, Russell Lee, and other agency photographers into the plantation South with specific instructions to find visual evidence to document eviction by machine. Among the photographs are Lee's of the Southern Tenant Farmers' Union members camped along Highway 61 in Missouri. In the autumn of 1940, Post Wolcott took a series of photographs of the Hopson Planting Company near Clarksdale, where tractors and the early International Harvester and Rust mechanical cotton pickers were tested and where day laborers were used in large numbers (fig. 4.6).

Wallace thought that deeper problems in the American economy, rather than the AAA or agricultural mechanization, actually underlay the eviction of tenant farmers. He also believed that mechanization was necessary to a viable future for the nation's agriculture and was part of a logical economic process. According to Wallace, the Department of Agriculture had "never taken the position that . . . mechanization, with its obvious effect on labor, should be stifled or ignored." Wallace wrote the following to Eleanor Roosevelt in the spring of 1939 when she requested more information on the condition of the displaced sharecroppers in Missouri:

Many thousands of needy farm families in other sections of the country are living under equally deplorable conditions. . . . Many of these families, like those in Southeastern Missouri, were sharecroppers or tenants who have been forced off the land by mechanization and other technological developments, or represent the historical and continuing off-throwing of population from the farms. The farms always have produced a much larger population than they support. . . .

When the Nation was expanding, this increase in population had opportunity either on new land or in the growing cities. The closing of opportunity in those two directions has resulted in damming up on the farms millions of people who normally would have been taken care of elsewhere. . . . There are few facts so fundamental to our whole economic problem of today.[55]

Paradoxically, after the Second World War the displacement of tenants from the land by mechanization and the flight of blacks into growing ghettos of the nation's metropolises continued with little notice by the media or the public until the confrontational phase of the civil rights movement in the 1960s (figs. 4.14, 4.15). Counter to the idea that tenant farmers were displaced from Southern plantations by machines, the concept emerged after the Second World War that planters were forced to mechanize because labor migrated. By the mid-1950s the theory of a "labor vacuum," which held that "machines do not displace labor, but merely come in to fill the vacuum left by labor which has already departed," was prevalent throughout the Southern cotton regions.[56] From the beginning of mechanization in the 1930s until the process was completed in the late 1960s, planters repeatedly denied that tenants, especially blacks, were displaced by machines. Rather, after the Second World War they complained of labor shortages, and many who purchased tractors and mechanical harvesters stated that they were forced to do so. Oscar Johnson, manager of the 36,000-acre Delta and Pine Land Company at Scott, Mississippi, wrote that although between 1935 and 1947 the number of tenant families on Deltapine dropped from 850 to 550, mechanization was "not the cause, but the result of economic change." Dorothy Dickins, concerned over the number of workers who might be evicted by agricultural mechanization in Mississippi, concluded in 1949, "At the present time there is no problem of a displaced labor supply; in fact, there is a scarcity of labor on some plantations." Arthur Raper and Harald Pedersen observed in 1954 in the Yazoo Delta, "Planter after planter . . . feel impelled to reiterate and re-affirm the defense which became the byword of the area in the late 30's: 'Not one family, not one person has been displaced by machines on this plantation.'"[57]

In his study of the Southern plantation economy, Mandle supports the idea of mechanization as a replacement for labor that departed as part of his thesis that changes in the plantation system during the twentieth century were initiated principally by blacks. Blacks began migrating from plantations during the First World War. The unleashing of blacks from the plantation system, however, occurred primarily during the Second World War. Blacks were confined to the Southern plantation regions by a com-

FIG. 4.14. A 1947 cartoon illustrating the feared impact of mechanical cotton pickers on black agricultural tenants and laborers. "Is this a new opportunity for your business, or will it become an economic Frankenstein?" (National Cotton Council 1947)

FIG. 4.15. An obsolete, abandoned one-row mechanical picker mounted on a Model-M Farmall Tractor in 1988, four decades after the tide of history had washed over the cotton plantation regions, their people, and the "economic Frankenstein." Charles S. Aiken

bination of racist hiring practices in Northern industry and immigration of cheap labor from abroad. With the advent of the Second World War, blacks were urgently needed by the military and by American industry. More than one million blacks migrated from the six Deep South plantation states during the 1940s, and "the vacating of the estates caused by the war accelerated trends towards mechanization of southern agriculture."[58]

WERE BLACK TENANT FARMERS DISPLACED BY MECHANIZATION?

Assessment of the relationship of mechanization of cotton production to migration of blacks from the Southern plantation regions must consider several factors. First, mechanization of cotton production did not play a role in black migration until the depression of the 1930s. The emigration of blacks from the rural South during the first three decades of the twentieth century was produced by other factors, including the "push" given by collapsing plantation systems on the lower Georgia Piedmont, the Alabama Black Belt, the Natchez district, and certain other regions. Migration from regions of agricultural demise was not just to cities but to viable Southern agricultural regions. While the black populations declined on the lower Georgia Piedmont, in the Alabama Black Belt, and in the Natchez district between 1910 and 1930, they increased in the Yazoo Delta and on the northern Mississippi Loess Plains. The "push" given by disintegration of plantation agriculture after 1900 coincided closely with the "pull" of new employment opportunities for blacks in the North. That the opening of jobs to blacks and the recruitment of black labor from the South during the First World War and the 1920s coincided with the impact of the boll weevil and the disintegration of the plantation system in particular regions was fortuitous. Without migration, even larger numbers of blacks would have been in destitute circumstances in the rural South during the 1920s and 1930s. Racial oppression in the South also was a major "push" of blacks to cities of the North and West, and the greater freedom that blacks found became a principal "pull" on relatives and friends who remained in the South. The de facto segregation in the North and West was not nearly as harsh as that mandated by law and by custom in the South. Many blacks who migrated thought that their children were in much better situations. Not only were schools in Northern and Western cities not legally segregated, but they were much better funded, had more comprehensive curricula, and extended through the twelfth grade.[59]

During the depression, the imposition of federal acreage allotments on cotton and the commencement of mechanization with tractors began to eliminate tenant farmers in viable as well as declining plantation regions. The number of black farmers evicted was relatively small compared with the number who remained on plantations. However, had agricultural mechanization continued under the economic environment of the depression, a horde of evicted tenants reduced to the status of day laborers attempting to eke out a living would have resulted. The Second World War intervened, and a "pull" was exerted by military personnel requirements and industrial labor demands. The expanding postwar economy continued to offer urban job opportunities to Southern rural blacks. Even if they did not find the number and types of jobs they sought, perception of a better life in the city drew blacks in increasing numbers from plantation shacks to the nation's cities. The number of blacks who migrated from Georgia, Alabama, and Mississippi was 773,000 between 1940 and 1950; 752,000 between 1950 and 1960; and 664,000 between 1960 and 1970.[60] That mechanization of cotton production and transition from the New South tenant plantation to the Modern South neoplantation occurred in phases kept even larger numbers of blacks from being suddenly "pushed" from the land between 1935 and 1965. Moreover, planters did not make decisions to adopt new technology strictly on the basis of economics.[61] Inertia and lack of confidence in machines and chemicals prevented immediate acceptance of new technology by some planters.

Despite the caution over mechanization, neither planters nor officials of the Department of Agriculture sought to retain the labor-intensive cotton production system. An important aspect of the relationship between mechanization and reduction in labor on plantations is that beginning in the 1930s wages for agricultural field labor were essentially inelastic. Planters who complained of labor scarcity during the 1940s and 1950s meant lack of abundant cheap labor, not the lack of workers. Except for skilled machinery operators, "competent labor," no attempt was made to hold or recruit agricultural workers by increasing wages or the percentage of the cotton crop given to tenants. Across the cotton regions, wages for weeding and picking cotton did not increase between 1949 and 1965. In Mississippi the wage for picking a pound of lint cotton actually declined from 8.6¢ in 1949 to 7¢ in 1965.[62] In 1965, the average worker could make only approximately $14 a day picking cotton.[63] When cheap black labor became scare, even reluctant planters adopted machines and chemicals rather than raising wages.

Another important aspect of the relationship between mechanization and labor is that the supply of agricultural workers was viewed by planters in a regional rather than in a local spatial context. Migration of tenant families and day laborers from plantations did not necessarily mean exodus from the region. A process Paul Taylor described in the late 1930s in which tenant families evicted from plantations moved to local municipalities continued through the 1960s. Growth in the regional rural-nonfarm and urban black populations escalated while planters complained of labor shortages and contended that mechanization did not displace agricultural workers but replaced those who left. Taylor thought that "the presence of this reserve, comprised principally of cotton workers recently swept from the land into the towns," was what made "tractor farming feasible." Blacks who did not migrate from the region were employed "only as a reserve to meet seasonal peak needs" for weeding and harvesting. Improvements in the rural infrastructure assisted planters in the elimination of tenant farmers. New types of spatial linkages developed between plantations and towns and cities. "Daily movement of labor between town and plantation," wrote Taylor, was "facilitated by very recent construction of graveled and hard-surfaced roads." Truckloads of day laborers from Greenville were hauled more than 35 miles over the new roads.[64]

One of the most perceptive and articulate expressions of mechanization and modernization of agriculture and their effects on those tied to the plantation was written by David Cohn just as the process began to accelerate. Having visited the Percy family's Trail Lake plantation in 1940 and returned a decade later, Cohn knew that a "mechanical revolution" was "changing . . . the shape and flavor of the plantation landscape" forever.

> When mechanization of the Delta agriculture has been completed, out into the unknown there will go about 70 percent of those presently upon the land; over 200,000 persons.
>
> As they go, the old-time plantation goes. Under mechanization the fields march from horizon to horizon unbroken by cabins, churches, gardens, and cowsheds. Scattered cabins and other structures give way to small villages as in European agriculture. Workers move to and from the fields on the self-propelled vehicles that are their tools.[65]

The only serious flaw in Cohn's prophecy is that he greatly overestimated the number of persons who would remain in the Delta's plantation agriculture.

CHAPTER FIVE

The World of Plantation Blacks

❦ ❦ ❦

We have already at least 10,000 Men of these descendants of Ham
fit to bear Arms, and their Numbers increase every day
as well by birth as by Importation. And in case there should arise
a Man of desperate courage amongst us . . . he might . . . kindle
a Servile War . . . and tinge our Rivers . . . with blood. . . .
It were therefore worth the consideration of a British Parliament,
My Lord, to put an end to this unchristian Traffick of making
Merchandize of Our Fellow Creatures. At least the farther Importation
of them in Our Colonys should be prohibited lest they prove . . .
troublesome and dangerous everywhere.

William Byrd to the earl of Egmont, 1736

A definite bi-racial problem underlies the economic and social order
of the Black Belt. Nowhere in the South does a sharper color line exist,
and nowhere do whites depend more upon blacks.

J. Sullivan Gibson, "The Western Black Belt," 1940

THE MEDIEVAL NATION

In many respects the Great Depression of the 1930s was the darkest and most severe decade that Southern rural blacks experienced between the Civil War and the Second World War. Not only were blacks adversely affected by the depression, but the decade was one of increasing uncertainty about the future of their historical role of providing cheap labor for plantation agriculture. The two decades preceding the depression were ones of great changes for the South's black population. The plantation system had begun to collapse in several of the densely populated older regions. The First World War and the opening of Northern industrial jobs contributed to increased migration and urbanization by rural blacks between 1910 and 1930. The great twentieth-century redistribution of the United States' black population had commenced. In 1910, 89 percent of the nation's blacks lived in the South, and 80 percent of the South's blacks were rural, concentrated in the plantation areas. By 1930, 21 percent of the nation's blacks lived outside the South, and the percentage of the black population in the South which was rural had declined to 73 percent.[1]

One of the most important distinguishing features of blacks in the plantation South at the time of the depression was poverty. If the South in the 1930s was the nation's number one economic problem, the plantation regions were among the cores of economic despair. Although a few black families in the rural South accumulated land, established businesses, and rose economically, the majority lived in the depths of poverty. The poverty was both relative and absolute. Southern rural blacks were among the poorest Americans. The real poverty lay beyond mere income statistics; it was overt in housing, diet, clothing, health, and education and conspic-uously reflected on the landscape. The cultural landscape of the Southern cotton regions was a "miserable panorama of unpainted shacks, rain-gullied fields, straggling fences, rattle-trap Fords, dirt, poverty, disease, drudgery, and monotony" which stretched for more than a thousand miles (fig. 5.1).[2] Perhaps nothing more blatantly revealed the depth of the poverty of rural blacks than housing. Most dwellings were poorly constructed and poorly maintained. One government study tersely stated, "The observing traveler in the deep South is rudely shocked when he sees for the first time the widespread evidences of rural poverty revealed in the farm homes."[3] Among the rudely shocked was Henry Wallace on his first visit to the lower Georgia Piedmont in 1932 to confer with Franklin Roosevelt at Warm Springs, his rural retreat and polio treatment center in Meriwether County. "On my way to Warm Springs," wrote Wallace, "I was utterly amazed and appalled at the red gashed hillsides, at the unkempt cabins, some of them without windows or doors. It was a situation, it seemed to me with an Iowa farm background, that was almost unbelievable."[4]

The abysmal condition of houses occupied by blacks was hardly a recent phenomenon. Eighty years earlier a planter complained that on the land-scape everything bore "the impress of progress and permanency, save plan-tation buildings, which continue[d] to wear the face of neglect and decay."[5] In architectural design and construction, plantation tenant houses were the embodiment of the dwellings built during the last decades of slavery. Most of the dwellings occupied by rural blacks at the time of the depression belonged to the second generation of post–Civil War housing. Log houses had largely vanished. In 1934, only 2.7 percent of black tenant farmers in

FIG. 5.1. The deteriorating mature New South Alabama Black Belt landscape between Montgomery and Camden in 1939. The landowner's dwelling is in the background, and several second-generation, one-room tenant cabins with windows that have shutters but no glass are visible. Marion Post Wolcott, FSA Collection, Library of Congress

the seven Southeastern cotton states lived in log dwellings, while 97 percent lived in frame ones. Like the slave dwellings they replaced, frame tenant houses had unsealed walls that were only one plank thick. Floors had cracks between the boards.[6] In part, the poor condition of dwellings demonstrated inertia of past beliefs about houses and disease prevention. However, characteristics promoted in the 1850s as benefits that would prevent disease, including unsealed walls and floors, by the 1930s were considered major structural defects.

Plantation tenant houses of the 1930s had certain features that indicated their evolution beyond basic slave dwellings. Unlike first-generation houses of the post–Civil War era, some of which had a detached kitchen, many second-generation houses had appended "lean-tos," an addition with a sloping roof attached to the rear (figs. 5.2, 5.3). Lean-tos converted a single pen into a two-room dwelling and a double pen into a three- or four-room house, depending on whether the addition extended halfway or all the way across the rear. Porches were also added across the front of single- and double-pen dwellings. Except for the coldest part of winter, the porch,

FIG. 5.2. The vanishing New South plantation landscape. An abandoned tenant house on the former Nolan plantation, Morgan County, Georgia, in 1965. The three-room dwelling is an example of the second generation of housing occupied by tenant farmers in the Southern plantation regions following emancipation. The house is constructed of unfinished planks and was built between 1900 and 1920. A lean-to kitchen, half the length of the house, is attached to the rear. The simulation-brick, tar-paper siding was a popular cheap insulation added to tenant houses after the Second World War. Charles S. Aiken

FIG. 5.3. An abandoned four-unit dwelling for black day laborers and Hispanic migrant workers on a plantation in Tunica County, Mississippi, in 1987. The house was constructed about 1940. Except for the front porch and lean-to rooms at the rear, the form of the dwelling is similar to multiunit slave houses built during the last decades of the Old South era. Charles S. Aiken

like the yard, functioned as an additional room where people worked, socialized, and even slept. The "shotgun" was another type of plantation tenant house (fig. 5.4). It was a variation of the double pen with the doors at the gable ends, two or three rooms in a line, and a front porch and occasionally a rear one. Large numbers of shotgun houses were built in the Mississippi Loess Plains and the Yazoo Delta, but they were not as common on the lower Georgia Piedmont and in the Alabama Black Belt. Most plantation tenant houses were unpainted. Eighty percent of the tenant houses in the seven Southeastern states in 1934 were unpainted frame.[7] A common practice on well-managed plantations was to paint all company houses the same color. Some plantations used cheap red or green lead paint, while others continued the antebellum practice of whitewashing houses, interior as well as exterior.

The houses of rural blacks were cold and drafty in winter and full of vermin in the summer. To lessen drafts through the cracks, newspapers and magazine pages, cardboard, and cheap garish wallpaper often were tacked to walls. Fireplaces were not intended to heat rooms comfortably but merely to provide places where people could warm themselves. Because of the danger of houses catching fire from sparks, fires usually were not allowed to burn after adults went to bed, even on the coldest winter nights. By the 1930s, "tin" (galvanized sheet steel) roofs had begun to replace wooden shingle ones. The steel roofs lasted longer and were easier to maintain than wooden shingles, and they did not catch fire from chimney sparks that fell on them. Walls and ceilings of tenant houses were lightly coated with soot from the fireplace and stove, and everything smelled of wood smoke and pork grease. Lack of adequate sanitary facilities was "the most critical problem related to southern rural housing" in the 1930s. Thirty-one percent of the dwellings occupied by black tenant families in the Southeastern cotton states in 1934 had no outdoor or indoor toilet. Fifty-six percent of those in Alabama had none.[8] In the absence of a privy, adults relieved themselves behind bushes and the house; children frequently squatted in the yard.

Shoddy materials, poor construction, and lack of maintenance meant that tenant houses deteriorated rapidly and that little variation in quality of dwellings existed either within or among plantation regions. Describing the Alabama Black Belt in 1934, Charles Johnson wrote, "With the exception of a few homes owned by Negro farmers, the dwellings available for the tenants present a dreary monotony of weather worn cabins. The physical pattern varies but slightly, and there are not enough houses of better model and construction about to incite interest in anything different."[9]

The Percy family's Trail Lake plantation in the Yazoo Delta southwest of Indianola depicted the foremost in tenant housing near the conclusion of the New South era. In 1937, black families occupied forty-nine dwellings on the 3,256-acre plantation. Eight of the dwellings had two rooms, twenty-five three rooms, sixteen four rooms, and one five rooms. The major difference between the houses on Trail Lake and those in Macon County was not architecture, building materials, or construction techniques. It was better maintenance. All Trail Lake houses were built of rough planks using board and batten construction methods, and all were unsealed. But the houses had screened doors and windows and fresh coats of green paint, and repairs were made annually.[10]

Planters' concepts of tenant houses as cheap ephemeral structures and the rapidity with which the dwellings deteriorated are illustrated by three plantations that the Farm Security Administration purchased in 1941 for the Tallahatchie Cooperative Leasing Association, a project in the Yazoo Delta to make small landowners of black tenant farmers. None of the twenty-nine tenant houses on the 1,200-acre Mahoney plantation was more than twenty years old. However, twenty-six were rated "poor" to "fair" by appraisers. The three that were in "good" condition were less than

FIG. 5.4. The mature New South landscape of Greenville, Mississippi, in 1939. This row of shotgun houses in the black residential territory was literally just beyond the railroad tracks. The dwellings lacked electric power and plumbing and are crowded together for reasons other than lack of space. Russell Lee, FSA Collection, Library of Congress

five years old. Little wonder that in describing the Yazoo Delta, which into the 1930s experienced devastating floods, William Faulkner wrote that "no man, millionaire though he be, would build more than a roof and walls to shelter the camping equipment he lived from when he knew that once each ten years or so his house would be flooded to the second storey and all within ruined."[11]

WITHIN AND WITHOUT THE SOMBER VEIL OF COLOR

Although they outnumbered whites across the plantation regions, blacks were members of a subordinate cultural subsystem. Economic and political motives only partly underlay whites' subversion of the optimistic goals of former slaves. It also was based on the belief that blacks were inferior and, therefore, fundamentally different from whites. From the beginning of importation of slaves in the seventeenth century, the fear of black domination prevailed among whites. The horror of slave revolts was superseded after the Civil War by the dread of black political domination. The beliefs American whites held at the end of the New South era concerning blacks had evolved over four centuries. They were the accumulation of decades of European and American ideas about race, class, intellect, morality, work ethics, and hygiene. From the time of initial contacts, Europeans tended to regard black Africans as inherently different. Not only were blacks less advanced technologically than Europeans, but they were significantly different in appearance and culture.[12] Slavery degraded blacks and widened the gulf between the races, and discrimination during the New South era perpetuated it. A vertical social order of cast and class established during slavery survived the Civil War and Reconstruction and prevailed throughout the New South era.[13] The consequences of discrimination, including low educational attainment by blacks, became for whites further evidence of their inferiority. Myrdal argued that such reasoning employed the "principal of cumulation," or the "vicious circle." He explained, "White prejudice and discrimination keep the Negro low in standards of living, health, education, manners and morals. This, in its turn, gives support to white prejudice. White prejudice and Negro standards thus mutually 'cause' each other." Myrdal thought that antiamalgamation underlay whites' belief that blacks were racially inferior. He found that both Northern and Southern whites employed the "anti-amalgamation doctrine" to defend cast distinctions between the races. Unlike members of white minorities, blacks were thought to be unassimilable. "Crossbreeding" was "undesirable" because the offspring was considered "inferior to the 'pure' white stock."[14] To prevent amalgamation, whites devised a set of de facto and de jure codes that segregated the races. W. E. B. Du Bois used the metaphor of a "veil" to characterize the separate, complex black culture that surrounded whites but was largely oblivious to them.[15]

The segregation codes had a rank order. Highest in the order were the ones opposing interracial marriage and sexual intercourse between black males and white females. Next was the social segregation of whites and blacks. Interracial dancing, bathing, eating, drinking, and other similar types of social relations were taboo. The third level of discrimination was segregation of public facilities including schools, hospitals, rest rooms, and railroad and bus station waiting rooms. The fourth order was political disfranchisement, and the fifth was discrimination against blacks by police, judges, and other public officials. The sixth level included discrimination in purchasing land and dwellings, in obtaining jobs, and in public relief and social welfare.[16]

In the South's plantation regions a set of rules governing relationships between blacks and whites began to evolve during the slave era. Some of the rules, including ones that banned interracial marriage and required segregated public schools, were enacted into law, but most were never legally codified. The rules were resilient and versatile. They survived emancipation and Reconstruction, and they were easily adapted to new innovations including the telephone, the automobile, and the movie theater. When telephones began to appear in municipalities and rural areas of the plantation South, even affluent blacks were often denied them. Once blacks obtained telephones, party lines were segregated. Movie theaters usually were constructed with a balcony and often a side entrance for blacks (fig. 5.5). The automobile presented a special challenge as to where a black who was not a chauffeur should ride relative to whites. Most whites did not want to be seen driving an automobile with a black riding in full equality as a passenger on the front seat, much less in the rear seat.

Adults of each race instructed their children in the rules of racial conduct both orally and by example. Cleveland Sellers came to realize the "subtle, but enormously effective, conditioning process" to which he was subjected as a child in a small South Carolina town. The older blacks taught the younger ones not "with words so much as . . . with attitudes and behavior."[17] The idea that blacks were distinct from and inferior to whites was continually reinforced in day-to-day activities. An intricate, excessive politeness evolved to ameliorate and restrain the harsh, rigid caste and class system. Throughout the plantation regions persons addressed one another by first names. Only whites who were not indigenous to a place were consistently called by their last. In addressing whites, blacks were expected to precede the first names of adults, and usually even children, with a title. Although "Mr." was the most common one for men, into the 1950s some older blacks still used "Master" when speaking to white men and boys. White females of all ages, whether married or not, were addressed using the title "Miss." Whites usually addressed blacks by just their first names. A few older blacks were called "Aunt," "Uncle," and "Mammy" by whites who had long known them. That blacks longed for the respect associated with titles is evidenced in their burial grounds. Rural cemeteries across the plantation regions are filled with commercial and homemade grave markers on which "Mr.," "Mrs.," and "Miss" are carved or painstakingly scrawled (fig. 5.6).

Black and white men and black and white women worked together, and black and white children played together. But in accordance with the segre-

FIG. 5.5. The mature New South landscape. Among the infamous photographs illustrating the age of segregation in the plantation South is Marion Post Wolcott's picture of the segregated Crescent movie theater in Belzoni in the Yazoo Delta in October 1939. Blacks purchased their tickets at the box office and then went around to the side of the theater and climbed the outside stairway to the balcony, where they had to sit. A segregated rest room for white men was outside the theater. The door to the right of the one marked "White Men Only" was to the rest room for black men and women. The white women's rest room was inside the theater. FSA Collection, Library of Congress

FIG. 5.6. Grave markers of blacks in New Salem Baptist Church cemetery in Wilkes County, Georgia, in 1992. The respectful titles "Mr.," "Mrs.," and "Miss," denied to blacks in life, could be had in death. Such markers are found in cemeteries across the plantation South. Charles S. Aiken

gation code, blacks and whites never sat down as equals to eat and never attended racially mixed social gatherings. Integrated recreation among men rarely extended beyond hunting, fishing, and shooting craps for modest stakes. White boys and girls, who played daily with the children of black sharecroppers, found that their birthday parties were attended only by white children, some of whom they hardly knew. Segregation extended even to eating utensils. In their cupboards some whites kept a special set of dishes, usually older ones or chipped and cracked ones. These were used only to serve food to blacks, even if the food for the white household was cooked daily by blacks. The unwritten code of discrimination required blacks to enter a white residence by the back door and, except for servants, never to venture from the kitchen to the other rooms unless specifically invited. If a planter had an office in his home, it was accessible to blacks by way of a rear or side door. On narrow rural roads, blacks were expected to pull their wagons and later their motor vehicles over to let those of whites pass. In plantation towns, blacks were supposed to move aside on sidewalks and let whites pass. A widely accepted custom that extended into the 1960s was that on Saturdays whites shopped in country and town stores during the morning and blacks in the afternoon (fig. 5.7). Although blacks could enter stores, banks, and post offices, places such as barber and beauty shops, restaurants, hotels, railway and bus depot waiting rooms, rest rooms, and drinking fountains were segregated.

Slight variations in the segregation codes existed across the plantation regions. At one extreme were the sexual taboos that were not to be violated anywhere. If a black violated a sexual taboo, the penalty frequently was death, but if a white violated one, the most severe penalty usually was ostracism by other whites. At the other extreme was racial etiquette that varied from region to region and even from place to place within a region. In Bolivar County, Mississippi, blacks were waited on throughout stores, but in Greene County, Georgia, and Macon County, Alabama, they were segregated to certain sections of stores. Blacks were allowed to try on gloves in Bolivar County stores but were not accorded this privilege in Greene County. In Coahoma County, Mississippi, blacks were served along with whites at drug store soda fountains, but the practice was not found generally across the plantation South.[18] The lack of more rigid segregation must be comprehended within the context of the small rural insular worlds of

FIG. 5.7. The mature New South plantation landscape. Belzoni, Mississippi, on a Saturday afternoon in October 1939. The period from October 1 through January 1 was the most gratifying of the year for blacks and whites in the plantation regions. Even a mediocre cotton crop meant a few weeks with money to spend while work in the fields decreased and ceased. This photograph places the solitary person in figure 5.5, who symbolizes the isolation of segregation, into the more supportive context of the group that was ostracized. Marion Post Wolcott, FSA Collection, Library of Congress

plantation blacks. A former Hale County, Alabama, black tenant farmer remembered, "Back then . . . there were no cars. There was no place to go to be segregated out of, 'cause you couldn' get away. You ride mules, you go only to the grist mill, haul cotton."[19]

In some planter families, children were taught that because blacks were inferior, they should be treated kindly and with tolerance. In part, this was self-serving, for whites who were fair and tolerant had fewer problems in recruiting and retaining tenants and day laborers. Slavery and paternalism led many whites to view blacks as a people who had to be continually watched over and cared for. Hortense Powdermaker thought that William Alexander Percy viewed Negroes as "happy Pan-like beings living only in the present, fundamentally and mysteriously different from white people."[20] Percy griped about his burden:

> A superabundance of sympathy has always been expended on the Negro, neither undeservedly nor helpfully, but no sympathy whatever, so far as I am aware, has ever been expended on the white man living among Negroes. Yet he, too, is worthy not only of sympathy but of pity, and for many reasons. To live habitually as a superior among inferiors, be the superiority intellectual or economic, is a temptation to dishonesty and hubris, inevitably deteriorating. To live among a people whom, because of their needs, one must in common decency protect and defend is a sore burden in a world where one's own troubles are about all any life can shoulder.[21]

Superficially, the plantation South was a place of informality and seeming affection between the races, especially as discerned by planter families. Throughout the New South era and into the period of the civil rights movement, whites continually wrote of the trust and harmony between the races. One of the most pretentious expressions of this mythical relationship, as perceived by whites, is in Ulrich Phillips' *American Negro Slavery*. Although his topic was slavery, Phillips confessed that his account of plantation homesteads was "drawn mainly from the writer's own observations in post-bellum times" (1877–1918):

> The lives of the whites and the blacks were partly segregate, partly intertwined. If any special link were needed, the children supplied it. The white ones, hardly knowing their mothers from their mammies or their uncles by blood from their "uncles" by courtesy, had the freedom of the kitchen and the cabins, and the black ones were their playmates in the shaded sandy yard the livelong day. Together they

were regaled with folklore in the quarters, with Bible and fairy stories in the "big house," with pastry in the kitchen, with grapes at the scuppernong arbor, with melons at the spring house and with peaches in the orchard. The half-grown boys were likewise almost as undiscriminating among themselves as the dogs with which they chased rabbits by day and 'possums by night. Indeed, when the fork in the road of life was reached, the white youths found something to envy in the freedom of their fellows' feet from the cramping weight of shoes and the freedom of their minds from the restraints of school.[22]

Arthur Raper, a sociologist who studied people of the plantation South more intimately than Phillips and who was not burdened by the racism or nostalgic agrarianism that hounded the historian, did not detect such a joyous relationship among children but, rather, the vast overt chasm of segregation which let the black child know from the time he was old enough to comprehend that he was inferior:

The white child and the Negro child go their separate ways to school and church and graveyard long before they are old enough to wonder why. They are already separated by the assumption of innate difference, already accept the dogmas which underlie cast distinctions. That the white man and the Negro are fundamentally and inevitably and unalterably different will scarcely be questioned so long as the two races go in opposite directions to recite their arithmetic tables, their reading lessons, their creeds about a loving Heavenly Father, their intentions to emulate a Jesus who called no man common or unclean.[23]

The seeming informality and friendliness between the races was actually a facade. Raper thought that the deceptive relationship rested, "to no small degree, upon the Negro's acceptance of a role in which he [was] . . . just an inferior man."[24] Rarely did whites and blacks engage in candid dialogue even on a one-on-one basis. Conversations between the races were confined to crops, animals, weather, illness, hunting, fishing, and routine field and household matters but almost never ventured into critical social, economic, and political topics. Complaints by members of one race against those of the other often were conveyed obliquely in the guise of stories. Beneath the superficiality lay a constant tension. Blacks griped to other blacks about whites, and whites complained to one another about blacks, not only in conventional contexts but even in formal intellectual ones. In his caustic review of *American Negro Slavery*, W. E. B. Du Bois charged that

"Mr Phillips . . . had done little with any Negro sources, most of which he regards as 'of dubious value'. . . . The result is a readable book but one curiously incomplete and unfortunately biased. The Negro as a responsible human being has no place in the book. . . . Nowhere is there any adequate conception of 'darkies,' 'niggers' and 'negroes' (words liberally used throughout the book) as making a living mass of humanity with all the usual human reactions."[25]

Despite a few brave outcries by persons such as Du Bois, the subordination of blacks was seldom questioned. Large portions of William Alexander Percy's autobiography are chatty renditions of beliefs that whites shared about blacks. "Apparently there is something peculiarly Negroid in the Negro's attitude toward, and aptitude for, crimes of violence," wrote Percy. "He seems to have resisted, except on the surface, our ethics and to have rejected our standards. Murder, thieving, lying, violence—I sometimes suspect the Negro doesn't regard these as crimes or sins, or even as regrettable occurrences." When John Dollard visited the Yazoo Delta to study Indianola for his classic *Cast and Class in a Southern Town* (1937), Will Percy asked him "for some explanation of the Negro's propensity to crimes of violence." Dollard replied that his tentative opinion was "that the frustrated hatred of the Negro for the white man, because of the frustration, [was] transferred to his own kind for fulfillment." Even though Dollard's answer was radical for the time, Percy would have disagreed with any answer that did not support his own closed-mind view. Not only did Dollard's answer fail to support Percy's definitive ideas about why blacks committed crimes, but it also attacked the cherished planter myth that harmony existed between the races. Percy's response was to scoff at the "oracle" from Yale.[26] Percy viewed himself as the paragon of the paternalistic planter and had what Jonathan Daniels called "the feudal lord's love for his serfs." Despite his frequent visits from his home in Greenville to Trail Lake, Mr. Will was detached from the plantation's people and the reality of its day-by-day operation. When LeRoy, Will Percy's nephew and Walker Percy's brother, assumed management of Trail Lake in 1939, he found that it was poorly supervised. A common form of punishment for aggressive and unruly black tenants, which LeRoy stopped, was to beat them with a two-by-four.[27]

Charles Johnson found that across the South at the time of the depression black youths were race conscious and believed that blacks were treated unfairly and suppressed economically by whites. Yet without a full comprehension of segregation, young black children thought that the discrimination they encountered was merely due to the failure of whites to understand what exclusion meant. "We don't have nothin'," a child told Johnson. "The white people don't even think about colored children. You can't go to the movies because there ain't none. We don't have no parks or nothin' at all." Segregation was more fully comprehended by adolescents. They knew that economic oppression, not just social inconvenience, was a product of discrimination. A young day laborer in the Yazoo Delta expressed what Johnson believed was a common perception of rural blacks as to economic relationships between the races: "Now the poor white folks is them that ain't got nothing, but thinks they is somebody. They like to pick on Negroes. They cheat them, 'buse them and meddle with them. An aristocratic white person is more decent about it. "Course they cheat Negroes too, but they's nice about the way they do it. Some rich people are just poor white folks too. Money don't change them none, its all in the way they act."[28]

By the 1940s, a few blacks in the plantation regions were beginning to convey to whites complaints about discrimination and segregation more forthrightly. On opening night of the new Greenland movie theater in Greensboro, Georgia, in January 1941, whites packed the 500 seats on the main floor and blacks the 160 in the balcony. During the dedication ceremonies, one of the white speakers stated that "what he had planned to say had already been said, and his predicament reminded him of 'an old nigger who . . . '" From the balcony someone shouted, "Callin' us 'niggers' again! Let's get out of here, and stay out!" For a few weeks local blacks boycotted the theater.[29] The incident revealed several things about race relations in the plantation regions at the end of the New South era. Southern rural blacks were beginning to emerge from reticence to challenge the status quo in race relations. However, they were not yet brave enough to do so openly. When whites sought to discover who initiated the walkout and planned the boycott, all blacks questioned denied being a leader or knowing who the leaders were. Most whites were amazed at what happened in the theater, for the incident was contrary to the dogma that blacks were well treated and were content. The segregated balcony of the new theater was a symbol of white paternalism and accommodation, and blacks' patronage on opening night evidenced their satisfaction with the racial system. The incident also challenged white authority. "Since when can't we call them 'nigger?' Why, we've done it all our lives!"[30]

Despite the sensation of the walkout, the boycott, and the challenge of white authority in a small plantation town, Greensboro's weekly reported nothing about the racial incident. The story of the opening of the new Greenland was upbeat and promotional.[31] The ignoring of blacks and racial incidents by local newspapers in plantation regions was another form of segregation. Even 25 years later as the segregated world came crashing down around them, most local newspaper editors failed to give their readers coverage of, much less insight into, racial reality.

Throughout the New South era some enlightened Southerners such as Thomas Dudley, Andrew Sledd, John Spencer Bassett, Will Alexander, and Lillian Smith comprehended the problems of blacks.[32] A few provincial whites came to understand the situation of blacks, even if to a limited degree, once they were made to see it. In 1934, Arthur Raper hired two young college-educated planters from Greene County to assist the state Federal Economic Relief Administration with tenant relocation projects on the lower Georgia Piedmont. When the two planters returned to the field office on the evening of the first day, they told Raper that "they didn't know there were any people anywhere who had so near nothing as the ones they had been investigating." Raper was especially pleased that the two were so enlightened by what they saw. He wrote of the incident: "I knew that on their plantations back in Greene County there were families on levels no higher, and in some instances I know lower. I knew then that these white college graduates of my own age [thirty-five] simply never had really seen their local situation. I was fascinated to know why. Did the pencil and a schedule make the difference? Or was it the fact of being away from home, and being paid to do it?"[33]

PUBLIC SCHOOLS: SPECIAL ICONS OF SEGREGATION

At the time of the depression, the contrast between the plantation South's dual segregated public education systems was at a peak (figs. 5.8, 5.9, 5.10). During the 1929–30 school year, Mississippi spent only 14 percent of its $10.4 million instructional budget on blacks even though they constituted 57 percent of the state's school-age children. In Georgia in 1934, $34.46 was spent on each white child but only $6.73 on each black one. The lowest per capita expenditures for black schools and the greatest inequities in funding were in plantation counties. In Greene and Macon Counties, the ratios were $30.87 to $1.84 and $47.10 to $1.82. Tunica County, Mississippi, spent $36.58 on each white child but only $1.96 on each black one.[34] A few plantation counties did not even have a public school system until well into the twentieth century. The little progress that rural blacks made in obtaining better facilities was largely through philanthropy. Between 1919 and 1931, the Julius Rosenwald Fund aided in the construction of new school buildings for blacks throughout the South, including 555 in Mississippi. Wilcox County, Alabama, did not begin to assume responsibility for the

FIG. 5.8. The mature New South plantation landscape. Children leaving the school for blacks at Gee's Bend, Alabama, in 1937. The school was housed in a church and had only one teacher for grades one through four. Arthur Rothstein, FSA Collection, Library of Congress

education of black children until the 1930s. Even in 1965, more than 40 percent of Wilcox County's black students were taught in schools owned and operated by the United Presbyterian Church of North America.[35] Not only did whites restrict funding for black education, but they also impeded the development of high schools for blacks. Early in the twentieth century, a movement to consolidate one- and two-teacher white elementary schools and to establish high schools swept the rural South.[36] Across the plantation regions, new rural and small-town consolidated schools for whites, many of which had twelve grades, began to replace modest elementary schools. By 1930, more than 90 percent of Mississippi's white children were enrolled in 975 consolidated schools, but there were only 16 consolidated schools for blacks in the state.[37]

What before consolidation of white schools had been statistical differences in funding of segregated systems by the 1930s had become overt landscape expressions. Rural plantation areas, such as eastern Tate County, Mississippi, with its Thyatira Consolidated, had twelve-year schools for whites, but numerous one- and two-room schools that did not go beyond the sixth grade still served blacks (map 5.1). The large new twelve-year

FIG. 5.9. The interior of a school at Gee's Bend, Alabama, in 1939. The room is freshly painted, and the tables, chairs, and books are better than those found at the time in most rural schools for blacks. According to Marion Post Wolcott, who took the photograph, the pupils are in the first grade, and the purpose is to show the extremes in ages. The tall young man in the background, who may appear to be a teacher, is actually a pupil learning to read. The scene is partly staged as a component of the Farm Security Administration's publicity for the agency's all-black Gee's Bend Cooperative Association farm-ownership project. Figures 5.8 and 5.9 are segments of before and after for the FSA promotion. FSA Collection, Library of Congress

school plants with their electric lights, central heat, gymnasiums, and orange buses contrasted sharply with the one-room shacks most black children attended (fig. 5.10). "One . . . can pick out the Negro schools by the appearance of their exterior while riding along the road at fifty miles an hour," wrote Raper (fig. 5.8).[38]

More than two decades after the consolidation movement began, almost all Southern high schools for blacks were located in cities.[39] Efforts by rural Mississippi blacks to secure high schools met with little success until after the Second World War and impending school desegregation. A black community leader in Carroll County, Mississippi, remembered the struggle to get a high school during the 1930s. "We tried to get a high school in Carroll County. They did not want us to have one. We had little schools all over the county. We finally got the [state-supported regional] agricultural high school in 1938 at the small community school in Summerfield. There was no transportation. The small school building was turned into a girls' dormitory; the attic of the school building was made [into a dormitory] for boys. The only way a child could go to high school was to go there and board."[40]

The facilities for black public schools were abysmal. Not only were most schools for blacks small and in disrepair, but many of the buildings were not even publicly owned. Black churches and lodge halls frequently housed them. Desks either were made of rough planks or were used ones from white

FIG. 5.10. The mature New South plantation landscape. Children's day at the Greene County Fair, Greensboro, Georgia, in October 1941. At the beginning of the Second World War, most counties in the plantation regions had consolidated twelve- and thirteen-grade schools and buses for whites. Most buses were small by late-twentieth-century standards, and they had wooden bodies. The worn, muddy tires and broken window of the bus indicate that although white schools were much better funded than black ones, by national standards the plantation regions lagged considerably in support for public education. The dress code of a planter-dominated society dictated Sunday dresses for girls and high-heeled shoes for teachers and chaperons, even at a county fair. Boys escaped with a Sunday shirt and a new pair of overalls, and they could go barefooted. Girls were expected to wear shoes. Fairs were segregated. One day of a three- or four-day fair was usually reserved for blacks. Jack Delano, FSA Collection, Library of Congress

schools. Most textbooks were tattered secondhand ones, and frequently there were not enough books for each child to have one. Schools, especially those housed in churches and lodge halls, did not have slate blackboards, bulletin boards, maps, and other basic equipment. Mount Pleasant in the Natchez district of Mississippi was typical of hundreds of black rural schools scattered across plantation regions a decade prior to the United States Supreme Court's 1954 landmark decision in *Brown v. the Board of Education of Topeka, Kansas*. Anne Moody remembered that the school to which she had to walk 4 miles was "a little one-room rotten wood building" located next to Mount Pleasant Baptist Church. "There were about fifteen of us who went there. We sat on big wooden benches just like the ones in the church, pulled close to the heater. But we were cold all day."[41]

In addition to the wretched facilities, the teachers were as poorly educated as they were paid. Most had not even completed high school. Only 7 percent of the 5,921 black teachers in Mississippi in 1930 had two or more

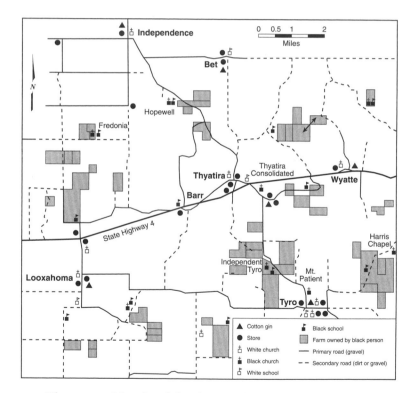

MAP 5.1. The mature New South landscape of eastern Tate County, Mississippi, in 1939. Thyatira Consolidated was a twelve-grade school created in 1920. Blacks had no local high school until after the Second World War (Tate County Property Assessment Map, 1939; Tate County Highway Map, 1939).

years of college. Thirty-seven of the 52 black teachers in Greene County, Georgia, in 1934 did not qualify for state certificates; only 3 had attended college. The average annual salary for the county's white teachers was $600; for black teachers it was $142. The median salary for a black elementary teacher in Mississippi in 1930 was $226, whereas a white one was paid $720.[42]

The contrasts that existed between black and white schools at the end of the Second World War are illustrated by Coahoma County in the Yazoo Delta. White students attended seven consolidated rural schools, five of which had twelve grades. Clarksdale, the county's largest municipality, had a public white school with grades kindergarten through twelve and numerous amenities, including a swimming pool. Although more than 70 percent of the Coahoma County's 48,333 inhabitants were black, no attempt had been made by county officials to consolidate elementary schools and create a public high school system for blacks.[43] Seventy-two black schools, forty-two of which had only one teacher, were scattered across Coahoma County plantations. Six of the rural black schools had junior high grades (grades seven and eight), and Clarksdale had a junior high school for blacks. However, there was only one high school for blacks in the county, a state regional facility located in the countryside 6 miles north of Clarksdale just beyond the border of the giant King and Anderson plantation. Public school buses were provided for white children but not for black ones. The schools for whites in Coahoma County had eight-month sessions, but those for blacks had six-month terms. All of the black schools and two of the white ones had split sessions. Both of the white schools with split sessions were located in areas of the county which had large numbers of white tenant farmers. Schools with split sessions opened in July and closed at the end of August, the beginning of the cotton harvest. The schools reopened in November and concluded the term in early spring just before cotton was planted.[44]

The abysmal state of schools had contributed to a fatalistic attitude among some blacks. An eighteen-year-old Coahoma County girl, who had married at fifteen, believed, "Education is all right for those who wants it for somethin'. . . . If you gonna farm all your life, though, you jus' wastin' your time goin' to school. And that's what most of "em gonna do. Dey gonna stay right here on some of these plantations all their life."[45] Despite the despicable buildings, the poorly trained teachers, the dearth of textbooks and teaching materials, the split school terms, the lack of public transportation, and the discrimination in funding, most blacks in the plantation regions remained optimistic toward education. Their confidence had not been quelled by discrimination or even by overt oppression. They still believed that the only hope to escape the plantation system lay in education.

SPATIAL CONSEQUENCES OF SEGREGATION

Although the dispersed settlement pattern of the New South era plantation landscape was one in which dwellings of blacks and whites were intermixed, the rural areas exhibited subtle and overt forms of residential segregation. One type of segregation prevailed in spatial patterns of tenants. Many planters either preferred black tenants to white ones or refused to employ tenants of both races on the same landholding. In the 1930s, 53 percent of the Southern cotton plantations were worked exclusively by black tenants, but only 5 percent were farmed solely by white renters. The white tenants on many of the plantations with racially mixed renter groups were actually children and other relatives of landowners.[46] On biracial plantations, an attempt was made to assign whites and blacks to houses and fields that were segregated from one another. Houses and fields, however, were not permanently segregated, for planters did not have tenant houses exclusively for blacks and whites. A white family might succeed a black one in a tenant shack.[47] If a plantation had a racially mixed group of sharecroppers and share tenants, the higher-status share tenants were usually whites and the sharecroppers were blacks.

Subtle spatial segregation on plantations included the deliberate placement of tenant houses, churches, and schools of blacks in relation to owners' and managers' dwellings (map 2.4). Although tenant houses might be within sight of owners' and managers' dwellings, they usually were not built close to them. Plantation schools and churches were even farther away than the dwellings of blacks, for the traffic and noise produced by black churches were undesirable to whites. It was not by chance that the small tract David Barrow gave to former slaves for a church and school was on a remote corner of Sylls Fork Plantation away from the headquarters house (map 1.4). For blacks, churches not only were symbols of freedom within the plantation system, but they also were the foci of the black cultural subsystem. Literally and figuratively a plantation had two headquarters, that of the white planter, the big house, and that of the black tenant farmers, the church. The spatial segregation and differences in the visual condition of the two symbolized the great economic, social, and political gulf that separated the races.

In the plantation regions, the social order carried over from the Old South era into the New was more a vertical class system than a spatial one.

But in the nation's growing, crowded cities blacks and whites were not as interdependent as in the rural South, and they became increasingly anonymous. A new spatial social order that segregated the races was advocated by many white urbanites. The nation's first laws that required residential segregation began to be passed by cities on the northern border of the South which had large influxes of blacks as the great rural-to-urban migration commenced at the beginning of the twentieth century. Baltimore and Louisville were among the first to pass ordinances that provided for all-black and all-white street blocks. Several Southern cities, including Norfolk, Richmond, Atlanta, and New Orleans, soon enacted residential segregation laws.[48] The newly organized National Association for the Advancement of Colored People fought the urban residential segregation movement, pursuing opposition to the Louisville ordinance all the way to the United States Supreme Court. In *Buchanan v. Warley* in 1917, the Court struck down the Louisville ordinance, holding that occupancy, purchase, and sale of property could not be inhibited by a state or municipality on the basis of race. Although property rights more than civil rights were paramount in the Court's ruling, the decision was a significant one. It ended the movement to legislate residential segregation almost as soon as it began.

Despite the lack of laws that decreed segregation and a few black servants scattered throughout affluent white neighborhoods, municipalities in plantation regions were rigidly segregated residentially. Blacks were not free to buy and sell property or live anyplace they wished. The black residential areas were actually territories to which blacks were confined by economic, social, and political constraints.[49] Most of the dwellings were similar to houses found on plantations. They were cheaply constructed two- and three-room dwellings built close together on small lots. Many were owned by white businessmen and planters, some of whom charged no rent to employees. The infrastructure of the black neighborhoods was minimal. There were no water or sewer lines, and even into the 1950s many of the dwellings did not have electric power, even though the municipalities had had electricity for many years.

Boundaries between the white and black residential territories were sharp. There was no transition zone with racial intermixture. The social and economic distinctions between the two populations were reinforced by physical barriers. Railroad tracks, the business district, a stream, or a major street usually formed part of the boundary between the white and black residential sections. Expansion of the black residential territory did not occur by blacks freely moving into the white residential area, even if vacant dwellings existed. If the demand for black housing grew, municipal leaders enlarged existing black residential territories or established new ones. Because all blacks, including the few affluent ones, were confined to the same territories, a significant range existed in the types and conditions of dwellings.

The black residential territories were both planned and unplanned. In older towns and cities some of the unplanned black residential areas had evolved from settlements created by former slaves who flocked to local urban areas during and immediately after the Civil War. Oxford, Mississippi, which is in the north Mississippi Loess Plains and was William Faulkner's model for his fictional Jefferson, had three major unplanned black residential areas: Freedman Town, the Hollow, and Saint Paul.[50] These "free towns," "liberias," or "quarters," as whites often labeled them, were filled with small shacks. Streets were narrow, unpaved trails, and there was no adequate provision for the disposal of garbage and sewage. Shallow wells and cisterns supplied water. Some municipalities eventually annexed unplanned black residential areas, but at the time of the depression and the Second World War many of them were still outside municipal boundaries. Municipalities created after the Civil War differed from antebellum ones in the spatial characteristics of black residential areas. New South municipalities, which were more carefully planned as segregated, biracial communities, had fewer blacks dispersed across white residential areas. Because the Yazoo Delta was settled primarily after the Civil War, most of its municipalities had planned black residential areas. Many of the blacks in Tunica, Mississippi, were concentrated in Tunica Colored Subdivision, which was created in 1912 just outside the municipal boundary.

Two classic studies of segregation and race relations in small Southern towns near the end of the New South era are John Dollard's *Caste and Class in a Southern Town* (1937) and Hortense Powdermaker's *After Freedom: A Cultural Study in the Deep South* (1939). Dollard and Powdermaker used the fictitious names Southerntown and Cottonville. Both, however, are actually the same place, Indianola, Mississippi, a Yazoo Delta town that was chartered in 1886 and had a spatially planned racial residential pattern characteristic of New South municipalities (map 5.2). Powdermaker and Dollard identified three principal spatial components of Indianola, white and black residential areas and the business district.[51] Powdermaker, who was not a Southerner, found the business district and the white residential area "typical of the small American town." The residences of Indianola's whites were "in orderly rows, set among well-groomed lawns and interrupted here and there by a church: Baptist, Methodist, Presbyterian, Catholic." She thought the "general effect [was] of trimness, newness, and middle-class suburbia, with the distinct impression of homogeneity that the term implies," and was especially taken by the "total absence" of white "slums." For her, the "most striking" geographical features of Indianola were "the segregation of Negro and white dwellings, and the contrast between the two sections."[52] Both Powdermaker and Dollard commented on the shoddy condition of the houses in the black residential section, the poor infrastructure, and the curious intermixture of a few substantial houses among the shacks. Dollard thought that the houses in the black residential area were "small and cheap." Further, "A well-cropped lawn is a rarity, as is a well-built house. At night one sees kerosene lamps gleaming through the windows; in a few of the houses, electric bulbs. Only two paved streets traverse this area where fifteen hundred people live. . . . Behind the houses the frequent privies testify to the fact that these people are not wholly included in our modern technology, as are those on the other side of the railroad tracks."[53]

Although for the majority of rural blacks the former slave's goal of

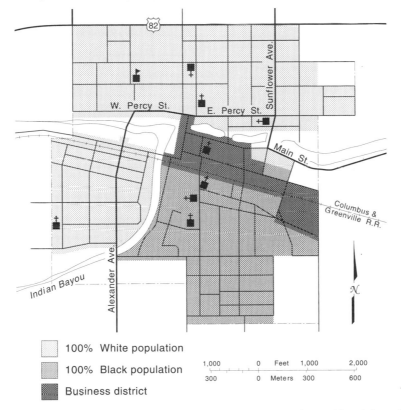

100% White population

100% Black population

Business district

1,000   0   Feet 1,000   2,000

300   0   Meters 300   600

MAP 5.2. Indianola, Mississippi, in 1940. The houses in the residential territories for blacks were on small lots, and some were crowded together like the dwellings in figure 5.4 (Aiken 1990). Courtesy of the Association of American Geographers

owning a farm was still unfulfilled at the time of the depression, across the plantation regions small numbers of blacks had managed to become land-owners. In Greene County, Georgia, the acreage owned by blacks increased from 412 in 1870 to more than 11,000 in 1934. Blacks acquired land in various ways. After the Civil War a few planters sold land to their former slaves. Among the examples of this practice is the sale of Hurricane Planta-tion in the Natchez district by Joseph Davis, Jefferson Davis's brother, to the freedmen who had farmed it as slaves.[54] In eastern Tate County, Mis-sissippi, the land owned by blacks around Harris Chapel was sold to former slaves by a white planter family (map 5.1). Planters who fathered mulatto children occasionally gave or willed land to their illegitimate offspring. The three largest landholdings at Independent Tyro Church were willed by a planter to his illegitimate mulatto children. In the 1930s and early 1940s, a few blacks acquired farms under New Deal programs of the Farm Security Administration. The four black landholdings at Hopewell Church in Tate County were obtained under the FSA's Tenant Purchase program.

Three geographical traits distinguished farms owned by blacks. First, as illustrated by eastern Tate County, the landholdings of blacks usually oc-curred in clusters of two or more, and the clusters often consisted of properties of extended families (map 5.1). Only rarely did single isolated tracts exist. Second, the landholdings of blacks were small by standards of plantation regions. Although the largest landholding in eastern Tate County in 1940 contained more than 400 acres, half of the properties of blacks were less than 100 acres. In Greene County, Georgia, 85 of the 211 black landowners in 1934 had less than 20 acres.[55] The third characteristic of land owned by blacks was that it frequently had undesirable physical traits. The landholdings had inferior soils or were hilly and badly eroded or poorly drained. Land that was physically undesirable was sold to blacks if no white buyers could be found. The landholdings of blacks in eastern Tate County between Thyatira and Wyatte had suffered severe sheet erosion and gullying.[56]

The only serious attempt to segregate blacks and whites in ownership of farmland was a campaign launched in 1913 by Clarence Poe, editor of the *Progressive Farmer*, a widely read Southern agricultural journal. Poe was quickly thwarted in his effort to create a spatial apartheid similar to that which was concocted in South Africa, for planters and affluent ur-banites, as well as influential blacks, opposed land segregation legislation. Not only would the laws hamper the purchase and sale of property, but in time they might also be used to dictate where black and white tenant farmers could live.[57]

Concentrations of black landholdings were enclaves within plantation-dominated areas. Churches usually were the foci of the enclaves, and they occasionally had small stores and other small businesses owned by blacks (map 5.1, fig. 5.11). Although socially autonomous, the enclaves were not independent economically, for black landowners needed the plantation infrastructure. Black enclaves usually did not have furnish stores, ginneries, or other enterprises that were integral to cotton farming. Blacks' lack of political and economic power meant that the enclaves did not receive the same degree of public services as other areas of a county. Unless the land-holdings of blacks happened to be on major public thoroughfares, roads usually were not well maintained. A vivid description of a black enclave in the 1940s is given by William Faulkner through the eyes of Mink Snopes, a poor white sharecropper:

> This type of road was familiar out of his long-ago tenant-farmer freedom . . . : a Negro road, a road marked with many wheels and traced with cotton wisps, yet dirt, not even gravel, since the people who lived on and used it had neither the voting power to compel nor the money to persuade the Beat supervisor to do more than scrap and grade it twice a year.
>
> So what he found was not only what he was hunting for but what he had expected: a weathered paintless dog-trot cabin enclosed and backed by a ramshackle of also-paintless weathered fences and out-houses—barns, cribs, sheds—on a rise of ground above a creek-bottom cotton patch.[58]

John Lewis, who rose to prominence as a young leader in the civil rights movement and was among the first Southern blacks elected to Congress in its wake, remembered a similar place, Dunn's Chapel, his boyhood home in the Alabama Black Belt where his father owned a farm. "We didn't have electric lights, we didn't have indoor plumbing. . . . We had unpaved roads, and for many years the county refused to pave the major road. They paved it up to where the black section of the county started."[59]

Despite the poverty, the enclaves were of immeasurable importance to blacks, for they were the type of places that J. B. Jackson termed "autonomous spaces." Not only did the areas permit greater freedom than plantations, but according to Woodson, "Negroes' best opportunities for struggling upward" in the plantation South were "in settlements and towns largely restricted to the black population."[60] The significance of the enclaves is revealed in the names, often symbolic ones, by which blacks referred to them, usually the same names as the churches that were foci:

FIG. 5.11. A mature New South black community on the lower Georgia Piedmont near Madison in 1939. Churches frequently were the foci of such enclaves. The two dwellings, which are similar to that in figure 5.2, were constructed between 1900 and 1920 and are representative of the second generation of housing occupied by blacks after the Civil War. Although utility lines passed through the enclave, neither the church nor the houses had electricity or telephones. The road through it was unimproved. Marion Post Wolcott, FSA Collection, Library of Congress

Independent Tyro, Hopewell, Fredonia (map 5.1). To whites, however, the enclaves were nonplaces. Whites rarely ventured into them and knew little about them. Names of large plantations and plantation hamlets were shown on county road maps and United States topographic maps, but black enclaves were identified only if they were incorporated towns or large rural hamlets.

A few black enclaves achieved prominence beyond the local area. Because they were concentrations of property owners and well-educated people, black colleges and all-black towns were chief among the enclaves. Black colleges were scattered across the plantation regions and included a few state-supported schools as well as a number of private ones. Tuskegee Normal and Industrial Institute, which Booker T. Washington founded in 1881, was one of the most famous centers of black higher education in the rural South. The school was located in Macon County at the edge of Tuskegee in the eastern Alabama Black Belt. Mound Bayou was the most prominent of the black municipalities in the plantation regions. Located in the midst of an area of small farms that had been carved from the Yazoo Delta wilderness on land sold to black settlers by the Louisville, New Orleans, and Texas Railroad after the Civil War, Mound Bayou was known throughout black America and was a source of pride in African American heritage. Although the population was less than one thousand at the time of the depression, Mound Bayou had grocery and general merchandise stores, a number of small shops, a cotton gin, a cotton oil mill, and a bank.[61] The Springfield Community in Hancock County on the lower Georgia Piedmont was also among the few rural black communities that attracted national attention. In 1934, the Association for Advancement of Negro Country Life at Springfield owned a community center, a canning factory, a store, a health center, a boys' camp, and a dairy farm.[62]

Near the end of the depression, the cooperative and tenant purchase programs of the Farm Security Administration created a number of new black rural enclaves, including the Hopewell community in eastern Tate County, Mississippi (map 5.1). A few of the larger cooperative projects, such as Gee's Bend in the Alabama Black Belt and Mileston Farms and Tallahatchie Farms in the Yazoo Delta, briefly achieved national recognition before they slipped into oblivion with the coming of the Second World War. Gee's Bend was a 7,500-acre plantation located in Wilcox County on a large meander of the Alabama River. About 1900, the owners left the plantation and ceased to manage it. The blacks, many of whom were the descendants of the plantation's slaves, continued to farm as cash tenants and made a "mildly profitable living." In 1937, the FSA purchased the plantation. The original plan was for approximately a hundred 80-acre farms on the original 7,500-acre tract and an adjoining 2,763-acre one. A cooperative, which owned a store, cotton gin, and gristmill, was organized. After an initial period of renting, the tenants were given the opportunity to buy their farms.[63] The Mileston Farms project was created on a 9,350-acre tract in the Delta section of Holmes County, Mississippi, in 1938. Regarded by local whites as land with poor agricultural potential, most of the tract was still in forest and was poorly drained. By 1943, Mileston had 106 small farms owned by former black tenants and a cooperative, which owned a store and a cotton gin. Tallahatchie Farms, established nearby in Tallahatchie County in 1941, was among the last of the cooperative associations created by the FSA before the agency's political adversaries terminated the programs designed to make landowners of tenant farmers. The FSA purchased three plantations, which were divided into tracts and sold to black tenant farmers.[64]

THE POLITICAL REALITY

Disfranchisement was a major victory for affluent whites in the plantation regions. Not only were blacks eliminated as a political threat, but the political power of the white yeoman farmers was also diminished. In the 1880s, approximately 64 percent of the Southern adult males voted; by the first decade of the twentieth century the percentage had declined to about 30 percent.[65] Enfranchisement of females did not substantially increase the percentage of Southern adults who voted. From 1920 through 1944 fewer than 20 percent of adults in Alabama, Georgia, and Mississippi voted in presidential elections. Significant variations, however, existed across the South in regional patterns of political participation. Concern for white dominance and sectional power resulted in high levels of political participation by whites in the plantation regions. In every Southern state the counties with the largest black populations were ones with high percentages of white voters. The relative number of voters was smaller among whites in counties in which they had numerical majority. "It is the whites of the black belts who have the deepest and most immediate concern about the maintenance of white supremacy," wrote Key. "Moreover, it is generally in these counties that large-scale plantation . . . agriculture prevails. Here are located most of the large agricultural operators who supervise the work of many tenants, sharecroppers, and laborers, most of whom are colored."[66]

The almost total absence of political participation by blacks and the relatively low participation by poor whites allowed affluent whites in the plantation regions to exercise political power far beyond what their numbers indicated. Under the banner of the Democratic Party, affluent whites in the plantation regions allied themselves with conservative business and industrial groups in the cities, which was a reemergence of the old pre–Civil War alliance that constituted the core of the Whig Party. Through the alliance, a small number of whites were able to subordinate entire states, and even the entire South, to their control. Throughout the New South era, as during the Old South one, plantation regions had a significance that was greater than their sizes or populations dictated. Spatially, into the post–Second World War period the plantation regions were "the hard core of the political South" and "a skeleton holding together the South." In 1949, Key wrote: "The black belts make up only a small part of the area of the South and . . . account for an even smaller part of the white population of the

FIG. 5.12. An unidentified woman, who was born a slave, hoeing cotton in 1937, near the end of the New South era. An ephemeral annual scene now long gone from the Southern landscape, the long dress, gloves, bandanna, hat, smoothly worn hoe handle, and intent gaze toward the minutest blade of grass and stem of a weed denote a professional with years of toil in cotton fields. Dorothea Lange, FSA Collection, Library of Congress

South. Yet if the politics of the South revolves around any single theme, it is that of the role of the black belts. Although the whites of the black belts are few in number, their unity and their political skill have enabled them to run a shoestring into decisive power at critical junctures in southern political history."[67]

The relative importance of the plantation regions was maintained by several devices in which racism was central. Although disfranchisement eliminated the political threat of blacks in Southern states, blacks had to be kept repressed. To sustain the South as a white man's land, no political division could be permitted. The plantation South became a one-party area. One-party rule changed general elections into trivial exercises, for the real election was the Democratic Party primary. Between 1896 and 1915 the

Democratic Party in every Southern state adopted a form of the direct primary to nominate candidates. By controlling the Democratic primary, the affluent whites of the plantation regions, with their affluent urban allies, could control a state. The "fundamental explanation" of Southern politics was a geographical one. The plantation region whites "succeeded in imposing their will on their states and thereby presented a solid regional front in national politics on the race issue."[68] Even with the surge of population in Southern cities during and after the Second World War, rural plantation whites were able to maintain significant political power by refusing to redistrict state legislatures and congressional districts fairly. Only after the Supreme Court's *Baker v. Carr* decision in 1962 did the plantation whites lose the rest of the political strength that they had maintained long after the majority of a state's population had shifted to cities.

The New South era was relatively brief, lasting less than a century in the cotton plantation regions. So short was the period that even persons who were born slaves lived through it (fig. 5.12). But near the conclusion of the New South, on the eve of the Second World War, restoration of voting rights to blacks and political realignment that redefined the political situation of the plantation regions were a quarter century in the future. Many changes, including sweeping spatial ones, occurred in the plantation regions during a tumultuous period that witnessed the end of the New South era and the transition to the Modern South. Like the Civil War and Reconstruction, the Second World War, the civil rights movement, and the War on Poverty shook the plantation regions to their social, political, economic, and geographical foundations.

THE
IMPACT OF THE CIVIL RIGHTS MOVEMENT
1954 TO 1998

II

CHAPTER SIX

Mobilization

✻ ✻ ✻

We speak now against the day when our Southern people
who will resist to the last these inevitable changes in social relations,
will, when they have been forced to accept what they at one time
might have accepted with dignity and goodwill, will say,
"Why didn't someone tell us this before? Tell us this in time."

William Faulkner

There is not a single legal reason for [the Brown *decision]. . . .*
It is immoral, infamous and illegal and . . . deserves no respect from us.
. . . We repudiate this decision; we will not abide by it. . . .
If a suit is filed to integrate the schools of this community,
don't obey it. Your future depends on your ability to resist.

United States Senator James O. Eastland

Superficially, the civil rights movement seems a happening, a haphazard undertaking filled with spontaneity and without significant thought given to design and strategy. The various national and local civil rights organizations differed in objectives, philosophy, and composition. Leaders of the different groups often were at odds with one another. The movement also progressed through different stages. Immediate objectives seemed to shift from desegregation of public schools, to registration of black voters, to desegregation of all public facilities, and then back to voter registration and school desegregation. Events often appeared to be more in control of civil rights leaders than the leaders in control of events.

The civil rights movement, however, involved much planned strategy. Even the Student Nonviolent Coordinating Committee (SNCC), an aggressive and impetuous organization, held numerous planning sessions and kept detailed minutes and records of meetings and projects. Strategy employed by civil rights organizations had important geographical dimensions. Because greater media attention was given to urban events, the basic geographical tactic was to focus on cities more than on small towns and rural areas. In addition, spatial concentrations of urban blacks made mass demonstrations easy to organize and created large confrontations between demonstrators and the police for the media to photograph, film, and televise. Not only were most rural blacks dispersed across the landscape, but they were less affluent than urban blacks. Under the plantation system, whites could closely control rural blacks by traditional economic, social, and political constraints. Although the primary emphasis was on metropolitan areas, the nonmetropolitan South was hardly overlooked. The plantation regions were historically the political nuclei of the Deep South and the cores of black oppression. The primary achievements of the civil rights movement, the 1964 Civil Rights Act and the 1965 Voting Rights Act, were not realized until the campaigns were extended into these regions. A number of critical events of the civil rights movement, including the 1964 Mississippi Freedom Summer and the 1965 march across the Alabama Black Belt from Selma to Montgomery, focused on plantation regions.

Although not intentionally part of a planned strategy, by necessity, the civil rights movement proceeded spatially from the metropolitan periphery to the nonmetropolitan plantation cores.

THE THREE STAGES OF THE CIVIL RIGHTS ERA

Distinct stages can be identified at both national and regional scales in the civil rights era, which extended from 1954 to 1972. Nationally, the stages were mobilization, confrontation, and acquiescence. The mobilization stage extended from the United States Supreme Court's *Brown* decision on school desegregation in 1954 into 1960. The *Brown* ruling encouraged a growing resentment among blacks of segregation and restrictions on voting. The Second World War produced swift and profound economic and social changes in the United States. Movement of blacks from the rural South to cities accelerated during the war. For many blacks, their world suddenly expanded from that of a small insular plantation society, the medieval nation, to a multifarious urban one free of many of the restrictions that they had known all their lives. In addition, the discovery of Nazi and Japanese concentration and death camps toward the war's end exposed racism taken to its ultimate.

During the first stage, older civil rights organizations expanded and new ones emerged. The National Association for the Advancement of Colored People (NAACP) was the principal organization that had fought for black rights during the twentieth century. It was the NAACP that took the five court cases that collectively became known as *Brown v. the Board of Education of Topeka, Kansas* from local courts all the way to the United States Supreme Court. Although the organization's headquarters was in New York, the NAACP's largest membership was in Southern cities. After the 1954 landmark *Brown* decision, the NAACP began an effort to increase its membership in the South, especially in small towns and rural areas. In 1957, the Congress of Racial Equality (CORE), which at the time was led primarily by white intellectuals, began to expand from its northern base into the South, primarily into South Carolina. The two most important new civil rights organizations were the Southern Christian Leadership Conference (SCLC), which was formed in 1957 with Martin Luther King Jr. as its president, and the Student Nonviolent Coordinating Committee, which emerged in 1960 at the beginning of the student "sit-in" movement.[1] The mobilization stage also witnessed the development of new methods and tactics that became weapons employed during the confrontational stage. The major strategy of the NAACP was legal. In addition to the *Brown* case, it scored Supreme Court victories in the other landmark decisions, among them the 1917 decree in *Buchanan v. Warley*, which ended residential segregation by law, and the 1944 ruling in *Smith v. Allwright*, which abolished the all-white primary election.[2] Despite the legal victories, by the late 1950s the NAACP's conservative approach had produced little impact on the segregated society and the voting restrictions that confronted blacks in the South.

The 1955–56 Montgomery, Alabama, bus boycott, which was led by Martin Luther King Jr., the new youthful minister of Dexter Avenue Baptist Church, was the most significant event during the mobilization stage. The SCLC grew out of the bus boycott. The success in Montgomery proved that direct action by blacks could bring about reforms in segregation. It also demonstrated the effectiveness of the boycott as a weapon and nonviolence as a tactic. The SCLC was linked to the NAACP in that most of its founders were members of the older organization, but the new organization advocated more aggressive efforts in desegregation and voter registration. At the organizational meeting in Atlanta in January 1957, discussion focused on three topics: future action that should be taken by blacks, the attitudinal framework for the actions, and the role that the federal government might play in reform. The SCLC also was closely tied to the black church. Not only were ministers the initial leaders of the SCLC, but their churches served as the local organizational devices.[3]

The Student Nonviolent Coordination Committee emerged at the end of the mobilization phase of the civil rights movement and helped launch the confrontation phase. SNCC grew out of the lunch counter "sit-ins" that were initiated by college students in Greensboro, North Carolina, in February 1960 and quickly spread to Nashville and other cities. The tactic was for students to take seats at lunch counters, ask to be waited on, and refuse to leave if they were not served. SNCC was founded at a conference held April 16–18 at Shaw University in Raleigh, North Carolina. The conference was organized by Ella Baker, executive director of the SCLC. Baker knew that the sit-ins were "concerned with something much bigger than a hamburger or even a giant-sized coke." The primary goal of the students was "to rid America of the scourge of racial discrimination—not only at lunch counters but in every aspect of life."[4] Baker's reason for holding the conference was not for the SCLC to take control of the sit-in movement but to try to establish more leadership and planning among the inexperienced students.[5] SNCC was a more impatient and aggressive organization than the SCLC and its leadership younger and more impulsive. Although initially the two organizations were closely linked and engaged in joint ventures, strong disagreements erupted over strategy and methods. The disagreements and jealousies eventually led SNCC leaders to charge that the SCLC had intruded into places to which it had established claim.

The vortex of the civil rights era was the confrontation phase. It began with the "sit-ins" in 1960 and extended into 1966 and was marked, not just by clashes between civil rights workers and local authorities, but also by ones between the federal and state governments. In 1961, CORE sponsored "freedom rides" to desegregate public facilities in bus stations across the South. SNCC initiated black voter registration campaigns in nonmetropolitan Mississippi and Georgia in the summer and fall of 1961 and moved into the Alabama Black Belt in the fall of 1962. For the summer of 1964, SNCC created and largely administered a massive assault by Northern college students on Mississippi to register black voters and test desegregation. The SCLC launched major campaigns in Albany, Georgia, in 1962, Birmingham in 1963, St. Augustine in 1964, and the Alabama Black Belt in 1965.

The persistent effort of civil rights organizations to involve the federal government more directly in the civil rights struggle began to be realized more fully during the Kennedy and Johnson administrations. Major federal-state confrontations occurred in September–October 1962 over the enrollment of a black student, James Meredith, at the University of Mississippi; in May 1963 over the effort of Governor George Wallace to prevent enrollment of blacks at the University of Alabama; and in January–March 1965 over Wallace's attempts to prevent a march by civil rights demonstrators across the Alabama Black Belt from Selma to Montgomery. The confrontation phase of the civil rights movement was marked at times by extreme violence. The brutality heaped on nonviolent workers and demonstrators spurred the federal government to action and contributed to the passage of the 1964 Civil Rights Act and the 1965 Voting Rights Act, which quickly and dramatically changed the South forever.

After the passage of the 1965 Voting Rights Act, the civil rights era entered the third phase, acquiescence. Not only did the effort to secure civil rights for blacks shift from the South to the North and West, but within the South desegregation of public schools and other public facilities and registration of black voters spread rapidly to the remotest rural places. This period, which in certain respects continued into the 1990s, was distinguished by the shift of the primary effort to secure civil rights for blacks back to the federal courts. To actually achieve desegregation of public facilities under the 1964 Civil Rights Act and effective political enfranchisement under the 1965 Voting Rights Act, numerous local court suits were filed. Some of the national, regional, and local organizations that emerged during the first two phases of the civil rights movement successfully made the transition to the third phase. Others, such as SNCC, floundered and vanished.

In retrospect, many of the demands of blacks during the civil rights era seem modest requests that they be granted basic American liberties. But for whites in the plantation South the civil rights movement challenged and sought to destroy the vertical social order, an anachronism that prevailed after it had begun to disappear from most segments of American life.

RESISTANCE TO BLACKS' GROWING DEMANDS
FOR CIVIL RIGHTS

To most white Southerners *Brown v. the Board of Education* appeared to be an obscure decision by nine old men who had no comprehension of blacks, segregation, and race relations in the South, especially in the plantation regions. "What the Supreme Court needs to do is to spend about thirty days in Georgia, Alabama, Louisiana, Mississippi, or East Texas and associate intimately with the average negro family," wrote Yale-educated Mississippi circuit judge Thomas P. Brady shortly after the *Brown* decision. "It should do this in August, mind you, and then the Supreme Court would have a passing understanding of the negro and his problems."[6] *Brown v. the Board of Education*, however, was a much larger and more important case than its name indicated. *Brown* was actually five separate cases that were argued as a body before the Supreme Court in December 1952 and reargued in December 1953. The other four suits were from Clarendon County, South Carolina; Prince Edward County, Virginia; Delaware; and the District of Columbia. *Briggs v. Elliott*, the oldest of the five cases, originated in Clarendon County, South Carolina, a stereotypical cotton plantation county on the Coastal Plain. Most of Clarendon's blacks were members of poverty-stricken tenant families on cotton and tobacco plantations. The county's dual school system was typical of those across the rural plantation South. In 1949–50, Clarendon's 6,531 black pupils were dispersed in sixty-one schools, most of which had only one or two teachers and half of which were housed in ramshackle buildings. The 2,375 white pupils attended twelve well-maintained schools. Only $43 was spent on every black child, while $179 was spent on every white one. The court suit initiated by black parents in Clarendon County did not seek school desegregation, only equity in funding. The case began in 1947 when parents asked the school board and the State Board of Education to provide public buses for black children as well as for white ones.[7]

As the five school cases moved from the lower courts to the nation's highest, Deep South judges were among those who heard and issued opinions. The judges were hardly homogeneous in their view of racial segregation and its effects on blacks. *Briggs v. Elliott* went through the United States Court for the Eastern District of South Carolina in Charleston on its way from Clarendon County to the Supreme Court. Of the three judges who heard the case in district court, one, George Bell Timmerman, was a rigid segregationist, and another, John J. Parker, was a capable justice who strictly followed the law. The third justice, Waties Waring, in lineage was the embodiment of the mythical Southern aristocracy. A member of a prosperous South Carolina family who had settled in Charleston in the seventeenth century, Waring was the descendant of slaveholders. As a child he was cared for by a mammy who had been born a slave. But over the years Waring came to realize that the discriminatory treatment of blacks was wrong, and he wrote a dissenting opinion to that of Parker and Timmerman, who denied the plaintiff's plea.[8]

In the first conference of the Supreme Court justices following the 1952 argument of the *Brown* case, the vote of one of the nine was clearly expressed and never in doubt, that of Hugo Black, the lone member from the Deep South. Black was born in Clay County, Alabama, and rose to prominence as a Birmingham lawyer and judge. To promote his political career he was briefly a member of the Ku Klux Klan. As a member of the United States Senate, Black had a mixed record on human rights. His appointment to the Supreme Court, however, freed him from his segregationist constituency, and increasingly he severed philosophical ties with his past. Black needed no lawyers and scholarly witnesses to explain to him the purpose of school segregation. He was an authority who knew both its intents and its results. School segregation was to subjugate blacks. Not only did Black believe that legally segregation was a violation of the Fourteenth Amendment to the United States Constitution, but he also thought that it was morally wrong. Although Black knew that difficulties lay ahead for the people of the South, he asked the other justices to join him in abolishing school segregation.[9]

The conflicting attitudes and opinions among Deep South judges reflected differing racial attitudes held by Southerners in general. In the 1950s, Martin Luther King Jr. was among the few who approached a full comprehension of the attitudinal differences among white Southerners with regard to race and equal rights for blacks. King believed that there were three basic types of white Southerners. First was a "minority" who "would use almost any means, including physical violence, to preserve segregation." At the opposite extreme was a "growing minority" whose members were "working courageously and conscientiously to implement the law of the land." This group believed "in the morality as well as the constitutionality of integration." Most white Southerners belonged to the third group. "Through tradition and custom" members of this group "sincerely believe[d] in segregation," but also stood "on the side of law and order." King optimistically thought that there were "in the white South millions of people of good will whose voices" were "yet unheard" and "whose course" was "yet unclear." They were silent because of "fear of social, political, and economic reprisals." He called upon members of the third group "to grid their courage, to speak out, to offer the leadership that . . . [was] needed."[10]

A study of Guilford County, North Carolina (Greensboro), directed by Melvin Tumin in the late 1950s, also concluded that white Southerners could be divided into three categories. Twenty-five percent were "hard-bitten, hard-core" segregationists. One-third were moderates who were willing to compromise on desegregation and liberals who supported it. The "large majority," about 40 percent of the population, were "habituated to traditional patterns of segregation." The ultimate success of massive resistance to desegregation depended upon the support of the last group, and many initially supported the outspoken defenders of white supremacy. Its members became a target of a "neobourbon" campaign designed to prevent enforcement of the Supreme Court's rulings. Ultimately, as Numan

Bartley observed, most members of the major group "were unwilling in the end to tear apart the fabric of southern society and commit the region to anarchy in defense of segregation." In the second Reconstruction, as in the first, the majority of bourbon and neobourbon political activists lived in the plantation South.[11] Into the 1960s and the critical 1962 "one man, one vote" decision by the Supreme Court in *Baker v. Carr*, which Jimmy Carter labeled a "turning point," the plantation regions were still critical parts of what Key called the "skeleton holding the South together."[12]

Martin Luther King Jr. also understood that acceptance and resistance by white Southerners of equal rights for blacks had a geographical dimension (map 6.1). As the civil rights movement progressed, King refined his geographical comprehension and strategically employed it at the local scale. In 1958, however, he delineated only the broad-scale spatial pattern of the three groups of Southerners. "Geographically speaking," there were three Souths. Oklahoma, Kentucky, Missouri, West Virginia, Delaware, and Kansas, the border states together with the District of Columbia, were the "South of compliance." Tennessee, Texas, North Carolina, Arkansas, and Florida composed the "wait-and-see South." The "South of resistance" was Georgia, Alabama, Mississippi, Louisiana, South Carolina, and Virginia,[13] the core states of the plantation South. Most of the defiant opposi-

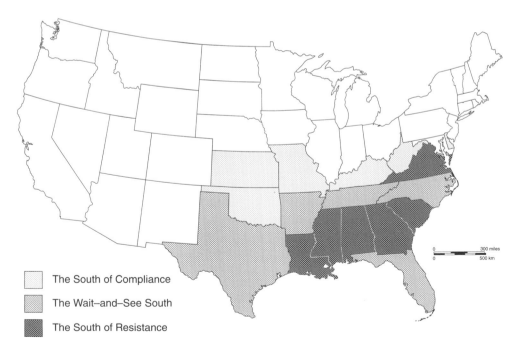

The South of Compliance

The Wait–and–See South

The South of Resistance

MAP 6.1. Martin Luther King Jr.'s three Souths (based on description in King 1958, 201)

tion to equal rights for blacks and most of the major confrontations during the civil rights era occurred in these six states.

One form of accelerating white resistance to the civil rights movement was organizations that fought desegregation and black voter registration. Among them was an old organization that was reborn, the Ku Klux Klan, and a new one that emerged, the Citizens' Councils.[14] Historically, there have been three Ku Klux Klans rather than one continuous organization. The original Ku Klux Klan emerged immediately after the Civil War. On Christmas Eve, 1865, six young former Confederate soldiers in Pulaski, Tennessee, amused themselves by inventing a mystic organization with absurd names, rules, and rituals. The name "kuklux" was corrupted from *kyklos*, the Greek word for circle. They soon discovered that their prank of riding through Pulaski and the surrounding countryside wearing white robes made from bedsheets especially frightened recently freed slaves. The organization spread, and amusing tricks played on blacks soon grew into intimidation by masked night riders.

Particular Southern leaders, including former Confederate general Nathan Bedford Forrest, saw the Klan as an organization with potential to fight political reconstruction of the South by Northerners and former slaves. At a meeting in Nashville in 1867, a plan was developed to create a structured organization. The name was changed from Kuklux Klan to Ku Klux Klan, and Forrest was elected Grand Wizard, the head of the organization. As the Klan grew in membership, it also grew more violent. Some of its members engaged in beatings, mutilations, and murder. In 1869, Forrest ordered the Klan to disband, but it persisted. Congress in 1870–71 passed three civil rights laws known as the Ku Klux Klan acts, which included giving the president the authority to use the military to enforce provisions of the Fourteenth Amendment to the Constitution. The new federal laws, growth of Southern opinion against the organization, and the overthrow of Reconstruction state governments contributed to the demise of the first Klan.[15]

The second Klan, the Knights of the Ku Klux Klan, was organized in 1915 as a fraternal society. William Joseph Simmons, an Atlanta promoter, got the idea for a new Klan from D. W. Griffith's epic motion picture *Birth of a Nation*, which romanticized the original Ku Klux Klan. Simmons borrowed from the Klan's constitution, bylaws, and ritual and promoted the new one as a patriotic, benevolent, and fraternal society. Simmons' motivation was purely monetary. As Imperial Wizard he received a percentage of each ten-dollar initiation fee, a monthly per capita tax, and profits from the sale of Klan paraphernalia. The organization flourished in the racism, xenophobia, and social upheaval that gripped the nation after the First

World War. By 1924, the Knights of the Ku Klux Klan was a national organization with a membership between three and five million. Indiana and Ohio had the most members. Many persons joined the Klan as they would a social club. Others were attracted because they thought membership would aid them in their jobs or, like Hugo Black, in local politics.

Like the original Klan, the Knights of the Ku Klux Klan grew increasingly violent. The organization sought not only to maintain white supremacy but also to fight Catholics, Jews, and other perceived menaces to America. The Klan also grew more political. In 1922, Louisiana governor John M. Parker asked President Harding for the federal government's help in breaking the Klan's grip on his state. The most common form of Klan violence was kidnapping, usually whites, who were then whipped and sometimes tarred and feathered. Bombings and murders also were committed by members of the Klan. The growing violence, together with anti-Klan laws and failure of the organization to achieve what it promised, contributed to the demise of the second Klan. By 1928, membership had declined to less than 300,000, and by 1934, it was less than 10,000. Members dissolved the bankrupt Knights of the Ku Klux Klan in 1944 at a national convention in Atlanta. Although small Klan klaverns continued to operate for several years in various parts of the nation, primarily in the Southern states, by 1952 most were extinct or dormant.[16]

A third Klan arose after the 1954 *Brown* decision. Unlike the two earlier Klans, which were largely united under one leader, the third Klan was composed of more than fifteen separate and independent factions, each of which had its own Imperial Wizard or Grand Dragon. In the mid-1960s, membership ranged from only 5 in the Mississippi Knights of the Ku Klux Klan to more than 15,000 in the United Klans of America. Combined membership of the third Klan was small compared with the memberships of the original and the second Klans. Despite such boasts as "These men are TENS of THOUSANDS STRONG," total membership in 1967 was estimated to be only 16,810 in 714 local klaverns, which included 56 ladies' auxiliaries.[17] Unlike the old Knights of the Ku Klux Klan, which sought to be a national organization, the third klan was located primarily in the South. North Carolina, Alabama, Georgia, Virginia, and Mississippi were the only states with 1,000 or more members. Almost half of the membership was located in North Carolina. United Klans of America, the largest group, had its headquarters in Tuscaloosa, Alabama, a small city just north of the Black Belt. Using white supremacy as a slogan, United Klans exploited the growing intensity of the civil rights movement to recruit members. By the mid-1960s, the organization had 556 klaverns in eighteen states.[18]

The membership of the third Klan was composed primarily of persons from lower socioeconomic ranks of Southern whites. Some members were farmers, but most were poorly educated laborers, tradesmen, and factory workers who resided in small towns and cities. A former officer of the United Klans of America described the "rank-and-file membership" as persons "drawn from uneducated elements of the population who never attained the social status they would [have liked] to achieve." Such people were "seeking comradeship but would not be at home in civic clubs such as the Rotary." Those who joined the Klan were also motivated by "hatred for Negroes, Jews and Catholics."[19]

The first Citizens' Council was organized in July 1954, only two months after the *Brown* decision. Ironically, the place was Indianola, Mississippi, the small Yazoo Delta municipality that two decades earlier both Hortense Powdermaker and John Dollard had selected as a typical Deep South town. The organizers and initial leaders were Indianola's white male civic leaders, including the mayor, the city attorney, and several prominent business-men. Herman Moore, an Indianola banker, was elected the first president, and Arthur B. Clark Jr., a Harvard-educated attorney, was chosen as the first vice president. Members of the Citizens' Council denounced secret ritual and violence and claimed that they wished to fight school and other types of desegregation openly and peacefully through legal methods to prevent amalgamation of the races. The Citizens' Council quickly spread to other Mississippi municipalities and to other states, even to ones outside the South, including California. The first Alabama Citizens' Council was organized with six hundred charter members at Selma in Dallas County in November 1954. Within a month additional councils were established in four other Black Belt counties. Local chapters formed state organizations, and in April 1956 representatives from eleven Southern states met in New Orleans and created the Citizens' Councils of America with headquarters in Jackson, Mississippi.[20]

The legal methods the Citizens' Councils espoused to fight segregation were based on outdated concepts of states' rights, including interposition of a state government between its citizens and the federal government and the power of a state government to nullify federal laws. The defeat of the Southern Confederacy should have put to rest forever the idea that state authority supersedes federal. The Citizens' Councils' literature and rallies were designed to appeal to a deep-seated feeling among many white South-erners of oppression and discrimination by the federal government. The Confederate battle flag was used by the organization more as a symbol of scornful defiance of federal authority than one of fanatic white racism. The fear of Communism and Red-baiting, which were so prevalent in the United States during the 1950s, were blatantly used by the Citizens' Councils. The cover of Thomas Brady's *Black Monday*, which was published and widely circulated by the Citizens' Councils, depicted the fist of Communism breaking through a states' rights banner with the caption "Lest We Forget Integration of the races and the destruction of White America is one of Communistic Russia's objectives."[21]

Just as black churches served as a network for the diffusion of the SCLC, so white male civic clubs, including the Lions and Rotary, facilitated the spread of Citizens' Councils. In 1955, the Mississippi Citizens' Councils claimed 75,000 members, and in 1957, those in Alabama had 80,000. Nationally, the membership in 1957 was estimated to be more than 225,000.[22] The Citizens' Council by no means appealed to all white Southerners. Poor whites were not aggressively recruited, and many of the better-educated and more affluent whites, especially persons who had traveled outside the South, found the Citizens' Council an extreme, shallow organization that espoused obsolete strategy and unrealistic goals. "Despite the efforts and claims of the Council," wrote Hodding Carter III in 1959, "many of Mississippi's college-educated youth do not accept all tenets of the segregationist faith."[23] A Northerner who studied the Citizens' Council at the peak of its influence found that "not every Southerner like[d] the Councils. . . . Some planters were always disdainful of them. . . . Despite their pretensions, the Councils [were] essentially a middle-class movement with a sprinkling of the top society—managers, not owners; merchants, not industrialists."[24] Hortense Powdermaker thought that whites in Indianola, where the Citizens' Council movement originated, were "almost entirely on the lower or middle rungs of the middle class." According to her, Will Percy, who considered himself an "aristocrat," had "no respect for Indianola's white citizens and actively disliked them."[25] A few Southerners who were regarded as moderate on racial issues were members of the Citizens' Council. A Hale County, Alabama, businessman, who characterized himself as an Adlai "Stevenson Democrat" in the age of Lyndon Johnson, saw no conflict between his council membership and his temperate stance that local biracial committees should administer federal antipoverty programs.[26] Able only to impede rather than to stop the rolling tide of change and not capable of fulfilling what it promised about states' rights and federal authority, the Citizens' Councils began to decline in the late 1950s. By 1961, the organization had essentially disappeared from the border South, and it was in demise in the Deep South.[27]

To Southern blacks and non-Southerners who participated in the civil rights movement, there was little difference between the Ku Klux Klan and the Citizens' Council. Both sought to deny blacks their rights; the Citizens' Council was but a white collar version of the Klan. In much of the interpretative literature of the civil rights movement, either the Ku Klux Klan and the Citizens' Council are lumped together and essentially treated as one organization, or attention is devoted to the Klan to the almost total exclusion of the council. Not only did the Ku Klux Klan and the Citizens' Council appeal to different socioeconomic groups, but they exhibited different spatial patterns in their memberships. In addition, despite the notoriety that the Klan received from infamous bombings and brutal murders, the Citizens' Council was more significant and effective in impeding desegregation and black voter registration. It was the council more than the Klan that helped to precipitate the critical events that led to crucial federal-state confrontations and to passage of major civil rights legislation. The Ku Klux Klan was a red herring during the civil rights era and continues to serve as one in interpretation of that period.[28]

Although the original Ku Klux Klan was organized primarily by members of the planter class, the second and third Klans were rejected by most members of the group. In the Deep South, the second Klan became a populist organization that the poorer whites sought to employ politically against planters and businessmen. Reflecting the planter perspective, Will Percy thought that the Knights of the Ku Klux Klan was a "monstrosity" that "was not even a bastard of the old organization which General Forrest had headed and disbanded."[29] The third Klan also found little support in the plantation regions. Although a few klaverns were in the plantation regions, most were in the surrounding white yeoman farmer areas and in blue-collar factory towns and cities. In Alabama, most Klan members resided north of the Black Belt, with large memberships in the industrial cities of Birmingham, Bessemer, and Anniston. The greatest Klan membership in Mississippi was in the southeastern part of the state, focused on Laurel, Hattiesburg, and Meridian.

Although the Ku Klux Klan was a vicious organization, its actual power and influence were much less than they might seem. The combined membership of all of the Klan groups was relatively small, and the organization had no prominent leaders and no substantive political power. In an opinion issued by the New Orleans Federal District Court against the original Knights of the Ku Klux Klan, Judge Minor Wisdom succinctly summarized the members' status: "None of the defendant klansmen is a leader in his community. As a group they do not appear to be representative of a cross-section of the community. Instead they appear to be ignorant bullies, callous of harm they know they are doing and lacking in sufficient understanding to comprehend the chasm between their own twisted Konstitution and the noble character of liberties under the law that is the American Constitution."[30]

Media reports and even official government documents of the 1960s often present the Klan as a major impediment to civil rights. In addition, some politicians exaggerated the importance of the Klan and their ability to control the organization as a means of insuring reelection. Powerful Mississippi senator James Eastland, who owned a plantation in Sunflower County and was an avid supporter of the Citizens' Councils, with FBI assistance, promoted the myth that he was personally responsible for keeping the Klan out of the Delta and northwestern Mississippi. The mayor of the small town of Jackson on the lower Georgia Piedmont "made it a point" to tell that although "the 'Klan' was very strong" in Butts County, he "had refused . . . a permit" for the organization "to hold a meeting at the . . . courthouse." However, United Klans of America Impala No. 42 was the only klavern in Butts County, and it had only a few members.[31] Knowledgeable civil rights workers advised rural blacks not to let the Klan panic them. In 1965, The "Benton County [Mississippi] Freedom Train" newsletter warned, "People are getting a little scarey [*sic*] because of the cross burnings and the Klan meeting. As a result, there have been many rumors about people being shot at. None of them so far are true. So please, let's not be like the boy who cried wolf."[32]

Most of the infamous acts of violence attributed to the Klan during the civil rights era either occurred in areas peripheral to the plantation regions or were carried out in the plantation regions by nonindigenous Klan members. The tragic highlight of the 1964 Mississippi Freedom Summer, the carefully planned murder of Andrew Goodman, Michael Schwerner, and James Chaney by members of the White Knights of the Ku Klux Klan, occurred in Neshoba County. Neshoba is a white yeoman farmer county located between the Loess Plains and Black Belt plantation regions. In 1960, only 22 percent of the county's population was black. The murder of Viola Liuzzo, a middle-aged white woman from Michigan who was helping transport civil rights activists participating in the 1965 march from Selma to Montgomery, happened in Lowndes County in the heart of the Alabama Black Belt. The three members of the United Klans of America who were indicted for her murder were from Bessemer, a Birmingham industrial suburb. Liuzzo's murder was a spontaneous act of violence rather than an execution that was carefully planned by the Klan.[33]

At the state level, the Citizens' Councils had their greatest membership and political influence in Mississippi, Alabama, Louisiana, South Carolina, Virginia, and Georgia, five of the six states that Martin Luther King Jr. identified as "the South of resistance." Reflecting its Indianola origin, the Citizens' Council was primarily a small-town and small-city organization with its greatest strength in plantation regions. In Mississippi, the council was prominent in the Yazoo Delta and in parts of the Loess Plains. The national headquarters of the Citizens' Councils of America was in Jackson, but the headquarters of the Mississippi Citizens' Council was in Greenwood, a small city in the Yazoo Delta. Most of the councils' Alabama membership was in the Black Belt and in the Birmingham area. The more militant and openly racist North Alabama Citizens' Councils, led by Asa Carter (a.k.a. Forrest Carter), an articulate troublemaker, split with the Black Belt–centered Citizens' Councils of Alabama, which had its headquarters at Selma.[34]

Although politically influential in Virginia, South Carolina, and Louisiana, the Citizens' Councils achieved their greatest political power in Alabama and Mississippi. James T. Folsom, who was elected the governor of Alabama in 1955, generally ignored the Citizens' Council, as did John Sparkman and Lester Hill, the state's two senators. The councils' political influence in Alabama increased under John M. Patterson, who followed Folsom as governor, and peaked under George C. Wallace, who was governor from 1962 to 1966, during the confrontation period of the civil rights era.[35] In Georgia, the Citizens' Council was not very strong, primarily because of the organization in 1955 of the politically powerful States' Rights Council of Georgia. However, the southwestern part of the state was a stronghold of the Citizens' Councils. Jimmy Carter was the only white man in Plains who did not join, despite considerable pressure and economic intimidation.[36]

Shortly after they were organized, the Mississippi Citizens' Councils initiated a strategy to control the state's government. Members of the council were behind the passage of two major amendments to the Mississippi constitution in 1954. One placed further restrictions on an already tightly controlled voter registration to insure that blacks remained disfranchised. The other authorized the state legislature to abolish public schools should they be desegregated. Although the Citizens' Councils quickly gained significant influence in the Mississippi legislature, Governors Hugh L. White and James P. Coleman, who held office during the 1950s, were not favorably disposed toward the organization. But Ross Barnett, who assumed office in 1960, was a member of the council, having joined the organization shortly after it was created. Barnett quickly proved that he was the right type of man at the right time, for civil rights leaders could not have cast a better person to play the segregationist foil. Under Barnett the Citizens' Councils gained control of the State Sovereignty Commission, a propaganda and investigative agency created by the legislature. Coleman used the commission primarily for public relations, but under Barnett it began to achieve its police state potential. Not only did the agency keep persons regarded as subversives under surveillance, but through it state funds were channeled to the Mississippi Citizens' Councils.[37]

The Citizens' Councils helped to create in the plantation South an atmosphere that encouraged and condoned white defiance of federal authority, intimidation and punishment of blacks who challenged and violated established codes of conduct, and retaliation against whites who wished to comply with desegregation and voter registration mandates. Thomas Brady's *Black Monday* specifically advocated that punitive economic measures be taken against blacks who sought change in the status quo. "Over ninety-five percent of the negroes of the South are employed by and work for white men or corporations controlled by white men," declared Brady. "A great many negro employees will be discharged, and, though it will work a grave hardship on many white employers, still it is better 'if our right eye offend us to pluck it out.' "[38]

The Citizens' Councils publicly denounced violence, and McMillen in his study of the organization found "no tangible evidence which suggests that it engaged in, or even overtly encouraged criminal acts."[39] A community leader in the Moundville area of Hale County, Alabama, actually believed that the Citizens' Council was a "force for moderation": "If it hadn't been for the Council there would have been a lot of killing around here. The Councils channel into rallies and meetings red-neck energies that might be expended on night-riding."[40] But if the Citizens' Councils did not officially condone acts of violence, the organization also did not attempt to prevent them. Some educated, white professionals and businessmen, the type that the Citizens' Councils boasted made up their membership, shared the view of brutality toward blacks expressed by J. J. Breland, a defense attorney in the 1955 Emmett Till murder case. Roy Bryant and J. W. Milam were charged with the brutal murder of Till, a fourteen-year-old black youth from Chicago who allegedly violated sexual taboo in a remark to Bryant's wife at a rural store in the Yazoo Delta. After an all-white Tallahatchie County jury acquitted the pair, Bryant and Milam sold their confession of the murder to William Bradford Huie, who published the story in *Look*. According to Huie, Breland, a Princeton graduate who acted as an instigator in the deal, wanted the Till slaying publicized as a warning.

"There ain't gonna be no integration. There ain't gonna be no nigger votin'. And the sooner everybody in this country realizes it the better." Breland considered Bryant and Milam low-class "rednecks." "But hell, we've got to have our Milams to fight our wars and keep the niggahs in line."[41]

The murder of Charles Evers, a Mississippi state NAACP official, by Byron de la Beckwith, a Greenwood fertilizer salesman, on June 11, 1963, left doubt in the minds of many about the relationship of the Citizens' Councils to racial violence. Beckwith, a member of the Mississippi Citizens' Council, had the organization's support in both of two mistrials for Evers' murder, and a White Citizens' Legal Fund was created to pay his trial expenses.[42] Years later, in 1990, the Evers case was reopened by Mississippi officials. Beckwith, who had moved to Tennessee, was reindicted by a Jackson grand jury, tried, and found guilty of Evers' murder.[43]

The efforts of the Citizens' Councils to dictate a racial dogma in Mississippi led James Silver, a history professor at the University of Mississippi, to pronounce the state a "closed society" that "imposes on all its people acceptance of and obedience to an official orthodoxy."[44] Even after the power of the Citizens' Councils began to wane, the white leadership in many plantation communities remained split between ultraconservatives, who were members of the council, and more moderate whites, who wanted to improve race relations. Following congressional passage of the 1964 Civil Rights Act and the 1965 Voting Rights Act, a federal investigator found among whites of Clarksdale, a small planter city in the Yazoo Delta, "a considerable element," including the president of the First National City Bank, who were "anxious to change the status quo." The moderate white leaders, however, wished to initiate improvements in race relations "without a head-on confrontation with the more conservative elements of the power structure" led by the local head of the Citizens' Council and president of the Bank of Clarksdale.[45]

THE FEDERAL GOVERNMENT AND BLACK RIGHTS

The mobilization period of the civil rights era (1954–60) corresponded with the Eisenhower administration. The period was one of significant progress in civil rights, especially if it is compared with the preceding eight decades. By the end of the Eisenhower administration, discrimination was officially terminated in the military and federal agencies; federal courts had mandated an end to segregation in the nation's public schools and in the public transportation system of Montgomery, Alabama; and the Interstate Commerce Commission had ordered an end to discrimination by interstate carriers. By sending troops to Little Rock during the 1957 school integration crisis, Eisenhower set a precedent for federal intervention to enforce court-ordered desegregation.[46] The first major civil rights bills since Reconstruction were passed by Congress in 1957 and in 1960. The Civil Rights Division within the Justice Department and the Commission on Civil Rights were among the provisions of the 1957 act.

Despite the accomplishments, the Eisenhower administration is often regarded as one of limited progress in civil rights for blacks. This interpretation exists, in part, because the administration displayed more hesitancy and political caution than the Kennedy and Johnson administrations.[47] Massive resistance of Southern whites to desegregation was aided by powerful Southern congressmen and senators. Shortly after John Kennedy assumed office in January 1961, he nominated his brother, Robert Kennedy, for attorney general. During the congressional hearings, Mississippi senator James Eastland, who chaired the Senate Judiciary Committee and had been an early counselor to Robert Kennedy, said to him, "Did you know that . . . [your predecessor] never brought a civil rights case in the state of Mississippi?"[48]

Although the Eisenhower administration took the initial steps to enforce federal court desegregation decrees, it attempted to avoid deep involvement in civil rights issues. Administration officials placed much emphasis on voting rights. Voting was the principal thrust of both the 1957 and 1960 Civil Rights Acts. The assumption was that once blacks were reenfranchised, they could use the ballot for self-protection and to achieve equal rights. The emphasis on voting resulted in the initial aggressive thrust of the civil rights movement into the South's plantation regions, the hard cores of white resistance, being taken by the Justice Department rather than by black rights organizations. Under the 1957 Civil Rights Act the Justice Department could bring two types of court suits. Section 1971(a) allowed the attorney general to file civil suits on behalf of persons who were deprived of the right to vote on the basis of race or color. Section 1971(b) gave the attorney general the authority for legal action against persons who interfered with the rights of others to vote by intimidation or coercion.[49]

During the closing months of the Eisenhower administration, from September 1958 to January 1961, only six voting rights cases were filed (map 6.2).[50] Although it might appear that the strategy was one of cautious delay, the newly created Civil Rights Division of the Justice Department faced several significant problems in enforcing the 1957 Civil Rights Act. Justice Department officials were handicapped in initiating and prosecuting suits under Section 1971(b) by a 1944 ruling of the United States Supreme Court in the *Screws* case.[51] For this reason, between 1958 and the passage of the Voting Rights Act in 1965, Justice Department voting litigation emphasized suits brought under Section 1971(a) of the 1957 Civil Rights Act. Even suits under Section 1971(a) were not easy to litigate. Geographically, the discovery of voter discrimination and litigation to eliminate it could not be conducted at the state level but had to proceed county by county, of which the South had more than one thousand. Federal prosecutors had to prove that racial discrimination was practiced by a county's voter registrars. This meant examination of hundreds of registration documents in each county. The task was further complicated by variation from county to county in the methods registrars employed to exclude black voters. According to John Doar, former assistant attorney general for civil rights, when he joined the Civil Rights Division in 1960, "None of us knew enough about registration records, nor the details of registration in Louisiana, Mississippi or Alabama, to direct [an investigation]. No registry

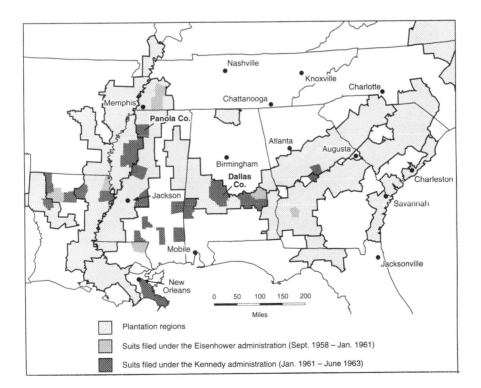

MAP 6.2. Voting rights suits filed by the United States Justice Department, September 1958–June 1963 (source of data: U.S. Department of Justice, Reports of the Attorney General, 1958, 1959, 1960, 1961, 1962, 1963)

offices had been inspected, and we were not familiar with many forms and books involved, nor the procedures or practices of the registrars."[52]

County officials in the plantation regions employed a variety of tactics to delay implementation of the 1957 Civil Rights Act. Court suits that challenged the constitutionality of the act were filed. Some county registrars refused to give the Justice Department access to voter files, resulting in the agency having to initiate court suits just to examine the records. Litigation over access to Bolivar County, Mississippi, records took almost four years. In Macon County, Alabama, the registrars purported to resign to keep Justice Department officials from seeing the voter files. "The defendants are making every conceivable effort to delay effective implementation of the intent of Congress as expressed in the Act," complained Assistant Attorney General Joseph M. F. Ryan Jr. in 1959.[53]

In addition to the legal problems encountered, the Justice Department moved slowly in bringing voting rights suits in the South because Attorney General William Rogers believed that close attention should be paid to Southern white sentiment. Rogers' idea was to bring court suits in counties where there were flagrant violations, to prepare the cases carefully, and to win them. He believed that after the Justice Department scored victories in several landmark cases, most white Southerners would permit blacks to vote, given the choice of obeying the law or facing overwhelming legal action.[54] Another factor that impeded implementation of the 1957 Civil Rights Act was that officials of the Justice Department's new Civil Rights Division had no sense of the geography of discrimination in black voter registration. By 1960, after more than a year of preliminary investigation, officials believed that racial discrimination was widespread in Alabama, Georgia, Louisiana, Mississippi, and South Carolina and that it existed in certain counties in Tennessee, North Carolina, and Florida. They also thought that no black voters were registered in at least twenty-seven Southern counties with large black populations. "After 4 years of hard work" investigating the geography of voter discrimination in the South, John Doar concluded, "We had underestimated the size of the problem."[55]

In 1958 the Justice Department selected Terrell County, Georgia, a rural county in the state's Inner Coastal Plain plantation region, for the first voting rights court suit. Less than 1 percent of Terrell's 5,036 eligible blacks were registered, compared with 83 percent of the eligible whites. Investigations revealed that the registrars maintained a set of color-coded records and required blacks to read and write a longer and more difficult paragraph from the state or the national Constitution than that given to whites.[56] The second suit was filed in February 1959 against registrars of

Macon County, Alabama. Although Macon was a rural Black Belt county, Tuskegee Institute, the college for blacks founded by Booker T. Washington, and a Veterans Administration Hospital for blacks created a relatively large, well-educated black middle class. Macon County was more than 80 percent black, but only 8 percent of the eligible blacks were registered to vote, compared with 97 percent of the eligible whites. To prevent blacks from voting, the county registrars kept irregular hours and employed more stringent standards for blacks than for whites. Only 510 of the 1,585 blacks who took the voter registration test between 1951 and 1958 were accepted. Some of the ones who were denied the right to vote held college degrees.[57]

Most of the Macon County blacks who were registered lived in Tuskegee, and despite the restrictions on registration, by 1955 blacks constituted 40 percent of the municipality's voters. In May 1957 Samuel M. Engelhard, a Macon County planter and an official of the Citizens' Councils of Alabama, with the approval of Tuskegee's white leaders, introduced in the state legislature a bill that deannexed all but 10 of the town's 420 black residents who voted. Tuskegee's boundaries were changed from that of a square to a twenty-eight-sided gerrymander, which a reporter termed a stylized seahorse and United States Supreme Court justice Felix Frankfurter called an interesting "essay in geometry and geography." All that remained of Tuskegee was the town square, the white residential core, and the streets leading from the square, along which were white residences and white-owned businesses.[58] For a number of years Macon County blacks had sought action by the Justice Department to end their voting problems. Hearings by the newly created Civil Rights Commission in Montgomery, Alabama, in December 1958 publicly exposed the deannexation of the black residents of Tuskegee and the blatant defiance by county and state election officials. The hearings led to the Justice Department's filing the Macon County suit.[59]

The third voting rights suit was filed by the Justice Department in June 1959 against the Citizens' Council and the registrars of Washington Parish, Louisiana. Louisiana law provided that two or more registered voters could challenge the right of a person to remain on the registration rolls. Members of the local Citizens' Council had challenged the qualifications of almost all of Washington Parish's 1,517 registered blacks but only a few of the 11,444 registered whites. The challenges were based on minor misspellings, petty deviations from printed instructions, and illegible handwriting. As a result, the parish registrars had removed 8 whites and 1,218 blacks from the rolls.[60]

Although in the 1960 presidential election blacks perceived John Kennedy as more favorably disposed toward civil rights than Richard Nixon, once Kennedy assumed office, controversial civil rights issues were not given high priority. The initial civil rights strategy was to appoint more blacks to high-level federal positions, make the president accessible to black leaders, and continue the Eisenhower administration's emphasis on black voting rights. According to Theodore Sorensen, Kennedy thought that with the right to vote blacks "could in time dramatically alter the intransigence of Southern political leaders on all other civil rights measures, shift the balance of political power in several states, and immunize Southern politics from the demagogue." Only later did he realize "that gaining the vote could not go far enough fast enough to remove a century of accumulated wrongs."[61]

The number of voting rights cases brought in Southern counties increased under Attorney General Robert Kennedy. The Kennedy administration also encouraged the creation of projects to help blacks to register to vote. Several philanthropic foundations, including the Taconic, provided funds for the Voter Education Project, which was created in 1962 and administered by the Southern Regional Council. Through the Voter Education Project funds were channeled to direct-action organizations, including the SCLC and SNCC, which worked at the grassroots level teaching blacks the elementary knowledge they needed to pass registration tests and giving them moral support when they went to register. Burke Marshall, assistant attorney general for civil rights under Robert Kennedy, made two important alterations in the operation of the Civil Rights Division. Attorneys who argued civil rights cases were assigned directly to the division, and attorneys were sent from Washington to participate in civil rights inquiries by the Federal Bureau of Investigation. The attorneys were to try to persuade local officials to comply with the law and to take them to court if they failed to do so.[62]

Although only six voting rights court suits were filed between September 1958 and January 20, 1961, when the Kennedy administration assumed office, six additional ones were brought between January 21 and June 30. Fifteen cases were filed during 1962 and twelve in fiscal 1963. The increase was due in large part to the more aggressive civil rights program of the Kennedy administration, which included a larger budget and more personnel for the Civil Rights Division. By the early 1960s, major legal challenges to the 1957 Civil Rights Act had been defeated in significant Supreme Court decisions that established the constitutionality of the act. Court suits were becoming easier to prepare and try. Suit preparation was facilitated by the passage of the 1960 Civil Rights Act. A major provision of the legislation permitted federal officials to inspect voter registration records and required county officials to keep the records for at least twenty-two months.[63]

John Doar was a Republican, but he was retained in the Civil Rights Division by Robert Kennedy. By the spring of 1961, Doar realized that in order to comprehend the spatial magnitude of black voter discrimination, he and other Justice Department officials would have to go into the field. In April 1961 Doar flew incognito to Jackson, Mississippi, to consult with Medgar Evers, field secretary for the Mississippi NAACP. If anyone knew the geography of voter discrimination in Mississippi, it was Evers. With large-scale maps of several counties that were under investigation spread out on Evers' kitchen table, Doar took the names and marked the locations of residences of blacks who Evers knew had been denied the right to vote. Using the maps, Doar drove to Natchez, Hattiesburg, and remote towns and rural places in central and southern Mississippi interviewing rejected black voters and learning the names of others who had been denied the right to vote. During his initial reconnaissance across Mississippi, Doar developed the field research techniques that were employed throughout the Deep South by Justice Department officials, primarily FBI agents, who followed. A map showing counties that offered favorable opportunities for voter discrimination court suits began to emerge. Large rural counties with sizable enclaves of black landowners had the most potential. Independent farmers who were parties to suits or were witnesses had less fear of economic reprisal by whites. Their independence was even greater than that of many middle-class blacks, such as public school teachers, whose jobs were controlled by whites.[64]

SPATIAL VARIATION IN WHITE RESISTANCE

All of the thirty-one counties in which the early voting rights suits were brought by the Civil Rights Division were in the Deep South, and most were nonmetropolitan ones in the plantation regions (map 6.2). As measured by the number and location of voters' rights cases that were filed, the plantation regions of Alabama, Mississippi, Louisiana, and southwestern Georgia were the hard cores of voter discrimination and resistance to federal authority. However, within the plantation regions spatial variation existed in the types and degrees of resistance to the registration of black voters. In some counties, the resistance by local officials proved to be only bluff, which gave way to compliance, at least token compliance, when the Justice Department threatened. In other counties, defiance could not be breached by dialogue, and an endless stock of legal and nonlegal stonewalling tactics were employed by local officials. Dallas County, Alabama, in the Black Belt, and Panola County, Mississippi, located partly in the Yazoo

Delta and partly in the Loess Plains, were among the first counties in which voter discrimination suits were brought by the Kennedy administration (map 6.2). In both counties, Justice Department officials found "over-whelming proof" of voter discrimination against blacks.[65] After attempts to obtain voluntary compliance with federal law, the Justice Department filed court suits against Dallas and Panola County registrars in April and October 1961.

Although blacks constituted more than half of the population in Dallas County, fewer than 2 percent of the registered voters were black. The situation in Panola County was even worse. Justice Department officials found only one black voter, an elderly man whose registration in 1892 predated disfranchisement.[66] The initial reactions by the registrars in Dallas and Panola to the Justice Department court suits were similar, but the tactics differed. In Panola County, registrar Leonard Duke denied allegations that he discriminated against blacks and challenged the constitutionality of 1957 and 1960 civil rights laws. In September 1962 Duke was succeeded by his assistant, Ike Shankle. Dallas County registrar J. P. Majors resigned seven weeks after the Justice Department filed the court suit against him. Majors was replaced by Victor B. Atkins, who registered 71 of the 114 blacks who had applied after the Justice Department suit was filed.[67]

The Panola County case was argued in the Northern Mississippi District Federal Court in March 1963 and appealed by the Justice Department in the Fifth Circuit Court in New Orleans. The outcome was a decree by the circuit court in May 1964 declaring that both Shankle and Duke "deprived Negro citizens of Panola County of their right to register to vote." Special registration for blacks was to be administrated by Shankle for one year, and a monthly report was to be made to the court. One stipulation was that an applicant's ability to read and write a section of the Mississippi constitution could not be tested with a passage of more than four lines.[68] Civil rights workers made several complaints to the Justice Department about continuation of voter discrimination in Panola County during the 1964 Mississippi Freedom Summer. However, for the first two months of the special registration period, 360 blacks and 118 whites were accepted as voters, while 1 white and 72 blacks were rejected. A report on voter registration in Sardis, one of Panola County's two county seats, during August 1964 by a civil rights worker declared: "There has been virtually no resistance to our work from local whites. No one has even circled our house regularly or followed us around town consistently. The police have left us alone, and when I received a legitimate traffic ticket from a highway patrolman, they jacked my fine up to $102, but my short stay in jail was completely peaceful and no police officer said a nasty word. There is a real possibility that some contacts with moderate whites could be made."[69] Blacks continued to be accepted as voters in Panola County over the next eight months. With expiration of the one-year special registration period approaching in May 1965, Ike Shankle's attorney told Justice Department officials that the registrar was willing to continue the special provisions informally.[70]

Unlike Panola County, what appeared to be the end of voter discrimination in Dallas County in 1961 proved to be an illusion. As in the Panola County case, the Justice Department had to appeal an adverse decision by the district federal court in the Dallas County suit to the Fifth Circuit Court. While the case was in process, SNCC made exploratory contacts with blacks in Dallas County during the summer of 1962. In February 1963 two SNCC voter registration workers, Bernard and Colia Lafayette, moved to Dallas County and established headquarters at Selma. The Lafayettes immediately began to encounter opposition from local Citizens' Council members who sought to prevent blacks from registering. Dallas County sheriff James G. Clark Jr. and his deputies were particularly aggressive and brutal in efforts to enforce segregation and keep blacks from registering to vote. In June 1963, the Justice Department filed a second Dallas County suit. This suit under Section 1971(b) of the 1957 Civil Rights Act (interference with the rights of others to vote) asked the federal district court to issue orders that would prevent Sheriff Clark and other county officials from intimidation of black voters.[71]

In the summer of 1963, SNCC intensified its "One Man, One Vote" drive to increase the number of registered blacks. The effort was marked by the continued harassment of SNCC workers and blacks who attempted to register. In September, SNCC and the Dallas County Voters League also launched an effort to end segregation in Selma. Sit-ins at downtown restaurants and lunch counters and voter rights demonstrations caused Sheriff Clark to organize a special "posse" of 300 men, many of whom began dressing in army-type uniforms. Clark also requested assistance from Colonel Al Lingo, Alabama public safety director, who sent approximately 150 state troopers to Selma. Despite the court suit the Justice Department had filed against Clark, he, together with local and state police, continued to harass and intimidate blacks who attended civil rights meetings and participated in demonstrations. A series of demonstrations were held during the third and fourth weeks of September. By the end of the month many of the demonstrators and their leaders, including John Lewis, had been arrested, sometimes brutally. The Justice Department responded by filing two additional suits under Section 1971(b) of the 1957 Civil Rights

Act. One sought to enjoin the misuse of state law enforcement devices, including prosecutors' offices and judicial processes, for the purpose of denying blacks the right to vote. The other was brought against the Dallas County Citizens' Council and sought to prevent the organization's members from taking punitive economic measures against blacks who tried to register to vote. Not only did members of the Citizens' Council fire employees who attempted to register and have students who participated in demonstrations expelled from school, but the organization even went so far as to place advertisements in the Selma newspaper which sought to recruit new white members and to intimidate blacks (fig. 6.1).[72]

Despite the Justice Department's having won its appeal of the suit

ASK YOURSELF THIS IMPORTANT QUESTION:
What have I personally done to
Maintain Segregation?

If the answer disturbs you, probe deeper and decide what you are willing to do to preserve racial harmony in Selma and Dallas County.

Is it worth four dollars to prevent a "Birmingham" here? That's what it costs to be a member of your Citizens Council, whose efforts are not thwarted by courts which give sit-in demonstrators legal immunity, prevent school boards from expelling students who participate in mob activities and would place federal referees at the board of voter registrars.

Law enforcement can be called only after these things occur, but your Citizens Council prevents them from happening.

Why else did only 350 Negroes attend a so-called mass voter registration meeting that outside agitators worked 60 days to organize in Selma?

Gov. Wallace told a state meeting of the council three weeks ago: "You are doing a wonderful job, but you should speak with the united voice of 100,000 persons. Go back home and get more members."

Gov. Wallace stands in the University doorway next Tuesday facing possible ten years imprisonment for violating a federal injunction.

Is it worth four dollars to you to prevent sit-ins, mob marches and wholesale Negro voter registration efforts in Selma?

If so, prove your dedication by joining and supporting the work of the Dallas County Citizens Council today. Six dollars will make both you and your wife members of an organization which has already given Selma nine years of Racial Harmony since "Black Monday."

Send Your Check To
THE DALLAS COUNTY
Citizens Council
SELMA, ALABAMA
YOUR MEMBERSHIP IS GOOD FOR 12 MONTHS

FIG. 6.1. An advertisement by the Dallas County Citizens' Council in the *Selma Times-Journal*, June 9, 1963

against the Dallas County registrar in September 1963, unlike blacks in Panola County, Mississippi, those in Dallas County were still restricted in their attempts to register. A massive registration effort on October 7, "Freedom Monday," resulted in only 14 of the more than 450 blacks who stood in line at the Dallas County Courthouse for eight hours being permitted to take the test to qualify as voters. Most of the 14 were not accepted.[73]

In 1965, John Doar characterized Panola County as "a locale where surprisingly good advances in voter registration had been made with relative calm." However, he described Dallas County as a place where "the litigation method of correction ha[d] been tried harder . . . than anywhere else in the South," but blacks were still denied "the most fundamental of their constitutional rights—the right to vote."[74] The principal difference between the two plantation counties was in the white leadership. Unlike the white leaders of Dallas County, those in Panola County did not allow resistance to black voter registration to advance to the stage of open hostility and blatant defiance of federal court decrees. The Citizens' Council was not strong in Panola County, and there were no Ku Klux Klan klaverns. In his study of the impact of the civil rights movement on Panola, Wirt concluded that "the relatively peaceful adjustment which Panola County enjoyed must be attributed to the white leadership's insistence on restraint and to the historical experience which made possible acceptance of that insistence." For Wirt, a critical element in the leadership was Panola County sheriff Earl Hubbard. Hubbard was a "parochial, prejudiced" man with "limited education." But unlike Sheriff James Clark, Hubbard was a law officer "who wanted no trouble in his county and who did not want the law broken—by either race."[75]

Neither Sheriff Hubbard nor Sheriff Clark was as powerful as Wirt and other civil rights movement scholars have interpreted them. Although the sheriffs and other elected officials of Southern plantation counties are conspicuous, they usually are not the actual leaders. Leadership historically has rested in the few powerful planter and business families who own much of the land and control the economic, social, political, and legal infrastructures. Sheriffs and other elected officials administer only at the pleasure of these frequently obscure rulers. Racial problems continued in Dallas County for years after James Clark lost his elected office in 1966 and disappeared from the public scene. According to J. L. Chestnut Jr., a black Selma attorney, circuit judge James A. Hare, not Clark, was the principal white leader:

Though big, burly Jim Clark became the symbol of white resistance Selma, Judge Hare was the power behind the scenes. Clark looked the part and played it well. He was a hothead. But he wasn't his own boss. He was subservient to Judge Hare both by law and by social class. Clark wasn't out there on his own authority ordering the marchers away from the courthouse and arresting them en masse. . . . The orders came directly from Hare . . . the commander in chief of the forces of white resistance in Selma. . . .

Judge Hare was a sort of 1960s version of an 1860s plantation owner. Jim Clark was his overseer, the lower-class white man who ran the fields and controlled the slaves. Hare told Clark many times, in my presence, that he, Judge Hare, was in charge.[76]

Dallas and Panola Counties illustrate that at the time of the civil rights movement the quality of local white leadership varied significantly across the plantation regions. As explained in chapter 3, counties in regions of agricultural demise such as the Alabama Black Belt tended to lose many of their astute white leaders, whereas counties in the viable agricultural regions such as the Yazoo Delta and the north Mississippi Loess Plains retained more of them. As early as the 1955 Emmett Till murder trial, the first event of the civil rights era to attract significant national and international media coverage in the rural plantation South, local white leaders in Talla-hatchie County, Mississippi, initiated efforts to negate the embarrassing image that the wrong type of sheriff projected. Two-hundred-seventy-pound Sheriff Clarence Strider, who during the trial repeatedly harassed black journalists, once bragging, "There ain't going to be any nigger reporters in my courtroom," was replaced by the more moderate Harry Dogan.[77]

In counties such as Panola, which initially resisted voting rights for blacks and then relented, the primary factor that motivated white leaders was not a sudden awakening to the condition of blacks followed by conversion to equal rights for all persons. Historically, blacks in the plantation regions were controlled by whites through a combination of intimidation, paternalism, and accommodation. During the early 1960s, as the civil rights era moved from the mobilization to the confrontation stage, white leaders in a number of counties across the plantation regions began to realize that for them the central issue was not blacks voting or even the preservation of segregated public facilities but continuation of white domination. In ceasing to resist federal laws, white leaders were prompted especially by a desire to retain political control, but not control of all elected offices. Allowing blacks to register to vote, and even supporting a few blacks for elected offices, became but another type of accommodation. Black voting was essentially co-opted by white leaders and incorporated into a new form of paternalism that included the use of federal social and economic programs and funds that began to trickle ever more swiftly into poor rural counties, especially after the launch of the War on Poverty by the Johnson administration in 1965.

During the mid-1960s, from counties across the plantation regions began to come reports of the collapse of white opposition to black voter registration. As voter rights court cases were won by the Justice Department, the strategy of example suits initiated by the Eisenhower administration and continued under the Kennedy administration began to succeed. Kennedy offered to negotiate compliance with local officials to avoid the charge that Southern civil rights suits were part of Kennedy's political strategy in the North. By 1964, Attorney General Robert Kennedy could report "voluntary compliance to Federal law forbidding racial discrimination in registration for voting in parts of Georgia, Mississippi and South Carolina."[78] There were even incidents of local officials assisting blacks with registration. In 1964, the sheriff of Tate County, Mississippi, which adjoins Panola County, held informal meetings at black churches and "encouraged . . . all Negroes that wanted to vote" to register and "he would offer them the protection of the law."[79]

Lack of open defiance of federal law resulted in anonymity for counties and municipalities. Despite the initial defiance by its white leaders, Panola County never attracted significant attention during the civil rights era. But counties such as Dallas, where white leaders were uncompromising in resistance to black rights, became notorious. Because of Dallas County's strong opposition to SNCC's attempts at voter registration during 1963, James Forman thought that Selma was an excellent place for an aggressive campaign "aimed at exposing to the nation the inadequacy of the 1957–60 Civil Rights Acts." If SNCC could "force the federal government to be involved . . . when voter registration [was] involved, then this" also would "be applicable in Mississippi and Georgia."[80]

Martin Luther King Jr. also was aware of the potential of uncompromising white resistance in Selma and Dallas County. Not only had King begun his ministerial career at Dexter Avenue Baptist Church in Montgomery in the heart of the Black Belt, but Coretta, his wife, was from Perry County near Selma, and Obie Scott, his father-in-law, still lived there. White thugs came close to attacking King in Selma in December 1962, during one of the SCLC's People to People promotional tours, and in the fall of 1963, he returned to give a public address sponsored by the Dallas County Voters League.[81] Soon a potent combination of King's strategy and unplanned events would make Selma the climax of the civil rights movement. Selma would cast a long and enduring shadow, not just across the plantation South, but across the nation. That the climax came in an old cotton town in the heart of the plantation South was fittingly appropriate.

CHAPTER SEVEN

Confrontation

❈ ❈ ❈

Go down, Moses, 'Way down in Egypt land;
Tell ole Pha-roh, Let my people go!

African American folk hymn

God knows how this race has been treated. And there's
a certain element that's workin to please God and overturn
this southern way of life. How many people is it today
that it needs and it requires to carry out this movement?
How many is it knows just what it's goin to take?
It's taken time, untold time, and more time it'll take before it's
finished. Who's to do it? It's the best people of the United States
to do it, in the defense of the uneducated, unknowledged ones that's
livin here in this country. They goin to win! They goin to win!

Nate Shaw

GO DOWN, MOSES

The sit-ins launched by a group of college students in February 1960 at a
Woolworth's store in Greensboro, North Carolina, marked the transition
from the mobilization to the confrontation stage of the civil rights movement. By April, sit-ins, accompanied by protest marches and picketing of
segregated businesses, had developed in many of the South's cities from
Richmond, Virginia, to Austin, Texas. Protest had even spread to a few
small plantation towns with black colleges, including Orangeburg and
Sumter, South Carolina, and Tuskegee, Alabama.[1] The plantation regions,
however, were largely untouched by the racial turmoil that began to rock
some of the South's cities, and they remained the bastions of the segregated
way of life. Most whites took reassurance from the belief that their total
control of the economic, social, political, and legal systems was sufficient to

suppress racial agitation, resist federal authority, and maintain the status quo. Six years later across the plantation South, civil rights leaders could declare what the SCLC's Hosea Williams told a crowd in Grenada, Mississippi, about continued white resistance to equal rights for blacks. "It's all over in Grenada, Miss. The Sheriff might as well know, the mayor might as well know, [Governor] Paul Johnson might as well know, the Ku Klux Klan and the White Citizens' Council might as well know—it is all over in Grenada."[2]

The major push by civil rights organizations into the plantation regions began in the summer of 1961. Although desegregation of public facilities was an objective of civil rights workers, the principal emphasis was on the restoration of voting rights to blacks. In part, the emphasis on voting was in response to the civil rights strategy of the Eisenhower and Kennedy administrations. But civil rights workers also realized that desegregation of restaurants and bus station and train depot waiting rooms in a small Mississippi or Alabama town meant little to poverty-stricken rural blacks. In addition, the dispersion of most of the black population across farms and plantations meant that mass demonstrations to support sit-ins would be difficult to organize and would not attract as much media attention as in cities.

By 1961 the NAACP had made some headway in recruiting members and establishing local chapters in the rural South. The NAACP, however, did not believe in confrontational, direct-action strategy. CORE was established in a few small plantation towns in South Carolina, but its local chapters resembled those of the NAACP.[3] Although the SCLC was more openly aggressive and Martin Luther King Jr. and other leaders believed in direct action, the organization was largely an urban one. When in 1958 the SCLC sponsored a voters' conference in Clarksdale, one of the Yazoo Delta's largest municipalities, fewer than two hundred persons attended a public rally to hear King speak. SCLC leaders did not begin serious consideration of the rural South until 1963.[4]

SNCC took the initiative in bringing aggressive campaigns for civil rights to the plantation regions. In the spring of 1961, SNCC was a new, poorly structured organization in the shadow of the SCLC. According to Cleveland Sellers, the reason that SNCC quickly rose to prominence was because "it was the only action-oriented civil rights organization in the South prepared to absorb the brash young militants who joined the movement because of the [Freedom] Rides."[5] "The brash young militants" were primarily young Southern and Northern blacks between ages eighteen and thirty, but from the beginning, SNCC's membership also included whites.

Most members were college students. Southern plantation regions were perceived as the most difficult and most dangerous areas by civil rights workers. In part, the symbolism that the regions held for the young workers is what made them enticing. The plantation regions both represented the shameful past of slavery and portrayed the worst in the contemporary denial of civil rights. According to Zinn, young educated blacks "wanted to return . . . to the source of their people's agony, to that area which was the heart of the slave plantation system, in order to cleanse it once and for all time."[6]

During 1961–63, SNCC launched voter registration projects in the non-metropolitan areas of Mississippi, Georgia, and Alabama. The focus was on specific areas rather than on entire states. Small cities in which field offices were located were the nuclei of the areas. McComb and Greenwood, Mississippi; Albany, Georgia; and Selma, Alabama, became significant landmarks in the civil rights struggle. A series of unplanned events in 1960 and 1961 made the McComb area the first outpost of SNCC's plunge into the nonmetropolitan South. Greenwood, Albany, and Selma, however, were deliberately chosen. The initial investigations of the Justice Department into the geography of voter discrimination were important to the selection of places. At a 1961 meeting with Justice Department officials, SNCC leaders saw a map on John Doar's office wall on which he had identified counties that offered the best possibilities for court suits under the 1957 and 1960 Civil Rights Acts. Among them were Terrell County, Georgia (the Albany area); Leflore County, Mississippi (Greenwood); and Dallas County, Alabama (Selma).[7]

Rural counties with small towns quickly proved to be poor bases of operation. Charles Sherrod and Cordell Reagon moved to Terrell County in the Inner Coastal Plain plantation region of southwest Georgia in the summer of 1961 and began a voter registration project. By the fall they had retreated to larger, but nearby, Albany in Dougherty County.[8] McComb, Albany, Greenwood, and Selma were small backwater cities with populations that ranged from twelve thousand to sixty thousand. Although all four had many of the provincial characteristics of small Southern towns, their larger size gave them attributes that made them more desirable for SNCC field offices. Each of the four cities had a black middle class composed of business and professional persons who were better educated than the majority of nonmetropolitan Southern blacks. The middle-class blacks had a degree of dialogue with the white leaders, and because some of them also were economically independent, they were not easily intimidated by whites. The larger populations of small cities also meant better chances for white moderates than in a small town. Finally, the larger black populations of small cities meant that black enclaves were larger. The larger the enclave, the more isolated and protected was a SNCC field office.

Robert P. Moses, a New York schoolteacher, was one of the brash young black militants who joined SNCC. Moses came to SNCC's Atlanta headquarters in June 1960 as a summer volunteer. He found a fledgling organization with only one salaried employee and no plans for volunteers. Moses created a place for himself. He helped in the office and participated in picketing Atlanta A & P supermarkets that had no black employees. In July, he traveled across Alabama and Mississippi recruiting ministers and NAACP chapter heads for SNCC's October conclave in Atlanta. Among the persons whom Moses contacted in Mississippi was Amzie Moore, head of the NAACP in Cleveland, one of the larger municipalities in the Yazoo Delta. Moore told Moses that he thought integration of schools and other public facilities was not central to the black struggle in the rural South. The primary issue was the right for blacks to vote.[9] Amzie Moore's belief that the emphasis should be placed on voting was hardly an innovative idea. Not only had the NAACP and the Eisenhower administration pushed the concept, but even the SCLC had attempted an unsuccessful voter registration project in 1958 called "Crusade for Citizenship."[10] What was new was an NAACP leader who was willing to welcome SNCC workers and their aggressive confrontational approach to voter registration. Moore thought that the conservative legal strategies of the NAACP were accomplishing little in the Yazoo Delta.

Moore and Moses planned a voter registration project in the central Delta for the summer of 1961. Because rural blacks had little formal education, they would have to be taught to read and to interpret the United States and Mississippi constitutions before they went to register. When Moses returned to Mississippi in June 1961, he found that Moore had not been able to obtain an adequate facility for a voter school. Curtis C. Bryant, head of the NAACP in Pike County, Mississippi, had read about the proposed voter project in *Jet* magazine. Bryant invited Moses to come to McComb, a city of twelve thousand in the southern part of the state, where facilities for a school would be provided by local blacks. McComb became the center of a three-county voter registration project on the eastern edge of the Natchez district (map 7.1).[11] Of the three counties, Amite historically was the most important for plantation agriculture. Although the plantation system had declined, nearly half of the county's population was still black in 1960. Pike County, in which McComb is located, and Walthall County, to the east of Pike, were on the fringe of the plantation region and had relatively small black populations compared with Amite. On August 3, Moses wrote SCLC headquarters in Atlanta that the McComb project was "getting under way": "We will open the school August 7 and continue for at least three weeks if possible." He asked for additional money, including a guarantee of $32.50 per week for room and board for ten volunteer workers if local funds were not forthcoming.[12]

A voter education school was established within a black enclave in each

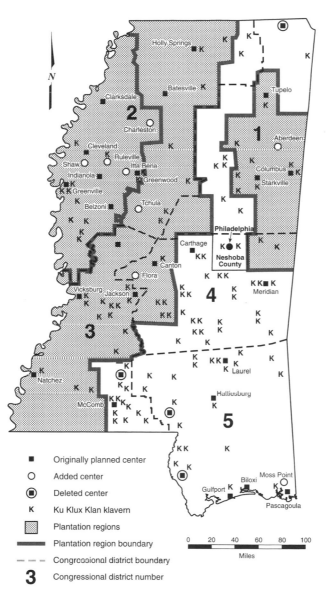

MAP 7.1. Mississippi 1964 Summer Project districts and centers and Ku Klux Klan klaverns (sources: SNCC 1963; Sutherland 1965; U.S. House 1967)

of the three counties. The first school opened in August in a Masonic hall above a grocery store in a black residential section of McComb. The school in Amite County was on the farm of E. W. Steptoe, head of the Amite County NAACP, and the one in Walthall County was at Mt. Moriah, a rural Baptist church.[13] Moses' first attempt to register new black voters was at the Amite County courthouse in Liberty. His arrest by a highway patrolman after he left the registrar's office caused SNCC headquarters in Atlanta to send additional workers, including Marion Barry. A second attempt to register black voters in Amite County resulted in an attack on Moses by a cousin of the sheriff. As events in the McComb area escalated during the late summer, local black high school students were drawn in. What began as a voter registration project led to desegregation efforts, including ones at the McComb Woolworth's lunch counter and the bus depot. Violence increased. John Hardy, a voter registration worker in Walthall County, was pistol-whipped by the registrar and then arrested for disorderly conduct.

The escalation of violence slowed but did not stop the voter registration project. On September 25, E. H. Hurst, a member of the Mississippi legislature, shot and killed Herbert Lee, an Amite County black farmer who had helped Robert Moses. Lee's death brought voter registration to a halt, but protest demonstrations by local high school students and SNCC workers continued. The McComb protest finally ended with the arrest of local leaders together with Moses and other SNCC workers.[14]

Although SNCC was defeated in its first foray into the nonmetropolitan South, valuable lessons were learned concerning strategy and techniques, and the national publicity given to McComb increased the organization's membership. Early in 1962, Robert Moses left McComb for Jackson, Mississippi, finding "the movement from the rural to the urban . . . irresistible."[15] He quickly organized a network of SNCC field offices throughout Mississippi, staffed in part by workers who participated in the McComb effort. Moses began to place special emphasis on the Yazoo Delta and the northern Loess Plains, the areas with the state's largest black populations. Greenwood, a city of 20,436 in Leflore County at the edge of the Delta, became the location of the principal state office.[16]

By the summer of 1962, SNCC's voter registration effort in Mississippi was better organized and began to accelerate. A principal reason for the improvement was the creation of the Council of Federated Organizations (COFO) in 1962 to coordinate the efforts of the NAACP, SCLC, SNCC, and CORE in Mississippi. The COFO was placed under the Southern Regional Council as part of the organization's Voter Education Project. Robert Moses was appointed director of the COFO's voter program. The cooperative project was one in which voter education schools were established in black churches, lodges, and schools to teach rural and small-town blacks the basic reading skills needed to pass literacy tests. The schools also sought to instill confidence through sermonlike inspirational talks and the singing of freedom songs. Many of the teachers were local blacks with leadership abilities who were selected for special instruction at training centers operated by the SCLC. One of the centers was in Atlanta and the other at Dorchester, an abandoned Congregationalist academy near Savannah in the extinct Sea Island rice and long-staple cotton plantation region. Although rural, the old Southern Puritan settlement was considered a safe place, illustrative of the diversity in the plantation South during the civil rights movement. By the spring of 1963, more than two hundred voter-school teachers had been trained. The methods and materials used in the training centers and voter schools were developed with the aid of the Highlander Folk School operated by Myles Horton in Tennessee.[17]

Despite the better organization, the number of blacks who attended voter education schools and attempted to register was relatively small. Interpretations of the general failure of voter registration efforts in Mississippi during the early 1960s stress the role of violence directed at civil rights workers and local blacks. But other factors were significant in preventing large numbers of blacks from attempting to register. One was the physical and social isolation of the plantation regions from the mainstream of contemporary American affairs. The stories of the major acts of violence have been told and retold, but almost no analysis has been conducted of the geographical context of voter registration. The relative isolation of plantation region blacks is revealed in Unita Blackwell's description of the arrival of the first civil rights workers in Mayersville, the small Yazoo Delta town of which she later became mayor:

> In 1963, we had heard that there were supposed to be these Freedom Riders . . . coming to Mississippi. Nobody thought they would ever show up. . . . Everybody talked about it—maybe they would come to Jackson or someplace like that. But I didn't think they would ever show up in Mayersville, Mississippi. . . .
>
> And then they showed up. It was two guys, two black fellows. They came walking down the road and I knew they were different. . . .
>
> Two Freedom Riders came to Sunday school that morning, and they were pointing the finger at me, saying, "Just like that lady talking back there in the Sunday school class says that God help those who help themselves, you can help yourself by trying to register to vote. . . ."

The white people knew what it meant. The black folks didn't know that much what it meant. I was only told when I started off that if I registered to vote, I would have food to eat and a better house to stay in. . . . We didn't know there was such a thing as a board of supervisors [the county governing body] and what they did, and we didn't know about school board members and what they did.[18]

Because of the lack of success in registering blacks, the directors of the Southern Regional Council's Voter Education Project threatened in November 1963 to cease financial support of the COFO effort in Mississippi. Robert Moses brought two proposals for salvaging the voter project to the SNCC executive committee meeting that was held in Atlanta on December 27–31, 1963. One was to develop a more intensive voter education program for Mississippi, perhaps, by the National Council of Churches. His alternative plan, "pushed by Al Lowenstein," was to "pour in thousands of students and force a showdown between local and federal governments in an election year." A confrontation project during the summer of 1964 would draw national publicity to racial oppression under "a totalitarian government, a closed society." Federal intervention would not give the Mississippi blacks their civil rights immediately, but it would be "an opening wedge for future pressure." John Lewis, Marion Barry, James Forman, and other principal SNCC leaders favored the confrontation proposal. Lewis thought that Lyndon Johnson could not "fail to respond" to "a physical showdown between the federal and local governments."[19] Moses believed that SNCC's mission was to "bring about just such a confrontation . . . to change the power structure": "Only when metal has been brought to white heat, can it be shaped and molded. This is what we intend to do to the South and the country, bring them to white heat and then remold them."[20]

SNCC's underlying strategy for the 1964 Mississippi Summer Project to produce confrontation was hardly new to the civil rights movement. Forced confrontation between civil rights demonstrators and local authorities was part of the SCLC's plan in the Birmingham campaign during the spring of 1963. From the unsuccessful Albany campaign in 1962, Martin Luther King Jr. and other SCLC leaders learned to select with great care the places in which to stage major civil rights efforts. Birmingham was "a deliberately engineered crisis intended to revitalize the nonviolent movement after Albany and to strike a damaging blow to the whole structure of southern segregation."[21] The code name during the planning stage of the Birmingham campaign was "Project C"—"C" for confrontation.

The violence in Birmingham, including that of the police led by the city's commissioner of public safety, Theophilus Eugene "Bull" Connor, helped to convince the Kennedy administration that new civil rights legislation was needed. On June 19, Kennedy sent a comprehensive civil rights bill to Congress.[22] Following Kennedy's assassination in November 1963, President Lyndon Johnson told Congress that "no memorial oration or eulogy could more eloquently honor President Kennedy's memory than the earliest possible passage of the civil rights bill for which he fought."[23] In a long, fierce congressional battle Johnson pushed for passage of the legislation, and on July 2, 1964, he signed the new potent Civil Rights Act into law. The act prohibited discrimination in public facilities, authorized the Justice Department to file suits to desegregate public schools, and made a sixth-grade education minimum proof of literacy for voting purposes. In what would prove to be the most far reaching section, the bill authorized termination of funding in federally financed programs.[24] The lack of more forceful guarantees of voting rights, however, caused civil rights organizations to continue their campaign for greater black enfranchisement.

SNCC's Mississippi Summer Project was relatively well organized and handled considering the large number of volunteers, the minimal amount of preparation they were given, and the idealism with which many of them entered rural Mississippi. Most of the more than one thousand volunteers were white Northern college students. After a week-long orientation session at the Western College for Women in Oxford, Ohio, the workers left for Mississippi by bus, van, and automobile. The Summer Project technically was under the oversight of the COFO, but SNCC and CORE were actually in direct charge. For organizational purposes, Mississippi was divided into five areas based on the state's five congressional districts (map 7.1). CORE was in charge of the Fourth District, and SNCC had responsibility for the remainder.[25]

Emphasis on voting during the civil rights movement resulted in the frequent use of congressional districts as regions. Employment of political units as cultural-social regions for the Mississippi Summer Project and a naivety that failed to comprehend the geographical diversity of the rural South soon proved fatal. Each of the five Mississippi congressional districts was rated for its violence potential. The southern part of the Fifth District, the Gulf coast, was considered safe because of tourism and the lack of white resistance to blacks voting. The southern part of the Third District, the McComb area, where in 1961 SNCC workers encountered violence and where the Ku Klux Klan had become active, was regarded to be dangerous, "with a long record of brutality and terrorism." The Fourth District, the one overseen by CORE, was perceived to be the "easiest" and safest of the five areas.[26] Most of the Fourth District was a white yeoman farmer area, a section of Mississippi where the small but vicious White Knights of the Ku Klux Klan was active. Here in Neshoba County, Andrew Goodman, Michael Schwerner, and James Chaney were murdered in June 1964.

The Mississippi Summer Project consisted of three programs: voter registration, freedom schools, and community centers. Voter registration and freedom schools proved to be more important than the community centers. The original idea for voter registration, the project that enlisted the greatest number of workers, was to register as many blacks as possible. Because the state's Democratic Party was not open to blacks, a new political organization called the Mississippi Freedom Democratic Party was created. The new organization held a mock election to demonstrate that unregistered blacks wished to vote, and it challenged the seating of the regular Mississippi delegates at the Democratic National Convention held at Atlantic City in August 1964. Freedom schools emphasized remedial education, contemporary issues, and black history. Approximately one thousand students were anticipated, but more than three thousand participated.[27]

The Mississippi Summer Project was successful in that national and international attention was focused throughout much of the summer on the conditions of blacks in Mississippi. If, however, success is measured in terms of the principal goal of creating a confrontation between federal and Mississippi officials, the project was a failure. The confrontation did not develop for three principal reasons. A tragic clash between federal and state officials had occurred almost two years earlier over the admission of James Meredith to the all-white University of Mississippi at Oxford, a small city in the north Mississippi Loess Plains. The potential for confrontation had been building in Mississippi since the *Brown* decision. Exaltation of defiance of federal authority by demagogues who exploited the race issue was permitted to go too far by Governor Ross Barnett in the fall of 1962. A riot resulted on the university campus in which a group of students and a mob of nonstudents, led in part by retired Major General Edwin A. Walker, who had commanded federal troops at Little Rock in the 1958 desegregation crisis and thought that he was on the "wrong side," confronted more than five hundred federal marshals, border patrolmen, and prison guards. Two persons were killed and more than two hundred injured in the confrontation.[28] After the riot, the tactic of Mississippi officials changed from one of defiance more to one that emphasized federal persecution. Never again did state officials permit civil rights issues to develop into open bloody confrontation between Mississippi and the federal government. When the Mississippi officials were confronted with strong school desegregation guidelines by the Office of Education in April 1965, Paul Johnson, who succeeded Ross Barnett as governor, told reporters, "Obedience to our laws is not optional."[29]

A second reason that a major confrontation between federal and state officials did not develop was the negative publicity Mississippi received over the brutal murders of Goodman, Schwerner, and Chaney by members of the White Knights of the Ku Klux Klan at the beginning of the Summer Project. The unfolding drama of the three's disappearance in June and the discovery of their bodies in August produced a flow of bad publicity concerning Mississippi throughout the summer. State leaders sought to prevent, rather than encourage, additional incidents that would reflect badly on the state.

The murders by members of the White Knights also provided the federal government with a red herring. Though the murders were horrible and sensational, as has been discussed, Klan activity was tangential to basic civil rights issues in the South. The murders, together with other violent acts by Klan members, resulted in a major federal effort, primarily by the Federal Bureau of Investigation, to destroy the organization. Despite the importance that some writers have ascribed to the Klan and to the FBI's effort to annihilate it,[30] neither was central to the civil rights story. Under the directorship of J. Edgar Hoover, the publicity strategy was to have the FBI continually fighting a subversive group thought to threaten the nation. The war against the Ku Klux Klan in the 1960s was merely the continuation of the battle against gangsters in the 1930s, saboteurs in the 1940s, and Communists in the 1950s. The fight against the Klan placed federal and Mississippi authorities on the same side rather than on a collision course over basic issues in the civil rights struggle. It also diverted federal attention away from the plantation regions, where the Klan was not very strong, to rural areas and cities in which the organization had its greatest strength but the black populations were relatively small (map 7.1).

THE GEOGRAPHY OF BARRIERS TO BLACKS' VOTING

In more than half of the South, primarily King's "South of compliance" and "wait-and-see South," there were no restrictions on voting. In the remainder of the South, potential black voters confronted barriers at two geographical scales. One group was local. In particular counties, such as Dallas, Alabama, whites kept blacks from voting by complicated and unfairly administered registration procedures and by economic and physical intimidation. These counties were in overt violation of the 1957, 1960, and

TABLE 7.1 Black Voter Registration, 1958 and March 1965

	1958	March 1965 % of 1960 Nonwhite Voting-Age Population	March 1965	% of 1960 Nonwhite Voting-Age Population
Alabama	70,000	14.5	92,737	19.3
Arkansas	64,000	33.2	77,714	40.4
Florida	144,810	30.8	240,616	51.2
Georgia	161,958	26.4	167,663	27.4
Louisiana	131,068	25.2	164,601	36.6
Mississippi	20,000	4.7	28,500	6.7
North Carolina	150,000	27.2	258,000	46.8
South Carolina	57,978	15.6	138,544	37.3
Tennessee	185,000	58.9	218,000	69.5
Texas	226,818	34.9	—	—
Virginia	92,172	21.1	144,259	38.8

Source: Price 1959; U.S. Commission on Civil Rights 1968b. Bureau of the Census 1963a.

1964 Civil Rights Acts, and eventually legal action by the Justice Department would have brought them in compliance with federal legislation. The other group of barriers was at the state level. In the plantation South states, laws dating from the post-Reconstruction era of disfranchisement required voters to pay poll taxes and pass literacy tests. The states of the Deep South, King's "South of resistance," together with Texas, had statewide impediments to minority voting such as literacy tests. Although the state laws were major barriers for poor illiterate blacks, they were not in violation of the United States Constitution or any federal statutes. In March 1965 the number of blacks registered to vote in the eleven former Confederate states was considerably larger than in 1958, just after the 1957 Civil Rights Act was passed. In North Carolina, South Carolina, Florida, and Virginia, significant increases occurred in black voter registration, but in Alabama, Georgia, and Mississippi they were slight (table 7.1). In all of the eleven states except Alabama and Mississippi, at least 25 percent of the black voting-age population was registered, and in Arkansas, Florida, North Carolina, and Tennessee it surpassed 40 percent.

By 1964, research by the Justice Department, which eventually surveyed more than five hundred Southern counties, and probing by civil rights organizations had revealed the detailed geography of local white resistance to black voter registration. Significant spatial variation existed. With a few exceptions such as Jefferson County, Alabama (Birmingham), whites in metropolitan counties offered little opposition to black voter registration, and registration was well above the average for the Southern states. In addition, whites in nonmetropolitan counties with small black populations usually did not attempt to keep blacks from voting. In Fulton County, Georgia (Atlanta), 31 percent of nonwhites were registered, compared with 44 percent of whites. In Charleston County, South Carolina (Charleston), and Orleans Parish, Louisiana (New Orleans), the figures were 43 and 89 percent for whites and 28 and 63 percent for nonwhites. Even in Jefferson County, Alabama, 21 percent of the nonwhites were registered, compared with 51 percent of the whites.[31] The principal areas of white resistance were counties in plantation regions where the black population surpassed 50 percent. Almost all of the thirty-one voting rights court suits that the Justice Department filed from September 1958 through June 1963 were against nonmetropolitan counties within or on the fringe of plantation regions in Alabama, Georgia, Louisiana, Mississippi, and Tennessee (map 6.2). Even among plantation counties the type and degree of white opposition varied, as illustrated by Dallas County, Alabama, and Panola County, Mississippi. The most comprehensive record of places of white resistance to black voter registration was compiled by the Justice Department for the purpose of sending federal registrars once such legislation authorizing

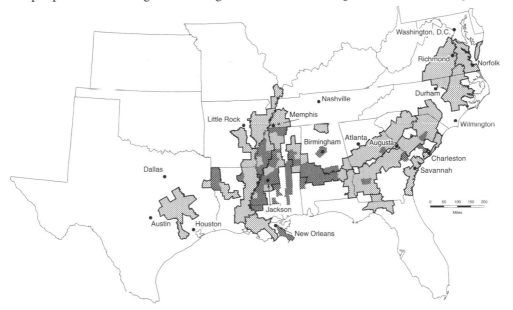

MAP 7.2. Counties to which the Justice Department sent federal examiners under the 1965 Voting Rights Act, August 1965–August 1967 (source of data: U.S. Department of Justice, Report of the Attorney General 1968)

them was passed by Congress.[32] Most were nonmetropolitan counties in Alabama and Mississippi, concentrated in the Black Belt, the Loess Plains, and the Yazoo Delta. Some of the defiant counties soon complied with federal law, but others continued to resist. Two years after the 1965 Voting Rights Act went into effect, federal registrars had been sent into sixty-two counties (map 7.2). Almost all of the sixty-two were nonmetropolitan ones in the plantation regions of Alabama, Mississippi, and Louisiana. Jefferson County, Alabama (Birmingham), and Hinds County, Mississippi (Jackson), were the only ones with large urban populations.

By 1964, civil rights leaders, as well as Justice Department officials, had a comprehensive knowledge of the detailed geography of white resistance to blacks' voting. The geography actually caused disputes over strategy which erupted among civil rights leaders. In 1962, Roy Wilkins, head of the NAACP, opposed SNCC's efforts to work in the rural plantation South because he believed that white opposition was so ruthless that few blacks would be registered. Much greater success could be realized by concentration on large cities where apathy was the main impediment to blacks' voting. But the fierce and violent white repression of blacks in the plantation regions was exactly what SNCC leaders wanted to expose. James Forman explained:

> We were interested in trying to register voters so as to expose the dirt of the United States. . . .
>
> Had we been content to register voters in the cities of the Deep South, we would have fallen victim to the intent of the federal government. For, in the cities, there was an absence of the kind of repression that existed in rural areas and our activities in the cities would have primarily centered around fighting apathy. . . . But in the rural areas we knew that the fear of the sheriff and the Ku Klux Klan, together with the desire of the whites to hold onto their power by any means, would surface.[33]

The plans for the SCLC's Project SCOPE also reveal a detailed spatial comprehension of white resistance by civil rights leaders. SCOPE (Summer Organization and Political Education Program) was a major voter registration drive during the summer of 1965 in 120 counties with 40 percent or more black population across the South from Alabama through Virginia. Planning for the project began in the fall of 1964. Ralph Blackwell, the director of voter registration and political education, was instructed by the SCLC's executive committee to include among the counties "ten . . . containing large cities where the Negro vote [might] be doubled and ten . . .

that consistently show[ed] strenuous resistance to Negro registration and hence Title I of the 1964 Civil Rights Bill [could] be applied."[34] Most of the 120 counties were nonmetropolitan ones in the plantation regions. That ten counties where whites vigorously opposed black voter registration had to be deliberately chosen indicates that by 1964 open defiance of federal law had shrunken to a relatively small number of places.

WHY SELMA?

The SCLC had little direct participation in the 1964 Mississippi Summer Project. After the successful 1963 Birmingham campaign, the SCLC spent much of 1964 involved with desegregation efforts in St. Augustine, Florida. Although the SCLC had announced a statewide voter registration project in Alabama during 1964, by April officials realized that not enough advance preparation had been done to launch it that year. During the summer of 1964, SCLC leaders began planning a six-month voting rights campaign in Alabama for 1965.[35] As originally conceived, the Alabama campaign had as its objective the same general goal as other efforts to register black voters in the South, "enfranchisement of the Negro people." In his plan for the project, James Bevel wrote: "We must keep in mind that unless we can in fact get Negroes registered, we cannot stop bombings of churches, unjust court proceedings, police brutality, etc. We must also keep in mind that unless large numbers of Negroes get registered, there will not be the climate for peaceful, large-scale school integration, integration of public accommendations [sic] and employment of Negroes on many city, county and state jobs."[36]

King's success in the 1963 Birmingham campaign leading to passage of the potent 1964 Civil Rights Act and his receiving the 1964 Nobel Prize for Peace gave him the confidence to seek a specific objective in the proposed Alabama campaign. The 1964 Civil Rights Act did not remove the legal state impediments to black voter registration and did not provide for sending federal registrars into counties, such as Dallas, which continued to resist orders of federal courts. King, at the crest of his influence, concluded that the Alabama campaign could be used to achieve a comprehensive voting rights bill that eliminated all barriers to blacks' voting. He devised a strategy in which the local illegal barriers that blacks confronted in a few counties were used to bring down the whole system of impediments—state-local, legal-illegal.

The primary question that confronted King in the fall of 1964 was where to focus the barrier-smashing campaign. In part, the answer to this ques-

212 • **Impact of the Civil Rights Movement**

tion lay in how King viewed the strategy of nonviolence. King's apparent adoption of India's Mohandas Gandhi's nonviolence was not what it superficially appeared. Taylor Branch explained that for tactical reasons King never attempted to deny the numerous interpretations of him as a Gandhian. King, however, was influenced more by the German theologian and philosopher Reinhold Niebuhr than by Gandhi. Niebuhr believed that injustice could not be eliminated without reformers involving themselves in power conflicts. King's nonviolence, what he termed "realistic pacifism," was actually a "Niebuhrian stratagem of power."[37] As the civil rights movement progressed, King's concept of the nonviolent stratagem evolved. According to Garrow, in the late 1950s and early 1960s, King's nonviolence was nonviolent persuasion. The tactic to "awaken a sense of moral shame in the opponent."[38] By 1963, nonviolence had become more a method than a moral commitment. Coercive nonviolence had replaced nonviolent persuasion. Violent confrontations attracted the attention of the national and international media and caused "moral pressure to bear on the federal government."[39]

The 1963 Birmingham campaign proved the success that could be achieved by prudent place selection and by meticulous planning. The best places for confrontations were ones in which state, as well as local, police would brutally attack protestors. For the Birmingham campaign, the principal tactic became one of forced confrontation in which the brutality of the oppressor was revealed by law enforcement officials attacking nonviolent demonstrators. In the final stages when it appeared that the campaign might fail, King gambled and sent surge after surge of schoolchildren through the city's streets as fodder for police dogs, fire hoses, and jails.

By the fall of 1964, the type of violent confrontation needed to place enough pressure on the federal government to insure the passage of additional voting rights legislation could have happened in a relatively small number of places in the South. A place had to have two attributes that were essential for potential confrontation. One was the ingrained commitment of local whites to continue to deny blacks their rights. Civil rights legislation, federal enforcement of the legislation, and the unrelenting effort of civil rights organizations had significantly reduced the number of places that overtly practiced voter discrimination. The second was a local leadership that would permit civil rights demonstrations to lead to police retaliation and racial violence. A number of hard-core counties that denied blacks voting rights, including Grenada and Leflore, Mississippi; Terrell, Georgia; Dallas and Perry, Alabama; and Plaquemines Parish, Louisiana, were among the possible candidates for violence initiated by local police.

If the sixty-two counties and parishes into which federal examiners were sent during the two years following Lyndon Johnson's signing the Voting Rights Bill into law are assumed to be those from which King could have picked the focus of his campaign for stronger voting rights legislation, why did he pick Dallas County, Alabama (map 7.2)? The question of moving the crusade to another state was never really debated. Had other states been considered, Alabama still would have been the logical choice. Clashes between state and federal governments over civil rights issues had caused most Southern politicians by 1964 to take less defiant stands. Moderate political leaders had even begun to be elected in some Deep South states. Although Carl Sanders, who was elected governor of Georgia in 1962, was described by one reporter as a man who had "no interest at all in the plight of the Negro," Sanders was pragmatic enough to foresee the political revolution that was about to sweep the South. Some SCLC leaders thought that Sanders recognized the future role of a black electorate and would "work to make county registrars cease their coercion of would-be Negro voters."[40] In 1964, there were still a number of counties across the "South of resistance" where local police would have gleefully attacked civil rights demonstrators, but only Alabama had a governor who was defiant to the point of allowing state police to join such attacks.

Martin Luther King Jr. was drawn to Alabama for the campaign for new voting rights legislation for reasons other than George Wallace. Although the SCLC headquarters was in Atlanta, Georgia, the organization had its greatest strength and influence among blacks in Alabama. The SCLC grew out of the successful 1955–56 Montgomery bus boycott and by 1964 had achieved its greatest victory in Birmingham. As Mississippi belonged to SNCC, so Alabama belonged to the SCLC. The primary question that confronted King in the fall of 1964 was where in Alabama the campaign for stronger federal voting rights legislation should be focused. James Bevel's preliminary plan for the Alabama project evaluated the pros and cons of staging demonstrations in several cities simultaneously or focusing on one city. Among the advantages of concentration on one place was that "the news media could cover more thoroughly and effectively one city than many." Moreover, "it would be more dramatic to have 5 or 6 thousand people in jail in one city rather than in many cities across the state."[41] Selma was not mentioned in the preliminary plan. In the proposed budget for the Alabama project, which was submitted on September 1, 1964, Montgomery, Gadsden, Tuscaloosa, Birmingham, and Mobile were the only places that were to have field secretaries and assistants. The only reference to Selma was buried in a section on "continued consciousness of specific problems . . . which [arose] locally, but with which a state-wide co-opera-tion could help."[42]

Shortly after he returned from a trip to Europe in October 1964, King learned that he had been awarded the 1964 Nobel Prize for Peace. A few days later when he made SCLC staff assignments for the coming months, the planning for the upcoming Alabama campaign shifted away from the state's large urban areas toward the Black Belt. Selma was given a promi-nent role. One assignment was for additional "research on Alabama— Selma in particular," and another called for an "exploratory meeting" with black leaders "in Selma." King also asked for information on "harassment of voters in small rural towns."[43]

Why did King shift the focus of the Alabama campaign from the large urban areas to the Black Belt? More factors would seem to be against such a change in geographical strategy than to favor it. The SCLC was primarily a metropolitan organization, but Selma, with a population of 28,385, was hardly a metropolitan place. When King was approached in the summer of 1964 by attorney J. L. Chestnut and other Selma black leaders about focusing the Alabama campaign on their city and the surrounding area, he expressed reservation about "the difficulty of achieving togetherness in what he called the 'open spaces' of the Black Belt."[44] The SCLC's only major venture into a small city in the plantation South, the 1962 Albany campaign, failed. The project was plagued by disputes between local and outside blacks and between SNCC and the SCLC (i.e., "the difficulty of achieving togeth-erness"), the problem of attracting substantive media attention, and the inability to win strong support from federal authorities. The underlying reason for the failure of the Albany campaign is usually ascribed to police chief Laurie Pritchett, who had read King's *Stride toward Freedom* and reasoned that the way to counter nonviolence was with nonviolence.[45] The ability of a police chief in a small city of 56,000 to checkmate his strategy gave King reservations about another campaign in a similar place. The Albany failure seemed to prove a frequent boast of law enforcement officials across the plantation South: "Here we know how to handle Nigras."

When the planned campaign had as its goal the general objective of "enfranchisement of Negro people in Alabama," the primary geographical strategy was to target the state more than a specific place. Birmingham, Montgomery, Mobile, and smaller cities could serve as foci. Once King decided to use the campaign to achieve the specific objective of new voting rights legislation, the geographical strategy shifted toward focusing on the right type of place within the state. The place had to have a well-developed local civil rights organization structure. The black community had to understand nonviolence strategy and had to have demonstrated that it was prepared to support a nonviolent campaign that could extend for several months.[46] The place also needed a local white leadership that would eventually permit violence in the effort to end the civil rights campaign. As early as December 1963, SNCC's James Forman thought that Selma was a place where "direct action" could "get local officials to violate federal laws thereby forcing a confrontation" that would cause the federal government to become involved.[47]

The metropolitan areas in Alabama offered King the black community infrastructure required for a major campaign, but two of them, Montgomery and Birmingham, had been used in previous major civil rights efforts. The campaigns in both cities ended with bargains struck between municipal and civil rights leaders. King told the group of black leaders from Selma, who met with him in the summer of 1964, that "Birmingham was compact and . . . organized" but "SCLC had already been there."[48] By 1964, Birmingham was the only major city in Alabama where blacks encountered significant local barriers in attempts to register to vote, and even in Jefferson County, 20.7 percent of the black voting-age population was registered. In addition, the relatively large numbers of blacks who were registered in Mobile (25.4%) and the belief of civil rights leaders that apathy was the major impediment to urban voter registration helped to rule it out. If major Alabama urban areas were eliminated, then King had only the state's small nonmetropolitan cities from which to select the place to focus the campaign for new voting rights legislation. Despite his reservations about another campaign in a place similar to Albany, there was no other choice but the "open spaces" of the Black Belt, and in the Black Belt there was no alternative to Selma. Perhaps nothing underscores King's predicament in the selection of Selma more than the conflict that the choice created between SNCC and the SCLC. Alabama might have belonged to the SCLC, but SNCC workers claimed Selma as their turf. Strife raged between members of the two organizations throughout the SCLC's Selma campaign.[49]

Superficially, in the fall of 1964 Selma did not seem unique in the civil rights struggle. It was thought to be "one of those lonely outposts in the civil rights movement where a handful of dedicated workers" were "pioneering against seemingly impossible odds."[50] However, other than Montgomery, Selma was the only city in the Black Belt with a longstanding local civil rights organization of any magnitude. Further, the two-year effort by SNCC to register black voters had helped prepare the black community for a sustained nonviolent civil rights campaign by the SCLC. The regional situation of Selma also was important. If a major confrontation between civil rights demonstrators and law enforcement officials failed to develop in Dallas County, then the campaign could be expanded into the small towns of Wilcox, Perry, and seven other Black Belt counties that also offered potential for violence and police brutality. In retrospect, it was obvious to Joseph Smitherman, Selma's young mayor who assumed office in October 1964, that Martin Luther King Jr. and other SCLC officials "picked Selma just like a movie producer would pick a set. You had all the right ingredients." Sheriff Jim Clark "had a helmet like General Patton, he had the clothes, the Eisenhower jacket and the swagger stick. . . . I was a young mayor with no background or experience, one hundred and forty-five pounds with a crew cut and big ears."[51]

THE SELMA CAMPAIGN

Four days after he was awarded the Nobel Peace Prize, Martin Luther King Jr. arrived in Selma to launch what an SCLC news release called "an intensive state wide Negro voter registration drive." But a few weeks later, Andrew Young, an SCLC spokesman, announced that additional voting rights legislation was the primary goal of the campaign.[52] The initial demonstrations, including marches to the Dallas County Courthouse, were countered with arrests by Sheriff Clark, but no major confrontations developed. Six weeks after the campaign began, *New York Times* journalist John Herbers wrote that "Selma was not likely to become another Birmingham."[53]

On February 15, as planned, SCLC leaders spread the voting rights campaign to Wilcox, Perry, and other surrounding Black Belt counties. Marches were held in Camden and Marion as well as in Selma. A bloody encounter between demonstrators and police occurred in Marion on February 18. Perry County's white leaders asked that state troopers be sent to help control the demonstrators. An evening march from Zion's Chapel Methodist Church to the Perry County Courthouse was blocked by the troopers. The streetlights suddenly went off, and the troopers attacked the marchers. Jimmie Lee Jackson, a young black man, was fatally shot by a trooper.[54] King immediately telegraphed Attorney General Nicholas Katzenbach: "We cannot in good conscience relent in our efforts to gain the right to vote and we urgently request federal protection in the pursuit of this basic right."[55]

The Marion incident proved to be only the prologue to what followed. On February 22, King told a group of more than seven hundred assembled at Brown's Chapel AME Church that a motorcade should be organized to drive across the Black Belt from Selma to the state capitol at Montgomery to protest denial of the right to vote. "We are going . . . to Montgomery. We hope to have people from all over the state to march on the capitol . . . to tell Gov[ernor] W[allace] we arent going to take it any more."[56] The death of Jackson a few days later caused the motorcade idea to be changed to a march, which was to be led by King and was to start from Selma on Sunday, March 7. George Wallace issued an order prohibiting the march and told newsmen that state troopers were "to use whatever measures [were] necessary" to prevent it.[57]

Although a march was planned, the actual decision to commence it was hastily made late in the morning on Sunday, March 7. In a letter to Martin Luther King Jr. dated March 7, John Lewis and Silas Norman denounced the SCLC's methods in Selma and stated that "the Alabama SNCC staff basically disagreed with the march" because we "strongly believe that the objectives . . . do not justify the danger and the resources involved."[58] Ralph David Abernathy was at the West Hunter Baptist and King at the Ebenezer Baptist in Atlanta with the services at which they were to preach under way when Abernathy received a telephone call from Hosea Williams at Brown's Chapel in Selma. "We got to march. . . . Everything's right," said Williams. Abernathy called King, who after pausing briefly reluctantly said, "Tell him he can go."[59]

When about 525 marchers departed Brown's Chapel, at the front of the column signifying a unity of cause, but not of purpose, was SNCC's John Lewis and the SCLC's Hosea Williams. As Lewis and Williams reached the Edmund Pettus Bridge, which spans the Alabama River on the southern edge of Selma, they were met by approximately 50 state troopers led by Major John Cloud and a large mounted posse of several dozen men led by Sheriff Clark. After ordering the column to halt and disperse, the troopers and posse advanced and attacked using clubs, whips, and tear gas, with a group of white spectators cheering them on from the side of the highway. The mayhem was photographed and filmed by the large party of reporters who had gathered to cover the march. Sixty marchers were treated for injuries, and 17, including John Lewis, who had a fractured skull, were admitted to the local hospital.[60]

For a number of weeks the Johnson administration had been working on new voting rights legislation, which was in draft form before March 7.[61] The incident at the Edmund Pettus Bridge, what later became known as "Bloody Sunday," assured speedy passage of the potent legislation that Johnson sent to Congress. A few hours after the event, film of police attacking and even pursuing helpless marchers was shown to the nation on television. During the days that followed, newspapers and magazines were filled with stories about the bridge incident and the helpless state of blacks in Dallas County and the plantation South. The beating to death of the Reverend James Reeb on a Selma street the day after the bridge incident, together with the continued attempts of George Wallace to prevent the march, added to the drama and increased support for voting rights legislation. In a nationally televised speech to a joint session of Congress on March 15, Johnson announced his proposal for a sweeping voting rights bill.

The march from Selma to Montgomery commenced again on March 21. Although the event was an anticlimax to the SCLC's Selma campaign, it was, in retrospect, the apex of the civil rights movement. Approximately thirty-two hundred persons from all parts of the nation gathered for the departure from Brown's Chapel. King knew that the march had assumed symbolic meaning beyond its original purpose, and he addressed the crowd in effusive biblical allegory that he frequently evoked: "Walk together children, don't you get weary, and it will lead us to the promised land. And Alabama will be a new Alabama, and America will be a new America."[62] The marchers included blacks and whites, even some white Southerners. Unknowns mingled with the famous. The route along U.S. Highway 80 led across the heart of the Black Belt, which figuratively as well as geographically was the heart of the plantation South. On Thursday, March 25, a procession of twenty-five thousand entered Montgomery. Wal-

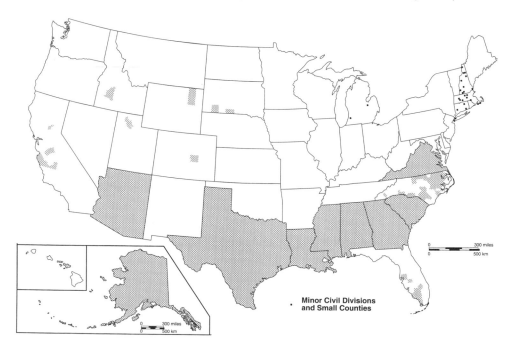

MAP 7.3. Jurisdictions under the Voting Rights Act as of 1985 (source of data: U.S. Department of Justice 1985)

lace refused to receive the petition that the marchers brought, but the entry into the city and the speeches to a crowd of seventy thousand at the state capitol were carried live to the nation by television and radio networks.

The Voting Rights Act that Lyndon Johnson signed into law on August 6, 1965, was the second piece of critical civil rights legislation of the era. The act abolished literacy tests, poll taxes, and other devices that were employed to prevent blacks from voting in seven Southern states. Because of legal devices and long histories of voter discrimination, certain states and smaller jurisdictions were placed under the provisions of the act by the attorney general (map 7.3). The act also provided for the attorney general to send federal examiners (registrars) to counties that continued to discriminate in voter enrollment and observers to those in which election fraud might occur. In what eventually proved to be one of the most powerful, far reaching, and controversial provisions of the act, the attorney general or the United States Court for the District of Columbia was required to approve any changes in voting laws. A few years after the act was passed, the United States Supreme Court ruled that changes in voting districts, polling places, and municipal boundaries came under the purview of the law.[63]

Immediately after Johnson signed the 1965 Voting Rights Act, the attorney general sent federal examiners to nine counties, including Dallas, Perry, and Wilcox, Alabama, to register black voters (map 7.2). By July 1966, examiners had been sent to thirty-two counties in Alabama, Mississippi, Louisiana, and South Carolina, where they had registered 117,017 black voters. The number of blacks registered in Dallas County increased from 1,516 in August 1965 to more than 10,000 nine months later.[64] During the two years following passage of the Voting Rights Act, federal examiners were sent to sixty-two counties, most of which were in the plantation regions of Alabama and Mississippi, the hard cores within King's "South of resistance." By August 1967, federal examiners had enrolled 160,417 persons, 7,383 of whom were white, in Alabama, Georgia, Louisiana, Mississippi, and South Carolina. An additional 1,962,588 voters were enrolled in the five states by local registrars who voluntarily complied with the provisions of the Voting Rights Act. With the reenfranchisement of blacks, voting acquired new meaning for politically apathetic whites. Sixty-five percent of the 1,962,588 voters who were voluntarily enrolled were whites.[65]

THE LAST CRUSADE

Following passage of the Voting Rights Act, the plantation South became less important to the civil rights movement as the geographical emphasis shifted to metropolitan black ghettos in the North and the West. For a brief period following the murder of Viola Liuzzo on U.S. Highway 80 at the conclusion of the march from Selma to Montgomery, SNCC focused its attention on Lowndes County, where the killing occurred. Stokely Carmichael and other SNCC members quickly came to the conclusion that no political future existed for the county's blacks in either the Democratic or Republican Party. The Lowndes County Freedom Organization, which held its first convention in May 1966, was organized, and a snarling black panther was adopted as the party's aggressive symbol.[66] Soon, however, SNCC's leaders were embroiled in a controversy over leadership and objectives. After Carmichael was elected chairman in the place of John Lewis, SNCC became more militant, emphasized black power as a theme, and began to be viewed as a part of the New Left.[67] Several of SNCC's leaders went on to prominent careers, including Marion Barry, who later was elected mayor of Washington, D.C., and John Lewis, who was among the first blacks elected to Congress in the wake of the civil rights movement.

The last major event of the confrontational phase of the civil rights movement in the plantation regions began in June 1966, as what superficially seemed a frivolous one-man journey by James Meredith, who four years earlier had integrated the University of Mississippi. For a brief period in 1962 and 1963, Meredith was at the center of the movement, but the reserved, eccentric man increasingly found himself on the fringe. After attending a White House civil rights conference on June 1 and 2, at which the failure of the Johnson administration to protect field workers was aggressively pressed by a number of those invited, Meredith left for Memphis. Wearing a pith helmet and carrying an ivory-headed cane, he departed the city on June 6 walking south along U.S. Highway 51 toward Jackson, Mississippi, on what he called a "March Against Fear." Meredith's route traversed the Loess Plains plantation region, and he announced that the purpose of his 200-mile "divine mission" was to encourage Mississippi blacks to register to vote. The Meredith walk across Mississippi was better planned than it appeared. Not only was Meredith accompanied by several associates, but he wrote Governor Paul Johnson and the sheriffs of the counties through which his route passed and asked that he be protected. Johnson responded by assigning two highway patrolmen to the march.[68]

On the second day, a couple of miles south of Hernando, an unemployed white man ambushed Meredith with a sixteen-gauge shotgun. The assailant was immediately captured by Mississippi state troopers and DeSoto County sheriff's deputies. Initial news reports stated that Meredith, who was only injured, was dead. Seizing on the opportunity to reemphasize to federal officials the danger that civil rights workers faced in the South, major leaders, including the SCLC's Martin Luther King Jr., SNCC's Stokely Carmichael, the NAACP's Roy Wilkins, CORE's Floyd McKissick, and the National Urban League's Whitney Young, rushed to Memphis. Several of the leaders, including King and Carmichael, declared that they would continue Meredith's march.[69]

The three-week trek across Mississippi in June 1966 was the last great march of the civil rights era.[70] The route generally followed the one planned by Meredith, but there were excursions from Highway 51 into the Yazoo Delta to the west and Neshoba County to the east. The march was both a victory parade and a funeral procession. The crusade highlighted the civil rights accomplishments of the past decade, but it also exposed dissension that worked to destroy the movement from within. Euphoria broke out among blacks along the way as the column proceeded across the countryside and passed through remote hamlets and towns. Reaction was as though the marchers brought the first word of the new freedom given to blacks by the 1964 Civil Rights Act and the 1965 Voting Rights Act and was reminiscent of the response of slaves a century earlier upon being told of their emancipation. When the column reached the isolated Yazoo Delta communities of Midnight and Louise, "people came running out of the fields to see and wave. A lot of them started in marchin'. At school they tried to keep them in, but the kids ran out and got in the line with them and started marchin.'"[71]

Dissension, however, set in among the civil rights leaders even before they resumed Meredith's march. Roy Wilkins and Whitney Young refused to participate because Stokely Carmichael wanted the march to emphasize the need for black political action independent from white control. King failed in an attempt to mediate the dispute. King left and returned to the march several times. He went to Chicago to plan a new strategy for the civil rights movement, a thrust into the Northern metropolises. During one absence, Carmichael made a speech at Greenwood, SNCC's most important stronghold in Mississippi, and unleashed his call for "black power." Although Carmichael's speech received only routine mention by the press, it ignited a controversy over exactly what the term meant and fueled the disputes over direction and leadership. Black power, coupled to statements such as Carmichael's "every courthouse in Mississippi ought to be burned down," also led to erosion of support for the civil rights movement, especially among whites, and contributed to its demise.[72] According to one civil rights activist, "the movement" became "defined out of context as bearded beatniks dedicated to overthrow the government."[73] The call for black power, however, helped in the transformation of the civil rights movement into an effort to use the newly restored black political rights to address historical problems and grievances.

IMPACT OF THE CIVIL RIGHTS MOVEMENT ON THE PLANTATION SYSTEM

The situation of blacks residing on plantations and in towns along the route of the Meredith march in June 1966 was, at best, ambiguous. Rapid mechanization and reorganization of plantations had been under way for more than two decades. Mechanization had eliminated thousands of jobs for tenant farmers and day laborers. Sharecroppers and traditional share tenants had almost disappeared from plantations. Since the Second World War, thousands of rural blacks had migrated to cities, but thousands still remained in the Loess Plains, the Yazoo Delta, and other plantation regions. Many blacks had moved to local municipalities, a continuation of the process that Dorothea Lange and Paul Taylor had witnessed during the 1930s. The growing local group of municipal blacks were under- and unemployed. Of the blacks who remained on plantations, the more skilled younger males found employment as machinery operators in the regions where agriculture was viable. Some elderly blacks and black females with young children also lived on plantations in old tenant shacks that had not been razed. This group, however, barely subsisted. Except for a few weeks weeding cotton in the spring and early summer and picking the crop for a few weeks in the fall, there was little work for them. They were trapped victims who were without jobs or hope whether they lived on the lower Georgia Piedmont and in the Alabama Black Belt, where the plantation system was in its dying gasps, or whether they resided in the Yazoo Delta, where the system remained viable in its neoplantation form.

By the mid-1960s only 4,595 black tenant farmers remained in the Yazoo Delta; in 1945 there had been 44,951. The old sharecropping system had been so decimated in the Delta and other plantation regions that the Bureau of the Census ceased to collect data on sharecroppers after 1959. Approximately 3,000 machinery operators were the mainstay of the Delta's reorganized plantation system. The jobs of the new labor elite were secure, but wages were low. Machinery operators worked from sunup to sundown for six dollars per day and rent-free houses, usually old tenant shacks that had been moved from the fields into the lines of dwellings along roads near plantation headquarters. Another 25,000 to 30,000 men, women, and children in the Delta were day laborers. They worked only two and one-half months, from May into July, weeding cotton. The pay was three dollars for a sunup-to-sundown day. Day laborers were recruited by black entrepreneurs who hauled the temporary workers to plantations in old trucks and buses for fifty cents per day.[74] Among the entrepreneurs were black preachers who secured day laborers from their congregations and transported them in church buses.

The mechanization of cotton plantations attracted significant media and scholarly attention at its genesis during the 1930s. Ironically, acceleration of mechanization after the Second World War and the replacement of thousands of men, women, and children by machines and chemicals went largely unnoticed and unstudied. Not until the final stages of the process, which coincided with the confrontation phase of the civil rights era, did the impact of mechanization on farm labor again become an issue. Labor displacement by machines reemerged as a national issue in the 1960s, in part, because planters could easily retaliate economically against blacks who attempted to register to vote, attended civil rights rallies, or in some other way challenged their subjugation. James Meredith's comprehension of this fear among blacks of his home state is what helped trigger his seemingly eccentric odyssey. The context in which many rural whites still viewed blacks, even after passage of the 1964 Civil Rights Act and the 1965 Voting Rights Act, is revealed by what the sheriff of DeSoto County replied when asked if he would protect Meredith. "We are going to treat James Meredith just like any other nigger chopping cotton in the fields."[75]

The solution to grass and weed control, the phase of traditional cotton production that seemed to defy all efforts to eliminate hand labor, was achieved with the introduction of the first effective herbicides in 1955, the year following the Supreme Court's landmark decision on school desegregation. Although in the mid-1950s planters had the means to eliminate all hand labor in cotton production, the availability of cheap black labor caused many not to do so. The herbicides marketed in the late 1950s and early 1960s were not as effective as chopping grass and weeds from cotton rows with hoes. Most planters who used herbicides employed day laborers to supplement the chemicals. Because hand-picked cotton was superior to that picked with machines, it was worth a few more cents per pound. This was an incentive for some planters to continue to pick all or part of their cotton by hand if inexpensive black labor was available.

The first large-scale eviction of black tenant farmers in retaliation for their civil rights activities which attracted significant media attention occurred in Fayette and Haywood Counties, Tennessee. In certain respects Tennessee would hardly seem to have been a state where black voter registration would have aroused white opposition. Unlike Alabama, Mississippi, and other Deep South states, which were parts of King's "South of resistance," Tennessee, a border state, belonged to the "wait-and-see South" (map 6.1). Tennessee had no major impediments to voting. The state had no poll tax and no literacy test; registration was simple and straightforward. Yet, in 1959 almost no blacks were registered in Fayette and Haywood Counties, and few had attempted to register during the twentieth century.[76] Unlike much of Tennessee, which is Upland South with a white yeoman farmer tradition, the southwestern corner of the state where Fayette and Haywood Counties are located has a plantation tradition. Physically, southwestern Tennessee is Loess Plains, and economically and culturally, it is an extension of the north Mississippi Loess Plains plantation region. In this area of wide fertile creek and river bottoms and gently rolling uplands, blacks outnumbered whites.

In 1959 Fayette County blacks complained to the Justice Department that they were excluded from voting in the local Democratic primary. The Justice Department filed a court suit against the Fayette County Democratic Party, and in April 1960 party officials agreed to end racial discrimination.[77] The discriminatory practices of the local Democrat Party caused blacks in Fayette County and adjoining Haywood County to organize Civil and Welfare Leagues, which launched voter registration drives. The local Citizens' Councils retaliated. During the summer of 1960, some of the black tenant farmers who had registered to vote were informed by their landlords that rental contracts would not be renewed and that they would have to move from the plantations by January 1, 1961. Whites began to subject members of the Civil and Welfare Leagues to other forms of economic reprisal, including refusal of white-owned wholesale firms to sell to black-owned grocery stores.[78]

To draw national attention to their plight, a tent city named "Freedom Village" was created for evicted sharecroppers on the farm of a black landowner. In January 1961 eight families lived in Freedom Village, including a thirty-six-year-old sharecropper with six children. The mass evictions were halted by a court order obtained by the Justice Department. In April, the Federal Court of Appeals held that evictions of sharecroppers who registered to vote violated the 1957 Civil Rights Act. By 1964, the number of blacks registered to vote in Fayette and Haywood Counties had surpassed the number of whites who were registered.[79] Although the eviction of black tenant farmers from plantations in the two Tennessee counties briefly disappeared as a public issue, the number of black tenant farmers expelled from landholdings actually increased. Mechanization, rather than civil rights activities, was given as the reason by landowners.[80]

There is no way to know the number of blacks in the South who were evicted from plantations because they attempted to register to vote or participated in other civil rights activities. Scattered through the documents of SNCC, the SCLC, the COFO, the Mississippi Freedom Democratic Party, and the Office of Economic Opportunity are numerous references from across the plantation regions to economic retaliation taken against blacks because of their civil rights activities. It mattered not whether the plantation system was viable or whether it was in demise. If machines did not replace tenants, the land was planted in pine trees or abandoned to sedge grass and scrub forest. From Sunflower County in the heart of the Yazoo Delta, a SNCC worker reported in 1965 that "a plantation worker [was] fired for being a freedom registrant and attending two voter registration rallies." She added: "This type of incident occurs often; it is seldom reported in detail."[81] That same year an Alabama SCLC official wrote Martin Luther King Jr., "In Wilcox County . . . there have been eighty-one reported cases of heads of families being told to either pay their bills by December 1st or be faced with eviction and arrest. This action is being taken by the white power structure as a retaliating measure against Negroes registering to vote. In the majority of these cases the families will be evicted from the farms on which they are sharecroppers or tenant farmers."[82]

Among the infamous incidents of economic retaliation because of civil rights activity was that taken against Fannie Lou Hamer. Hamer's landlord, W. D. Marlowe III, fired her as a timekeeper and sharecropper and evicted her from his plantation after she went to the Sunflower County Courthouse in Indianola in 1962 to register to vote. Although Hamer had little formal education, she became one of the most effective speakers and grassroots leaders of the civil rights movement. Sponsored by SNCC, she toured Northern university campuses and told her story to help raise money and recruit workers for the struggle in the South. Hamer had a major role in the Mississippi Freedom Democratic Party, the organization that challenged the seating of the state's regular delegates to the 1964 Democratic National Convention in Atlantic City, and she remained active in the civil rights struggle in Mississippi until her death in 1977.[83]

The last half of the 1960s witnessed the final confrontations over the displacement of men, women, and children from cotton fields by machines. The process that had begun three decades earlier in the depths of the Great Depression was swiftly completed during the waning years of the civil rights movement. In the Yazoo Delta the percentage of cotton harvested by machines jumped from 69 in 1964 to 95 in 1966.[84] Although attempts by plantation blacks to secure their rights dramatized the final phase of mechanization and accelerated it a few years, the replacement of humans by machines would have occurred in Southern agriculture if there had never been a civil rights movement. The primary factor that caused agricultural mechanization across Southern, as across Northern and Western, farmlands was economic, not social. William S. Hopkins, a Stanford University economist, in 1940 succinctly told a congressional committee investigating rural migration what was happening in American agriculture. Denouncing the image of the prosperous small family farm as a myth, Hopkins went on to say, "There are at work in American agriculture two forces which, by the logic inherent in them, press continually toward the reduction of the importance of the small farmer. These are, first, the efficiency of intensive mechanization; and second, the efficiency of large-scale operation."[85] In the cotton plantation South, the 1960s saw the fulfillment of Hopkins' predictions.

Many adult blacks who remained on Southern plantations in the 1960s were middle-aged and older.[86] Most knew little about tractors, mechanical cotton harvesters, or modern agricultural methods. They were not retrainable as agricultural machinery operators, nor did they have rudimentary skills for occupations other than farming. Fannie Lou Hamer, who was forty-four when evicted from the Marlowe plantation, was representative of a class of strong, middle-aged black females who composed a large part of the residual labor force. Even though famous as a professional civil rights activist, Hamer moved to Ruleville and joined scores of other poverty-stricken, middle-aged black women in the growing shantytowns located in and on the fringes of Yazoo Delta municipalities. There they barely subsisted on surplus food distributed by the Department of Agriculture and, eventually, on food stamps and other federal programs for the poor.

In a brave, naive effort to improve their deteriorating lot as plantation workers, a group of Yazoo Delta blacks with the help of a group of civil rights workers in 1965 organized the Mississippi Freedom Labor Union. Approximately one thousand persons, primarily day laborers, joined the union, and two hundred of them immediately declared that they were on strike. Although few tractor drivers participated in the strike, twelve on the Andrews plantation near Shaw joined the day laborers. The tractor drivers and their families were housed in a group of old army tents on a nearby black-owned farm. Called "Strike City," the camp was reminiscent of

"Freedom Village" in Fayette County, Tennessee, five years earlier. "Strike City" was advertised by a highway sign depicting a large black fist breaking a chain shackle. Union demands were modest, even by standards of the 1960s: a minimum wage of $1.25 per hour, an eight-hour workday, health and accident insurance, and equal employment practices.[87] Both the union and the strike, however, were folly. Though sympathetic for the condition of blacks, the president of the Mississippi AFL-CIO advised union organizers that the situation of agricultural workers in the Delta "was impossible to deal with from a trade union point of view": "How do you organize and produce benefits for some twenty-five thousand seasonal workers who are employed only two and a half months each year and who are largely unemployable elsewhere?"[88]

In the context of the time, the Mississippi Freedom Labor Union was a disappointing and unimportant effort to improve the tragic lives of a few rural blacks. Viewed in historical perspective, however, the union and its failure are significant for what they reveal about the changes in the relationship of blacks to the Southern plantation system. Following the Civil War, the former slaves were able to extract a few concessions from planters by bartering their labor as plantations made a transition from slave to emancipated workforces. By refusing to work under squad and other labor systems that they considered too oppressive and too reminiscent of slavery, blacks "won" share tenancy, which at the time they thought gave them greater freedom. A century later efforts by blacks again to barter their labor as the plantation system made another transition failed. The demise and the mechanization of plantation agriculture resulted in black labor no longer having value in the agricultural system to which American blacks had been intimately tied for almost four hundred years. When the twelve tractor drivers who struck the Andrews plantation said to the owner, "We ask you once more, Mr. Andrews, if we can have a raise [to $1.25 an hour]," Andrews replied, "Hell, no. I've told you no. Before I'd go up, I'd go down. You work for what I pay you, or get out." After he fired the strikers, Andrews renovated the plantation houses and hired white tractor drivers.[89]

The severing of American blacks from the plantation, their principal home and place of employment from the seventeenth into the twentieth century, occurred during the relatively brief period between 1910 and 1970. In the first decade of the twentieth century 71 percent of the nation's blacks were still concentrated in the rural South. Sixty years later only 18 percent of the black population remained there. History offers few comparable cases of the redistribution of such a large group of people in so brief a period.

Ironically, a federally mandated minimum wage for agricultural workers, which was intended to better the conditions of the blacks who were still dependent on the plantation, actually proved to be the *coup de grâce*. The 1966 Federal Minimum Wage and Hour Law required all farms that used five hundred or more hours of hired labor to pay at least $1.00 per hour in 1967, $1.15 in 1968, and $1.30 in 1969.[90] Sharecroppers came under the legislation because planter-led Southern legislatures had legally changed their status from tenant to laborer when sharecropping emerged on plantations following the Civil War. The minimum wage law caused owners of viable plantations to eliminate all nonessential workers, including day laborers and remaining sharecroppers, by taking full advantage of new agricultural technology and placing more emphasis on soybeans and other crops that could be grown with less labor than cotton.

In 1910, 89 percent of the nation's black population was in the South, primarily in the plantation regions. Sixty years later 53 percent of the blacks were in the South, 39.5 percent in the North, and 7.5 percent in the West. Although several million blacks emigrated from the rural South to the nation's metropolises between 1910 and 1970, at the end of the civil rights movement, a large residual population of several million remained in the plantation regions. Whether the plantation was viable as in the Yazoo Delta or whether it was essentially extinct as on the lower Georgia Piedmont and in the Alabama Black Belt, geographically and economically most blacks were now free from the agrarian system that had enslaved them on the North American continent for more than three centuries. But blacks who lived in the old plantation regions faced a myriad of new problems and new questions. How could the segregation barriers that remained in the face of the 1964 Civil Rights Act be destroyed? How was the political power restored by the 1965 Voting Rights Act to be used? More fundamental were problems and questions related to the desperate situation in which many blacks found themselves. Where would they live apart from the plantation, and how would they sustain themselves?

CHAPTER EIGHT

The War on Poverty

❧ ❧ ❧

Any realistic effort to deal with the problems of
rural poverty and rural housing would require large appropriations
over a long period of years.

Henry Wallace to Eleanor Roosevelt, 1939

The federal government don't care about poor folks.
Well, we got to find out what it does care about and then we got to raise
a lot of hell and disturb them in this thing they care about.

Unidentified Mississippi black woman, 1966

THE POOR SHALL NEVER CEASE OUT OF THE LAND

If a person who knew the plantation South in the Great Depression of the
1930s had left and returned three decades later, he or she would have found
profound changes with the exception of the deprivation. Several million
blacks and whites remained in the depths of poverty. The few federal
programs introduced to alleviate the suffering of the destitute had hardly
made an impression. In 1964, as in 1934, across the plantation regions
people were still ill fed, ill housed, and ill clothed. Because of the demise
and the mechanization of plantation agriculture, many were actually in
worse relative circumstances. Housing was still the most profound landscape expression of the poverty (fig. 8.1). Although plantation housing of
the mid-1960s appeared little changed since the depression, there were a
few subtle improvements. The Rural Electrification Administration created
by the New Deal had helped to bring electric power to all but the remotest
dwellings. Naked light bulbs now hung from the ceilings at the ends of
electric cords, exposing the dingy, greasy newspaper- and magazine-cov-
ered walls of tenant shacks. In some small crowded kitchens, an electric
refrigerator stood across from a wood-burning stove. Electric pumps on
wells had brought water to the shacks on many plantations. Hydrants in

yards had replaced wells and cisterns, but in the early 1960s, a cold water faucet over a rust-stained kitchen sink was a luxury. The affluent poor owned washing machines that often sat proudly on the front porches, placed there as status symbols as well as for the lack of other storage space. Dilapidated automobiles and pickups were parked in front of some shacks. To reduce maintenance costs more than to keep the wind from whistling through cracks between wall planks, some planters had covered their tenant houses with cheap tar paper that resembled brown and red brick veneer.

The civil rights workers who ventured into the plantation regions during the 1960s, whether they were black or white, Northern or Southern, were primarily from the urban world. Most had no experiences that prepared them for what they found in the rural South.[1] The urbanites encountered a destitute world that they not only had never seen but did not even know existed in the United States. The naivety of young civil rights workers, especially affluent Northern whites, resulted in letters to parents and friends filled with graphic descriptions. One student's deflowering encounter with a black family on a Mississippi plantation during 1964 Freedom Summer is hauntingly reminiscent of Raper and Reid's description of Seab and Kate Johnson a quarter of a century earlier. "Upon approaching the house, we were invited on the porch which was strewn with bean shellings, rotten cotton sacks, pieces of a broken stove, and other assorted bits of scrap. . . . On a drooping cot to our right as we came in the door lay a small child (six months old). The child's eyes, nose, and mouth were covered with flies. Not being able to stand such a sight, I tried to chase them away only to be met with the reply of the mother. . . . 'They will only come back again.' "[2]

The depravity among blacks across the plantation regions in the 1960s was merely the continuance of the poverty that had plagued them since emancipation. Although a few blacks became landowners and small-town businessmen, accumulated wealth, and climbed from the abyss of the plantation, most remained destitute tenant farmers or day laborers, who were severed from the agricultural system as it declined or was transformed. The rates of poverty among blacks were staggering across the plantation regions. In the Yazoo Delta an astounding 56 percent of the blacks were poor.[3] The distress of rural blacks would have been even greater if several modest federal programs to assist the poor had not been initiated during the depression and expanded after the Second World War. Two national social aid programs introduced by the New Deal and modified under the Truman and Eisenhower administrations gradually provided modest

FIG. 8.1. Dwellings of black households, Coahoma and Tunica Counties, Mississippi, 1987. Electricity, refrigerators, freezers, washing machines, televisions, propane gas for heating and cooking, and motor vehicles helped to revolutionize life after the Second World War, but for many blacks basic housing remained similar to that of the second generation of post–Civil War dwellings. Charles S. Aiken

amounts of assistance to more and more blacks in the plantation regions. In addition to old age and survivors and unemployment insurance programs, the Social Security Act of 1935 also created grant programs for the aged, the blind, and families with dependent children. The grant programs differed from the social insurance programs in that they required cooperative federal-state funding. By 1940, all states had enacted legislation to participate in the grant programs.[4]

Food assistance was another social aid program begun under the New Deal. In 1935 Congress passed the Potato Control Act, which included a provision that authorized the Agricultural Adjustment Administration of the Department of Agriculture to purchase surplus farm products for distribution to needy families. The primary purpose of the program was to reduce agricultural surpluses rather than to assist the poor. Criticism of the surplus commodities program by grocers and even by food recipients resulted in the introduction of a food stamps program that operated from 1939 until 1943. The food stamps program served only about 4 million persons nationally. The surplus commodity program declined with the introduction of food stamps and was limited during the years immediately following the Second World War by the export of surplus food to feed the hungry in war-torn countries. During the 1950s, renewed emphasis was placed on the domestic surplus commodity program, and the number of recipients increased from 248,000 in 1950 to more than 7 million in 1963. The surplus commodity program helped poor families to subsist but hardly provided an affluent diet. Cornmeal, grits, rice, flour, peanut butter, nonfat dry milk, and butter were major staples. Other foods, including cheese and canned meat, part of which was purchased by the Department of Agriculture, enhanced the diet. A poor family received generic brown bags and cans of food with the exact amount based on the size of household. Because the program was meant to supplement food that was grown and purchased, families could not live sufficiently on just surplus commodities. In 1964, Congress passed a new food stamp act administered by the Department of Agriculture. The surplus commodity program was continued, but counties could not participate in both.[5]

Throughout the South in general and in the plantation regions in particular, the federal assistance programs administered by state and local officials discriminated against blacks. Shortly after emergency relief measures were initiated by the New Deal, an investigation of the programs in Dallas County, Alabama, found: "Negroes predominate on the relief rolls. Nevertheless the discrimination against them is so sharp in the matter of money allotted that this gives an inaccurate picture of cost distribution. . . .

The average amount of relief given to the colored households for the month of June was $5.07 while for the white households the amount was $15.20. These allotments are based on budgetary needs and the Negro is assumed not to have as high needs as the white worker."[6]

Discrimination against blacks continued under New Deal assistance programs. Myrdal found the discrimination in Aid to Dependent Children was pronounced in seven of the Southern states where "the proportion of Negroes among those accepted for aid to dependent children was smaller even than the proportion of Negroes among all children under 16 years of age." He concluded that in "the urban South" blacks received "a larger share of the total relief benefits than correspond[ed] to" their "population ratio." They were "worse off in the rural South, where the most apparent racial discrimination [was] shown, at the same time as the general relief standards [were] very low."[7] When federal legislation to aid the poor was introduced, powerful Southern congressmen and senators saw to it that programs with the potential to assist blacks were greatly diluted and that there was significant state and local control. The 1935 Social Security Act did not include agricultural and domestic workers in the old age and survivors and unemployment insurance programs.[8] The Social Security grant and Department of Agriculture surplus food programs were administered through state and local agencies. In the plantation regions, federal assistance programs for the poor were co-opted and corrupted by planters. The programs became an extension of paternalism; planters largely determined to whom aid would be given and from whom it would be withheld.

As plantation agriculture was mechanized and black tenant farmers were eliminated, federal programs to aid the poor were manipulated to help maintain the supply of cheap day laborers. The programs allowed blacks to subsist when there was no work in the fields, but aid was withdrawn in late spring when cotton choppers were required to weed fields and was not resumed until the cotton harvest was almost finished late in the fall. In 1943, Louisiana introduced a requirement that Aid to Dependent Children be withheld from families in which the youngest child was seven years old if the mother was capable of working in the fields. An "employable mother" rule was adopted in Georgia in 1952 for families in which the youngest child was over three years of age. Under Georgia rules, a mother could not refuse any type of suitable employment, even if the wage was below what could be received from the Aid to Dependent Children program. A Yazoo Delta day laborer with seven children asked a SNCC worker, "What can a person do for his children when there's not enough food to eat?" To force her and her children into the fields, the administrator of the

Bolivar County USDA surplus food program did not issue rations to the family between April 1 and November 1. In 1961, 24 percent of black mothers in the South who received Aid to Dependent Children worked at least part time, compared with 8 percent in the remainder of the nation. In the plantation regions, the median monthly ADC payment was $11.70 for blacks and $21.90 for whites.[9]

Among the techniques used to manipulate blacks and poor whites were the speed with which applications for welfare were handled and the ways in which annual incomes of applicants were computed. Because most blacks worked irregularly and usually kept no records, they had no proof as to whether or not they qualified. Lack of proof left the decision to the discretion of county administrators. In Humphreys County, Mississippi, one family was denied participation in the surplus food program because "they had fixed up their house." Another family was removed from the program at the insistence of a planter who had "blacklisted" the father.[10] A frequently told story of the civil rights movement was the attempt by the white power structure of Leflore County, Mississippi, to use the federal surplus food program as a weapon to halt the voter registration efforts of SNCC. The county board of supervisors voted in October 1962 to cease participation in the program, which supplied food to several thousand blacks. SNCC retaliated with a large food and clothing drive among its affiliates. Despite the interception of the first shipment by Coahoma County law officers, SNCC workers eventually managed to deliver food and clothing to poor blacks in Leflore County.[11]

In part, the administration of programs to assist the poor in the Southern plantation regions was supported by how "welfare" was conceived and viewed at the national level. Aid to Families with Dependent Children was created by elitist Northern urban social workers who misperceived those whom they sought to help. The program was designed to assist a largely mythical and rapidly vanishing ideal family. The social workers failed to grasp that divorced and deserted women with children, especially blacks and members of other minority groups, increasingly headed families that needed welfare. Aid to Dependent Children, however, was devised to assist widows and orphans and low-income families with husbands who were temporarily unemployed. If single women with children were assisted to a too great extent, men's commitment to their families would be discouraged. By the 1950s, at the national level, sympathy for the poor who were not members of nuclear families, especially for blacks and other minorities, had almost disappeared. Aid to Families with Dependent Children, surplus food distribution, and other welfare programs were comprehended as en-

couraging immoral behavior by subsidizing wayward women to have illegitimate children.[12] Manipulation of welfare programs in the South to
force women and even children into the fields did not violate the national
conscience, much less the local ones. Consensus among affluent whites,
and even among lower-income ones, in the plantation regions was that the
welfare programs wasted money on undeserving immoral black women
and their illegitimate children. An oft repeated story was that a young
unmarried black woman supposedly told an older married white woman,
"Every time I gits another one, I gits more money."

THE ECONOMIC OPPORTUNITY ACT

The 1964 Civil Rights Act and the 1965 Voting Rights Act were not the only
major legislation passed during the Johnson administration which had a
profound impact on the plantation regions. Poverty, which had largely
gone unnoticed during the 1940s and 1950s, suddenly reemerged in the
1960s as a critical national issue. From the end of the depression into the
early 1960s, American poverty seemed to be on the wane and was largely
forgotten. John Kenneth Galbraith's *The Affluent Society*, published in 1958,
presented the idea that the poverty that had been so widespread across the
United States in the 1930s had largely been annihilated during the Second
World War and the booming postwar era. Galbraith's book, however, was
followed by studies that depicted a darker and bleaker picture. Robert
Lampman, in a study commissioned by Illinois senator Paul Douglas, who
questioned Galbraith's sanguine view, found that by the late 1950s the
movement of persons out of poverty had decreased significantly.[13] John
Kennedy called for a "war against poverty" during the 1960 presidential
campaign but made no major commitment to instigate it once he took
office. In 1962, a short provocative book by Michael Harrington, *The Other
America*, depicted the poor as an "invisible" mass of 40 to 50 million
persons who were "maimed in body and spirit, existing at levels beneath
those necessary for human decency." Harrington's book and "Our Invisible
Poor," an article by Dwight MacDonald published in the *New Yorker* in
1963, helped to renew the issue of poverty for Kennedy.[14]

Historically, poverty in America was a social and an economic issue. The
civil rights movement recast poverty and made it also a major political
concern. Although blacks were the focus of several critical political struggles in the nation's history, they were not major participants and were
ignored as a direct political force. At the beginning of the twentieth century
the nation's blacks were still concentrated in Southern plantation regions,

their historical homelands where they had been recently stripped of their political power. Interregional migration, however, began to change blacks' potential as a political force. By 1960, blacks were a political majority in several cities and congressional districts in the North where there were no voting restrictions. Despite seeming political inertia in the plantation regions, by 1960 the civil rights struggle, congressional passage of civil rights legislation, and federal court decisions indicated that the power of black votes would soon be unleashed in the South. Poverty was a critical issue to blacks. Not only was a large segment of the nation's black population poor, but because black poverty, in part, was a product of discrimination, it was an issue linked directly to the civil rights movement.[15]

Two events during the 1960 presidential campaign vividly demonstrated to Democrats the connection between racial issues and black voting. Although both the Republicans and the Democrats sought the endorsement of Martin Luther King Jr., he refused to back either Nixon or Kennedy. In October, King was arrested for participating in a sit-in at Rich's, an Atlanta department store. Atlanta mayor William Hartsfield, a racial moderate, had King released from jail, but he was quickly rearrested on order of a DeKalb County court judge for violation of probation on a charge of driving without a license. The judge sentenced King to four months at the state prison at Reidsville. Neither Richard Nixon nor John Kennedy sought to intervene directly. Kennedy, however, called Coretta, King's wife, to reassure her while Georgia governor Earnest Vandiver had the civil rights leader released. The media quickly spread news of the phone call, and the Democrats promoted the incident as evidence that Kennedy had more concern for blacks than Nixon. Kennedy received 68 percent of the black vote, whereas in 1956 Adlai Stevenson had received 61 percent. The outcome of the 1960 presidential election in Fayette County, Tennessee, reinforced for Democrats the relationship between black issues and voting. Under the Eisenhower administration the Justice Department had broken the hold of the all-white Democratic primary on the county. Counter to the national trend, most of the twelve hundred newly registered blacks voted for Nixon and caused Fayette County to go Republican for the first time since Reconstruction.[16]

Just prior to his assassination in November 1963, Kennedy instructed Walter Heller, his principal economic advisor, to include an antipoverty initiative among the new legislative proposals to Congress. Lyndon Johnson learned of the antipoverty proposal shortly after he became president and told Heller, "That's my kind of program."[17] In March 1964 Johnson sent his antipoverty bill to Congress, and on August 20 he signed the Economic Opportunity Act, which initiated the War on Poverty.

R. Sargent Shriver, a former Chicago businessman who was a brother-in-law of John Kennedy's and director of the Peace Corps, was selected by Lyndon Johnson to head the task force that drafted the Economic Opportunity Act. The legislation consisted of six interrelated programs: youth opportunity, community action, rural assistance, small loans, work experience, and volunteer workers.[18] As Congress debated the antipoverty bill, relatively little attention was paid to the potential consequences of the programs. Most of the opposition to the legislation was directed at the structure and administration of the proposed War on Poverty. A large part of the debate over the impact of the legislation concerned Title III, the part of the bill specifically directed toward rural poverty. Title III authorized loans and grants to low-income rural families and the creation of corporations that could purchase and develop land for resale to poor farmers, a program reminiscent of the thwarted New Deal resettlement and tenant purchase projects a quarter century earlier. Breaking up large holdings for resale to small farmers still reeked too much of land reform. The Senate removed a section that authorized grants to small farmers and accepted an amendment by Senator Frank Lausche of Ohio which deleted the part that gave poor rural families the opportunity to purchase small farms. However, a section designed to improve housing, sanitation, and education for migrant and seasonal agricultural workers was added.[19]

Title II, Community Action, was the program that proved to be the most controversial. The greatest impact of Community Action was in Northern cities, where it became a device that organized and unleashed the urban poor. Ironically, the Community Action section of the bill was designed primarily for poor blacks in the plantation regions of the South, where the program also had a significant and far-reaching impact. According to Adam Yarmolinsky, a principal member of the task force that drafted the Economic Opportunity Act, because the problems of black ghettos in Northern cities were not fully comprehended in 1964, "Negro poverty was thought about and talked about largely in the geographical context of the Deep South." Congressmen who represented Northern urban districts viewed the proposed antipoverty legislation "as a 'grits and greens' bill designed primarily to meet the problems of rural poverty in the South."[20] Passage of the Economic Opportunity Act, however, was aided by the rediscovery of poverty in southern Appalachia, which was brought to national attention, in part, by Harry Caudill's unforgettable *Night Comes to the Cumberlands*, published in 1962. Congress might not have enacted the bill that Johnson sent to Congress "had Appalachia not supplied poverty with a white face."[21]

The Community Action section of the Economic Opportunity Act contained six programs: Head Start, Upward Bound, Legal Services, Health Centers, Local Initiative, and Demonstration Projects. The Community Action program was quickly seized upon both by those who wanted to help poor rural blacks and by rural blacks who wanted to help themselves. Community Action provided a method by which local blacks could organize in an attempt to achieve what the 1964 Civil Rights Act and 1965 Voting Rights Act promised. Title VI of the Economic Opportunity Act, Volunteers in Service to America (VISTA), was a program through which civil rights workers could make a metamorphosis and become agents of economic, social, and political change with federal support.

What initially made the Community Action program innovative and controversial was that under the original bill local Community Action agencies could be organized and funded without any state or local government control. The task force that drafted the Economic Opportunity Act considered the provision to give aid directly to the poor through agencies that they controlled one of the most critical features of the bill. A principal concern was that Southern governors and state legislatures might deny poor blacks antipoverty programs. When the bill was drafted, mention of Alabama governor George Wallace's name immediately quelled any discussion that questioned the provision whereby local Community Action groups could bypass state and local governments in seeking funding.[22] During congressional hearings, Southerners were concerned that Community Action grants might be given to new local organizations involved in civil rights. Title II was modified to require that Community Action grants be given only to public agencies, private nonprofit agencies with an established record of concern for the poor, or new private nonprofit agencies created by existing ones with established records.[23]

Congressional efforts to restrict access of civil rights organizations to Office of Economic Opportunity programs were naive and meaningless, for the War on Poverty was an extension of the civil rights movement. Not only did civil rights workers organize Community Action agencies, but they also sought employment with the Office of Economic Opportunity. In San Francisco, the leaders of CORE filled the upper staff positions of OEO programs.[24] Within the OEO, the unquestioned belief prevailed that civil rights and the efforts to eliminate poverty were essentially one and the same. At a July 1966 White House meeting of federal civil rights coordinators held in the wake of Stokely Carmichael's call for black power during the James Meredith march across Mississippi, Samuel Yette, special assistant to Sargent Shriver, emphasized that there was "an unusually high correlation between the goals of the civil rights movement generally, and the goals of the OEO."[25]

The civil rights mission of the OEO sometimes overshadowed the primary objective of alleviating poverty. OEO executives often took special effort to make the civil rights mission of the agency clear to Southern local officials. In the rejection of the application of the newly organized Gainesville-Hall County [Georgia] Community Action Program, John Dean stated, "The application contains language that is paternalistic and racist. . . . While I am sure there are individuals with this kind of thinking who have mauvered [*sic*] into positions of authority in CAPs in many places, must we now accept so blatant an example of it in an application requesting funds to correct the evils that these stereotypes created?"[26] The efforts to have provincial Southerners, who often had known only a segregated society, comply with even simple civil rights mandates sometimes led to delay and confusion. The chairman of the Heart of Georgia Community Action Council complained to Congressman William Stuckey Jr. in the spring of 1966 that after "two years, three re-organizations, four . . . elections and two futile mass meetings" to meet civil rights guidelines, he was ready to "throw in the towel."[27]

VISTA provided a ready-made avenue for civil rights workers to make the transition to community development. Problems arising from the civil rights activities of VISTA workers caused Markham Ball, the program's director, to issue special guidelines. Ball acknowledged that "the problems of poverty and civil rights [were] closely connected" and that it was "natural and correct for . . . a VISTA Volunteer to see the battle for racial equality as one aspect of the War on Poverty." But the volunteers were advised that their jobs required they "bridge the gap between the poor and the rest of the community": "Where once you might have led a lawful protest, now . . . you will simply work quietly on the sidelines in ways that are less spectacular, but . . . more productive in light of the entire VISTA program."[28] Civil rights activists who helped to organize local Community Action programs were not under the constraints of VISTA workers. For them, civil rights and poverty were so intertwined that the transition from civil rights activists to representatives of the poor was a simple, logical process. Leaders such as Thomas Levin, a New Yorker who became an organizer and the first director of the Child Development Group of Mississippi, a Head Start program, thought that "the movement" was "the only channel of communication with the Negro poor and the only source of acceptable leadership." He believed that the OEO staff was compromised by dealings and bargains with the Mississippi white power structure, while the true civil rights workers, the "law-abiders," were those who had remained within "the movement."[29]

RESPONSES TO THE FEDERAL ANTIPOVERTY INITIATIVE

The War on Poverty was launched at the apex of the confrontational phase of the civil rights era. The initial responses of Southern state officials to the federal antipoverty initiative, especially to programs that could aid poor blacks in the plantation regions, varied. Officials astutely realized that substantial amounts of federal funds could be pumped into state and local economies, but they also sensed the civil rights mission in the OEO programs. A former school superintendent in Talbot County, Georgia, told an investigator for the agency that "numbers of people" were afraid that the program was a "vehicle to force integration."[30] It quickly became apparent that if state and local governments did not take control of the OEO programs, federal funding would bypass them and go directly to private Community Action groups. State officials in Georgia, Mississippi, and Alabama, however, differed in their initial responses to the War on Poverty. Of the three, Georgia was in the best situation to devise quickly a public system to manage programs and funds, and Mississippi was the least equipped to assume immediate government control. Alabama officials wanted the infusion of new federal money, but the opposition of George Wallace to programs that might aid civil rights efforts in any way impeded the creation of public agencies that could secure and manage the federal funds.

In Georgia, the State Office of Economic Opportunity, a technical assistance agency, was established in November 1964, a few months after congressional passage of the Economic Opportunity Act. Using existing state-sponsored Area Planning and Development Commissions, regional community action agencies and community action committees began to be created.[31] Some of the CAPs quickly became involved in a struggle over whether the new federal programs were to be extensions of the Area Planning and Development Commissions or independent agencies. The Area Planning and Development Commissions wanted to maintain control of the CAPs because of the federal funds they pumped into the agencies and because in some parts of Georgia the new planning agencies needed to show "*immediate* and *tangible* results." Results could be demonstrated more easily by OEO programs than through long-range planning and development.[32] Area Planning and Development Commissions, however, varied significantly in leadership abilities of the local boards that supervised and the managerial abilities of the persons who administered them. In some lower Piedmont counties, the local white leadership was so lethargic that no one showed any interest in the antipoverty programs, even though they could pump badly needed funds into depressed economies. Oglethorpe County was described to an OEO official as a place where "everyone . . . who had any get up and go got up and went—black and white." The county's agricultural extension agent thought that "Oglethorpe's big problem" was "apathy, . . . nobody care[d] about much of anything." When questioned about local racial issues, the elderly superintendent of county schools baited the OEO investigator with the reply that "he inherited lands back in 1913 and 1914 and 'it was as much my responsibility to look after the colored people as it was the mules.'"[33]

Although integrated, most of the initial county CAP boards organized by the Area Planning and Development Commissions were structured so that white community leaders were in control. More than a year after launch of the War on Poverty, an OEO evaluation of the six-county Oconee Planning and Development Commission on the lower Piedmont found that the major problems were "lack of programs, applications, planning, and community action." Only one of the six counties had an active CAP board, and it was "patterned after" the Atlanta CAP board, a "set up" that produced "control by power structure, with poor and Negroes relegated for the most part to [a] 'Citizens Participation Committee.'" At the root of the problem in the Oconee district was an ineffective planning and development director who was controlled by the white power structure.[34]

Ironically, the Planning and Development District that complied best with the civil rights guidelines of the OEO was located in the infamous Inner Coastal Plain plantation region of southwest Georgia, the scene of severe oppression of blacks. The eight-county West Central Georgia Planning and Development District at Montezuma initiated the OEO programs by hiring black and white codirectors and attempting to create local CAP boards that were racially integrated and democratically chosen. One "key man" was largely responsible for the district being "the best combination" of planning and economic opportunity in the state. Jimmy Carter, an Annapolis graduate and state senator, was chairman of the West Central Georgia Planning and Development Board. After the death of his father, Carter left the navy and returned to Plains, a small town in Sumter County near Albany, to manage the family's agribusinesses, which included a peanut warehouse. Carter worked to improve both race relations and the local rural economy. In 1965, a confident Carter told an OEO official, "We have received some criticisms regarding the racial aspects of EO but have ignored them. . . . Our commission . . . can stand on its own. It has stature and is respected in the area."[35]

The Chattahoochee Community Action for Improvement in the western part of the lower Georgia Piedmont plantation region was also among the more viable Georgia Community Action programs. The CAP's district contained seven counties with a total of 156,000 persons, 36 percent of whom were blacks. The initial endeavors included Head Start programs in three counties, Neighborhood Youth Corps and health programs in two, and a Farmers Cooperative in one. VISTA workers assisted with the Neighborhood Youth Corps. An OEO investigation found a considerable number of problems in the Community Action programs, primarily the use of funds to subsidize the Planning and Development Agency and selection of the county CAP boards by the white power structure.[36]

The commencement of the War on Poverty in Alabama corresponded with the climax of the civil rights movement. White hostility to anything that might assist blacks was at a peak, fanned by a race-baiting governor, the SCLC's Selma voting rights campaign and Project SCOPE, and the efforts of the Office of Education to break the school segregation barrier. In January 1965, the same month that Martin Luther King Jr. initiated the Selma campaign, George Wallace appointed a thirteen-member committee composed of the heads of various state agencies to organize and oversee the Community Action programs in Alabama. The committee had no black members. The original plan Wallace devised for controlling the OEO poverty programs also called for creation of local CAP committees composed entirely of mayors of municipalities within a county. After the OEO rejected this scheme in a confrontation with Wallace over the composition of the board in Jefferson County (Birmingham), the governor's next plan was to control the poverty programs by dividing the state into districts. District boundaries were gerrymandered to insure white majorities. Lowndes County, which was 81 percent black, was placed in a four-county district in which blacks constituted 37 percent of the population.[37] A year into the War on Poverty, only two of what OEO officials regarded as "positive examples" of Community Action programs had been created in Alabama. One was in Tuskegee, which had Tuskegee Institute, a majority black population, and newly elected black city officials. The other was in Huntsville, a north Alabama city that was a center for the federal space program. A white businessman had organized a biracial Community Action committee in the rapidly growing city which had a large number of persons who were not Alabama natives.[38]

The lack of aggressive grassroots black leaders in Alabama's plantation regions and the initial failure of state government to develop programs meant that representatives of the poor took the initiative to establish Community Action agencies for blacks. Field workers for the SCLC, SNCC, and the NAACP began to organize poor blacks and encourage them to demand their welfare rights as well as their civil ones. Macon, Lowndes, Dallas, Greene, Hale, and Wilcox Counties were areas of significant activity. Early in 1965, the NAACP began efforts to organize Community Action programs in Greene and Hale Counties. Orzell Billingsly, a Birmingham attorney, was the principal instigator and leader of the effort. Billingsly explained the plan to Sargent Shriver: "It is a fact of life in the South that in a given geographical area there are at least two communities—Negro and White. Our committee's initiative has arisen within the Negro community. However, we are moving presently to contact members of the white community in our county. . . . If the white community rejects our offer to cooperate, . . . we plan to move ahead and apply for a grant."[39]

To checkmate the nascent Community Action efforts of civil rights organizations in the Black Belt, George Wallace had the state Community Action Committee to begin the appointment of local CAP boards. The committee proposed for Greene County consisted of four whites and two blacks. The Greene County Community Action Program Committee organized by Orzell Billingsly and the local NAACP chapter protested to Shriver, alleging that "Negroes who ha[d] been giving their very lives to the cause of social change" were "eliminated" from the state-appointed committee. They also complained about the control of the proposed committee by the white power structure, including the probate judge, the chairman of the county board of commissioners, and the mayor of Eutaw, the largest town. For blacks, the proposed committee could "only . . . maintain the system of exploitation and Negro dependence on the white man."[40]

By the fall of 1965, more than a year after the creation of the OEO, Sargent Shriver and other high-level administrators were concerned about the failure of the agency's programs to reach effectively the plantation regions of the Deep South. A field investigation of the western Alabama Black Belt by Tersh Boasberg, an OEO official, revealed that federal programs to assist the poor were not operating. None of the nine counties surveyed participated in either the surplus food or the newly initiated food stamp programs. Only 12 of Alabama's 339 public housing projects and none of the state's urban renewal projects were in the nine-county area. Six of the nine counties had not submitted school desegregation plans to the Office of Education, and five did not receive any federal aid for public education. The area did not have a Neighborhood Youth Corps, an Adult Basic Education or Work Experience program, a VISTA worker, or a Community Action program other than Head Start. Six small Head Start programs and twenty-four rural loans were the extent of the War on Poverty in the area. At a recent hearing of the Commission on Civil Rights held in Demopolis, "Negroes in the audience laughed heartily and derisively at OEO's fumbling and inadequate efforts to bring the Dream of Opportunity to the Blackbelt."[41]

Boasberg found that three major barriers impeded establishment of Community Action programs in the Black Belt. One was the lack of black community organizations outside of black churches. "No Negro has ever had experience in assessing community needs or planning comprehensive community programs." The lack of expertise among poorly educated rural blacks to comprehend detailed instructions for submitting complicated OEO application forms also hindered organization of Community Action programs. The most important impediment to the OEO programs was the plantation South's vertical social structure with its rigid caste and class system that prevented the cooperation of blacks and whites on community projects. Stated tersely, "face to face contact around the discussion table between Negroes and whites . . . [was] absolutely unheard of." To retrieve the situation in the plantation regions, Boasberg recommended a project called "Operation Dixie." Contrary to the innovative idea of the Economic Opportunity Act that the poor could and should administer their assistance programs, Boasberg concluded that it was "absurdly naive . . . to assume that poor, uneducated and inexperienced rural Negroes and whites [would] suddenly meet together to plan and organize an effective CAP." "Operation Dixie" would be "a massive OEO effort to send into each county of the deep South, a person trained to organize and implement a comprehensive local Community Action Program," including organization of local black leadership and initiation of racial cooperation.[42]

Sargent Shriver and other top OEO officials had to be constantly aware of the federal agency's delicate political situation in handling its revolutionary social programs. The intention of Alabama state officials to organize local Community Action committees checkmated the proposed intervention of the OEO at the grassroots level. Shortly after he suggested his local intervention scheme, Boasberg wrote Samuel Yette, "The problems that the Negroes are having in these counties would be materially affected if we could implement a plan like Operation Dixie but I feel that things have proceeded to a stage beyond this in these five counties. . . . I feel it necessary we take some action here to resolve the questions and before we get into an outright hasole [*sic*] with Governor Wallace's Office."[43]

In part, factionalism and squabbling among blacks interfered with success in organizing Community Action agencies and with the smooth operation of several that were created. Conflict even developed within the Tuskegee Community Action Program in 1965 as the result of the aggressiveness and "almost pathological drive" of its associate director, who, though respected, was feared.[44] The schism that erupted between SNCC and the SCLC in January 1965, when the SCLC moved into Selma and the central Black Belt, which SNCC claimed as its territory, carried over into the War on Poverty. In 1966, the newly organized Dallas County and City of Selma Opportunities Board was ripped by a power struggle between members of the two civil rights organizations.[45]

It was considerably easier for whites to obtain black members for Community Action committees than for blacks to solicit white members for committees that they initiated. Through long-established mechanisms extending from the first Reconstruction, and even back to slavery, ruling plantation-region whites had employed the services of the more affluent, better-educated black elites—landowners, ministers, teachers, and businessmen. According to Myrdal, "an upper class position in the Negro community nearly automatically . . . [gave] a Negro the role of Negro leader. He [was] expected to act according to this role by both whites and Negroes. . . . The whites soon learned that they could find as many 'Uncle Toms' among Negroes of upper class status as among the old-time 'darkies,' and that educated persons often were much more capable of carrying out their tasks as white-appointed Negro leaders."[46] When told by OEO officials to appoint blacks to Community Action committees, the white leaders instinctively turned to the traditional group of black elites.

The officers of the Wilcox County SCLC chapter complained to the OEO that two state senators were attempting to organize a Community Action program for Wilcox and Marengo Counties, ignoring their efforts to create one. The SCLC had failed in its attempt "to get their brothers in the power structure to cooperate with them in bringing the war against poverty to Wilcox County." The board of the Community Action Program that whites organized was described as an example of "railroaded democracy." Although it had ten black and thirteen white members, the blacks had no authority. Office of Economic Opportunity officials had no objection to blacks organizing Community Action agencies that had boards with no white members, but they expected proof that an exhaustive effort had been made to solicit white participation.[47]

Progress of the War on Poverty in Alabama continued to be impeded, and even the Huntsville CAP proved that it was not the "positive example" it initially was thought to be. Closer inspection revealed that the CAP was dominated by its chairman rather than governed by a board. Its programs were "old-line welfare activities," including "a thinly disguised attempt to provide maids and gardeners for Huntsville's booming suburban middle-class."[48] By the summer of 1966, two years after the antipoverty crusade was launched, OEO programs still had not effectively penetrated the Black Belt. Half of Alabama's counties, including most of those in the Black Belt, did not have a functioning Community Action program, and some Black Belt counties still did not even participate in the Department of Agriculture's food programs. At the instigation of civil rights workers, pathetic letters painstakingly written in wretched scrawls began to be sent to the White House and the Commission on Civil Rights. "It was five family on . . . [this] place that I know about are not farming because he [the white owner] have took the cotton land from these poor color people," stated one letter. "All of the family that are living on his place have a house full of children. If you all can get a food program in Hale County, please hurry up because some of these poor people need one."[49]

One means of establishing Department of Agriculture food programs outside local county governments was for Community Action agencies to administer them. However, the failure of the Community Action efforts in the Black Belt impeded implementation of this method of providing food to the poor. In 1966, the OEO attempted to create "single purpose organizations" to distribute the federal food. One of the first such organizations was established in Hale County, but it was obstructed by the local probate judge, who initially refused to approve the food program.[50] Two years later, the federal food programs were still not well established in the Black Belt. The Commission on Civil Rights found that in 1968 approximately 75 percent of the public assistance recipients were not receiving food stamps.[51] The lack of initial success of the Community Action programs in the Alabama Black Belt demonstrated the vulnerability of the provision in the Economic Opportunity Act for the programs to bypass hostile state and local officials. Such officials could be effectively evaded only if there were capable and aggressive local leaders and nonprofit organizations that would serve as sponsors for Community Action programs.

Mississippi did not have the effective state agency structure for creation and administration of Community Action programs which Georgia had, nor did it have a governor who, like George Wallace, attempted to block or secure tight control of all antipoverty programs. Initially, public officials in Mississippi, the nation's poorest state, were apathetic to the War on Poverty. The first large OEO grant was not given to a state agency until more than a year after the commencement of the War on Poverty, when the Mississippi Department of Public Welfare received $1.6 million for Project HELP, a program to improve the distribution of USDA surplus food. Even Project HELP was an indirect result of private organizations. The state government proposed the food program to the OEO only after the Delta Ministry, an organization supported by the National Council of Churches, and the United States National Student Association sought to begin distribution of surplus food in Mississippi counties. Lack of interest by state and local officials meant that the largest share of initial OEO funds in Mississippi went to Community Action agencies sponsored by private organizations.[52]

Four agencies—Mid-State Opportunities, STAR, the Child Development Group of Mississippi, and Coahoma County Opportunities—controlled large shares of the thirty-one million dollars allocated to Mississippi OEO programs through January 1966. Mid-State Opportunities was created by a former member of the Mississippi House of Representatives and former director of the State Game and Fish Commission. STAR, which was closely affiliated with the Roman Catholic Church, was a program to improve the literacy and vocational skills of the unemployed poor.[53] The Child Development Group of Mississippi (CDGM) was the most controversial of the initial private CAPs. The CDGM was established by civil rights workers associated with SNCC, the Mississippi Freedom Democratic Party, and the Delta Ministry, who recognized the need for day-care centers to serve the poor. Rejected by poor whites whom they contacted, the civil rights workers found significant interest in child day care among poor blacks. An application to fund day-care centers in several Mississippi counties was made to the OEO. The Child Development Group of Mississippi was an agency of Mississippi Action for Community Education (MACE), an umbrella organization for various antipoverty groups. Although the CDGM did not seek to exclude white children, its outgrowth from civil rights organizations meant that its programs almost exclusively served blacks. The CDGM had to find a sponsor in order to meet the requirement of the Economic Opportunity Act that Community Action funds could only be given to private agencies with an established record of concern for the poor or to new nonprofit agencies created by existing ones. Mary

Holmes Junior College, a small obscure private school in Mississippi aided by the Presbyterian Church in the United States of America, was approached by CDGM organizers. The trustees agreed for the college to serve as the sponsor, lured by the opportunity to become part of a "mission with a vision" and twenty thousand dollars to administer the grant.[54]

Because the CDGM soon had Head Start projects in more than one-third of Mississippi's counties, concern developed among state and local officials that the private nonprofit agency would shortly organize the entire state. Most of the counties in which CDGM established Head Start programs were in the Yazoo Delta and the central and northern parts of the Loess Plains. Bolivar was among the Delta counties in which the CDGM operated a Head Start program in 1965. Several white and black leaders, including two county commissioners, several planters, and the mayor of all-black Mound Bayou, believed that the county should assume control of the local Head Start and initiate other War on Poverty programs. They formed the biracial Bolivar County Community Action Program.[55] The whites were motivated by the prospect of federal funds pumped into the local economy and by the belief that a radical group such as the CDGM, which promoted integration, should not have charge of education of the county's black children. Although the board of the new Bolivar County CAP did not include the poor and the black members were selected by the whites, the OEO funded the new agency. It was "crucially important" that the program be funded, an OEO official stated, because this was "the greatest stride in the history of the county in race relations" and would "be a catalyst in moving the entire community toward the removal of present patterns."[56] That the mayor of Mound Bayou and other local blacks wished local control of the War on Poverty programs illustrates that race alone does not unite a people on issues. The biracial Bolivar Community Action Program soon found that it had competition from another new local organization, the all-black Association of Communities of Bolivar County, which an OEO investigator termed "a self-appointed 'Committee of the Poor of Bolivar County.' "[57]

As in Bolivar, whites in many other plantation counties across the South showed interest in War on Poverty programs only after local blacks or outside organizations had begun to create Community Action groups. Whites' perception of blacks and their organizational capabilities continued to be strikingly naive. In the autumn of 1965, the Harris County, Georgia, ordinary (judge) told an OEO investigator, "As far as I know, local Negroes are not organized in any fashion. However, somebody is pushing Negro registration. I don't know who it is."[58] White leaders in Coahoma

County, Mississippi, became interested in the War on Poverty only after they were told that the Southern Education and Recreational Association, a private nonprofit corporation, proposed a large antipoverty project. The whites were amazed that such a venture could have progressed as far as it had without them even knowing of it.

Jessie Epps, an electrician who was born in Coahoma County and had been part of the great migration of rural blacks to Northern cities, was the impetus behind the Southern Education and Recreation Association. A resident of Syracuse, New York, Epps was chairman of the local chapter of the International Union of Electrical Workers civil rights committee. The influx of Southern blacks had created a number of racial problems in Syracuse. Epps came to the conclusion that if the blacks who migrated from the rural South were better educated, their economic and social integration would be easier. The AFL-CIO and the Ford Foundation expressed interest in Epps' idea to create an adult education project in the Yazoo Delta. The War on Poverty began while plans for the project were in an embryonic stage, and the OEO was approached for funding.

Epps was able to interest several blacks in Coahoma County in the adult education project, including Aaron Henry, a Clarksdale pharmacist who was a major Mississippi civil rights leader, and Benny Gooden, a former classmate at Coahoma Junior College, a state school for blacks. To Epps' surprise, several prominent whites also expressed interest in the project, including Oscar Carr, a wealthy planter who was a graduate of the United States Naval Academy. Carr stated that he joined the effort to assist poor illiterate blacks because "one has to have compassion for his fellow man." But, like most other prominent Mississippi whites who supported the War on Poverty, Carr also believed that lack of white leadership would mean black control of the antipoverty programs and more federal intervention in local affairs.[59]

Epps' idea for an adult education program led to the creation of the Southern Educational and Recreational Association, a nonprofit organization chartered in New York. In May 1965 the agency received a $345,000 OEO grant for its programs in Coahoma County. A few months later, Coahoma Opportunities, a Mississippi-chartered nonprofit organization with a biracial board of directors, replaced the Southern Education and Recreational Association as the agency through which OEO funds were channeled to the county.[60] Although Coahoma Opportunities was only "within the rough parameters" of OEO guidelines and the rules were "bent in the areas of Civil Rights, Parent Participation, and Program Administration," the organization was selected as a demonstration project. The OEO gave Coahoma Opportunities more than $2.5 million during its first six months. Among the agency's positive attributes were a group of moderate white community leaders who were willing to oppose conservative ones, a CAP board that had a "stormy past" but was "integrated 50/50" and seemed "to be working well," a program administration that was "not bad," and a black deputy director, Bennie Gooden, who was a "sharp enough" experienced administrator. By 1967, the Community Action programs had grown to include job development, legal aid, early childhood development, and medicare alert, in addition to adult education and Head Start.[61] Coahoma was among only a few nonmetropolitan counties where the OEO funded an intense attack in the War on Poverty.

EMERGENCE OF A REDEFINED PATERNALISM

Neither rank-and-file blacks nor whites in the plantation regions initially comprehended the objectives and potential of the War on Poverty and its programs. To whites, the antipoverty programs, at best, wasted money on undeserving blacks. At their worst, they aided the civil rights efforts and eventually would lead to a takeover by blacks. To blacks, the programs were unfairly administered to prevent them from getting their fair share and to keep the white power structure in control. "We face the dilemma of being criticized sometimes for funding, and at other times for not funding, and frequently for funding the groups we do fund," Samuel Yette lamented at the July 1966 White House meeting of federal civil rights coordinators. The superintendent of Harris County, Georgia, schools explained the segregated Head Start program to an investigator: "My hands are tied. If I had not moved nothing at all would have been done. I want to help these people, my conscience makes me do this, but I have already received criticism and this puts me in a bind. It is some accomplishment that white people here haven't stood in the way of Negroes being helped."[62]

The disputes over the composition of Community Action boards confirmed that most whites still were not willing to share political or economic decisions with blacks. As the historical geographical cores of black population, the plantation regions still constituted the medieval nation, the holdouts of what forty years earlier Ulrich Phillips termed "a common resolve indomitably maintained" that the South "shall be and remain a white man's country."[63] Expressed in the gutty, racist language of the sixties, this attitude of many whites is revealed in what William E. Young, president of the Coahoma County Board of Supervisors, allegedly said to Oscar Carr, a moderate, concerning the formation of a biracial Community Action board: "Mr. Oscar, I know a nigger better than anyone in the county. I'm older than you and I've dealt with them all my life. Give a nigger an inch and he will take a mile. Aaron Henry [local head of the NAACP] ain't coming to my office and he ain't being appointed to any board I got anything to do with. I ain't got education but I've experience dealing with niggers." An attorney who was a leader of conservative whites in Coahoma County but had worked to improve funding for segregated facilities for blacks stated that Clarksdale had never had a biracial committee and that he opposed the creation of one. When asked about the need to improve communications between the races, he responded with a conventional contrived answer that dialogue had always been open.[64] A few months after his trip to evaluate the OEO's problems in the Alabama Black Belt, Tersh Boasberg conducted a similar field investigation in the Yazoo Delta. He found that conditions in the Delta were as bad as, if not worse than, those in the Black Belt, and he confirmed the closed-mind attitude of most whites. "Out of the subtle, exceedingly complex tangle of Mississippi politics, economics and customs," wrote Boasberg, "at least two facts arc unmistakable: (1) the white man has no intention of giving thc Ncgro the franchise; and (2) the condition of the Delta Negro is abysmal. Any meaningful solution to the problem calls for massive Federal intervention, wholesale change in Mississippi politics, and huge programs planned and administered with Negro participation. While this is the goal, it is clearly unattainable at this time."[65]

Some planters, such as Billy Pearson, who owned the 1,500-acre Rainbow Plantation in Tallahatchie County, Mississippi, in retrospect declared that the socioeconomic consequences of mechanization were oblivious to them. But as early as 1955, J. J. Breland, an attorney for the defendants in the Emmett Till murder trial, boasted: "The whites own all the property in Tallahatchie County. We don't need the niggers no more."[66] However, one reason that some white leaders in the plantation regions resisted the War on Poverty or any type of program that might help to alleviate chronic poverty or assist blacks in any way was that they wished to encourage the migration of a people who increasingly were viewed as no longer having any economic purpose. More important, the 1964 Civil Rights Act and, especially, the 1965 Voting Rights Act made blacks a significant liability. Civil rights workers were "moving across barren and muddy fields from one clapboard plantation cabin to another, carrying out voter registration drives and forming community action groups."[67] Reenfranchisement and imminent desegregation threatened white political domination and the vertical caste-class social structure.

Federal programs that had been co-opted to manipulate black labor were now to be used to force blacks, not just from plantations, but from the plantation regions. A Coahoma County, Mississippi, planter candidly told OEO investigators that he opposed the agency's programs because their real purpose was to disrupt the laws of supply and demand for cheap labor and to impede the out-migration of blacks from the Yazoo Delta. That many white leaders in the plantation regions did not have a genuine desire to recruit factories or initiate any new type of economic development to provide new job opportunities for blacks severed from the plantation system is illustrated by what the white chairman of the Bolivar County, Mississippi, CAP asked an OEO investigator. "Since no industry will be coming into Bolivar County why train people for jobs?"[68] Another OEO official thought that "the important whites" in Hale County, Alabama, were "adamantly opposed to any anti-poverty program on the quaint grounds that they hope[d] to starve the Negroes out." The editor of the county weekly newspaper, who owned several thousand acres in the Black Belt, told him, "We like our Nigras, but we can't afford to keep 'em around."[69]

Some white leaders realized that all blacks would not migrate, and in counties with large numbers of blacks, their populations would remain relatively large. Furthermore, the tide of change rolling across the plantation South redefining traditional relationships between the races could not be stopped. Rather than resist the impending changes, these perceptive whites, like ones in Panola County, Mississippi, comprehended that only by acceptance of federal programs could they control them and continue their economic, political, and social domination. The War on Poverty was initiated in the midst of the rapid deterioration of the conventional relationship of rural blacks and whites. The demise of traditional intimate plantation paternalism was an important component of the redefinition of racial relationships. Although the old paternalism with its various types of favors and accommodation tied to the plantation system was outmoded, it could be replaced by a more anonymous form. Out of the turmoil of the 1960s emerged a new paternalism in which whites, together with some aggressive blacks, controlled and manipulated federal programs, especially the new antipoverty programs and transfer payments, to redefine and extend their domination.

At a biracial meeting held in race-torn Lowndes County, Alabama, to discuss a proposed comprehensive health program and a clinic to be built with OEO funds, a local white doctor asked, "If we, the Board of Health, do not take this program, could it inevitably come into the county anyway?" When an OEO official replied, "It could," the doctor responded, "We'll take it, then." According to John Hulett, the first black elected to public office in Lowndes County in the wake of the 1965 Voting Rights Act, the health program was an example "of black people's needs attracting lots of federal money for white officials to administer."[70] Although the OEO made a special effort to insure that blacks received at least some of the clerical jobs for the 1966 Project HELP food program in Mississippi, in Leflore County the venture had "the appearance of a program operated for Negroes by whites" because "a white staff certified plantation workers brought to them by white plantation owners."[71] In 1965, the NAACP cautioned that the War on Poverty offered a great opportunity to assist poor blacks, but in communities where they made up a large part of the population, which included central cities of the nation's metropolises as well as the plantation regions, the programs could disintegrate into a form of "white welfare paternalism." Two years later, Stokely Carmichael, SNCC's promoter of black power, together with Charles Hamilton, warned of "welfare colonialism."[72]

Once white leaders comprehended the relatively large sums to be spent in a local community by the OEO, together with the difficulty of continuing to resist the federal programs, they acquiesced. The initial checks paid to blacks working in the CDGM Head Start program in Holmes County, Mississippi, had to be cashed at a five-and-ten store in Lexington because the local banks and other white merchants would not honor them. When the bankers discovered that the local Head Start program had a budget of several hundred thousand dollars, the banks not only began cashing the checks but eagerly solicited the local program's account. The First National City Bank of Clarksdale, Mississippi, sought the accounts of Coahoma Opportunities and also assigned an accountant as comptroller.[73] By the early 1970s, the Lowndes County, Alabama, Comprehensive Health Program, which local white leaders reluctantly accepted, had an annual budget of $1.4 million and, with one hundred workers, was the county's largest employer.[74]

Bolstered by the civil rights movement and by new federal laws, a new, more aggressive black leadership did not always acquiesce to whites' assumption of new federal antipoverty and welfare programs. The programs often initiated local struggles over their control, not just among different white factions within a county or municipality but between whites and blacks. A white attorney in Taliaferro County on the lower Georgia Piedmont publicly charged that the two blacks who administered the local CAP were criminals. An investigator thought that "it was OEO's $212,724 grant that evoked the strong reaction . . . 'giving' the money to a group of local Negroes instead of channeling it through the 'proper people,' i.e., elected officials, that most incensed" the whites. The chairman of the Taliaferro Board of Commissioners candidly told the investigator, "It's not that we don't need the money, but it needs to be properly administered. . . . I feel that if the money comes here, I'm going to get my share of it, whether anyone else does or not."[75] In 1966, a white woman in Bacon County, Georgia, complained to Senator Herman Talmage that "the poor" were not hired in a local Community Action program. Although her family's income exceeded the three thousand dollar annual limit required to qualify for a job with the local CAP, wives of more affluent members of the white community were employed. The spouses of the county tax assessor and owner of a local apparel factory were among the employees, and the wife of a county commissioner was director of the program.[76] "Old City" and "Southerntown" were among the places where whites and blacks clashed over the control of Community Action programs. In Natchez, two whites who worked closely with blacks to create a Community Action agency were removed from the board when state officials insisted that established local white leaders be appointed. At Indianola, a nascent Community Action agency organized by members of the Mississippi Freedom Democratic Party charged that Sunflower County Progress, the agency funded by the OEO, was "hostile to . . . the very people it purport[ed] to serve" and had a biracial board whose black members were puppets of the white power structure.[77]

The belated efforts of Mississippi state officials to control OEO programs and the civil rights activism of the Child Development Group of Mississippi led to a major confrontation that made the federal agency appear to have been compromised. In 1965, Mississippi senator John Stennis launched an investigation of the CDGM. Both a Senate subcommittee and an OEO study found that the CDGM had problems in meeting certain standards for Community Action programs. A new agency, Mississippi Action for Progress, was established by state officials to replace the CDGM as the primary sponsor of Head Start programs in the state and included Aaron Henry, Hodding Carter III, and LeRoy Percy on the board.[78] As an alternative to Mississippi Action for Progress, civil rights workers organized Friends of the Children of Mississippi in 1966. During its first eighteen months the new Community Action Head Start program operated on a small grant from the Field Foundation. After the OEO began funding the agency in 1966, its Head Start programs spread to fifteen counties, most of which were in the Yazoo Delta.[79]

Two decades after the War on Poverty began, a journalist's description of the welfare office in Tunica County, Mississippi, could have been one from scores of counties across the plantation South. After passing by the desks of the staff, which included a cordial, well-dressed black woman, he reached the director's office. The wife of a former state senator, the director was "an immaculately dressed white woman behind a cluttered desk, with a beaded necklace, rosy cheeks and two diamond rings on her left hand." Her detachment from the plight of the poor blacks in her care was centuries old, stretching not only across the New South but reaching into the Old. Seventy-three percent of Tunica County's residents received some type of welfare. The director told the journalist, "I don't think there's a feeling of hopelessness here."[80]

Although in the 1990s the War on Poverty was generally regarded as a failure, not all of the programs that it initiated were abandoned, nor did they fail to produce important results. The Community Action programs, especially, had a significant impact in the plantation regions. Foremost, Office of Economic Opportunity guidelines that required boards of community action groups to include members of racial-minority groups challenged the vertical social structure of the plantation regions. Poor black day laborers found themselves sitting across the table from affluent white planters. Even though whites usually dominated them, the biracial boards were precedent-setting in symbolic equality. A black board member of Coahoma County Opportunities thought that the Community Action program gave moderate whites "a legitimate vehicle for association and communication with Negroes."[81] In clashes with blacks over the composition of boards and control of OEO programs, whites discovered a more competent leadership among blacks than what they traditionally perceived. Protected by federal laws and guidelines, blacks were not as afraid to confront whites. Once the dialogue between blacks and whites was opened, it was extended beyond matters of Community Action committees. The eyes of some moderate whites were opened for the first time, and they began to comprehend the destitute condition of many of the blacks among whom they lived. A white board member of Coahoma County Opportunities candidly admitted that he initially became involved to control the agency, but he later became "committed to the program because the need [was] so great."[82]

OEO programs also placed blacks in white-collar jobs where they dealt with whites as equals and were overt role models. At the initiation of Mississippi's Project HELP in 1966, the only jobs held by blacks in the Mississippi Welfare Department were as warehouse laborers. Although the OEO was not successful in efforts to get the department to hire black supervisors, almost seventy were employed in clerical positions in state and county offices. In a few Community Action programs, blacks were employed as assistant directors or codirectors. Although James Hamilton, the codirector of the West Georgia Planning and Development District, initially worked only with blacks, soon he and Robert Bailey, a Georgia native who was the white codirector, began appearing together at local meetings.[83]

Community Action programs also helped blacks to take advantage of other federal programs. In part, this came through dissemination of information about the federal programs from Community Action groups. Participation in Community Action meetings gave rural blacks more confidence in throwing off the constraints imposed by the plantation. Most important, Community Action programs permitted leadership of Southern rural blacks to move beyond the confines of the black church to which it had been largely limited since emancipation. Ironically, the longstanding role of churches in black communities is what made poor rural blacks better prepared than poor rural whites to take advantage of Community Action programs. Officials of the Chattahoochee Community Action for Improvement found that assistance to whites was limited because they were "much more widely dispersed throughout the region, . . . less organized, and . . . much less noticed than the Negro group."[84] Local civil rights activities had prepared the way for blacks to serve on Community Action boards and as directors of programs. The steps beyond community action boards were participation in political parties and then election to public offices.

CHAPTER NINE

School Desegregation

✠ ✠ ✠

I draw the line in the dust and toss the gauntlet
before the feet of tyranny and I say segregation now,
segregation tomorrow, segregation forever.

George C. Wallace, inauguration address, January 1963

Ol' Wallace, you never can jail us all.
Ol' Wallace, segregation is bound to fall.

African American freedom chant, 1963

The Supreme Court's ruling in *Brown v. the Board of Education* was a precedent-setting decision on segregation in all types of public businesses and facilities.[1] Shortly, the Court struck down de jure segregation of public golf courses, beaches, state parks, and athletic competitions by merely citing *Brown*.[2] However, across the South, especially the plantation regions, most facilities that served the public remained de facto segregated as the 1950s gave way to the 1960s. Although the direct action campaigns of SNCC and the SCLC in the plantation regions focused on the restoration of voting rights to blacks, there also were attempts to crack the segregation barriers. Direct action desegregation efforts were largely confined to privately owned businesses that served the public, including lunch counters, restaurants, movie theaters, and bus and train waiting rooms, and to publicly owned facilities such parks and libraries. The central symbol of segregation in the South, the all-white public school, was left essentially untouched by the direct action organizations. The effort to desegregate public schools continued to be pursued primarily by the NAACP through the courts. This did not mean that direct action had no relevance to school desegregation. The potent 1964 Civil Rights Act was passed by Congress, in part, as a result of violent white reaction to the SCLC's Birmingham campaign and other savage responses to direct action.

Although schools were public facilities, they were deeply interwoven into the fabric of the plantation regions' white society. They were more than just educational institutions. As public high schools emerged and large consolidated plants replaced small one- and two-room structures after the First World War, the institutions increasingly became the foci of white community social and recreational activities. A white community split religiously into Baptist, Methodist, Presbyterian, and Catholic factions was strongly united in its school. Unlike urban schools, small-town and rural ones often were attended by two or more generations of the same families. Rivalry in sports became a major source of entertainment, even for the middle-aged and the elderly. Schools that were not large enough to field football teams competed in basketball, with the girls' teams often generating as much fierce rivalry as the boys. Entire communities turned out for school plays, for Halloween, Thanksgiving, Christmas, and Valentine carnivals, and for graduations. Consolidation of small high schools into even larger ones after the Second World War was accompanied by a meteoric rise in the popularity of football in plantation communities. Introduction of marching bands, majorettes, and halftime shows not only perpetuated white community pride but caused it to ascend to new heights.

In the fiscally conservative rural South, public funding, abysmal for black schools, was not adequate even for the white ones. Education was continually subsidized in a variety of ways. Much of the subsidy came from parents, who donated supplies and labor to maintain buildings and athletic facilities and who purchased teaching materials, supplemental textbooks, athletic uniforms, and a plethora of other paraphernalia for their children. The numerous school community events were held for more than academic and recreational reasons, for the admission prices, though modest, were a critical part of the financial support. Planters and prosperous businessmen usually sent their children to the public schools and dominated their boards. The cores of the faculties were composed of the planters' and businessmen's wives, who had degrees from Southern state and private colleges and universities and, though provincial, were a cadre of fair to excellent teachers. At the beginning of the Great Depression, 76 percent of the teachers in rural and small-town white high schools in Mississippi had four-year college degrees; fewer than 1 percent did not have at least two years of college.[3] By 1950, almost all teachers in the white schools held four-year degrees. High schools emphasized basic subjects, including English and mathematics, and produced graduates who were relatively well educated, if hardly cosmopolitan. Some affluent planter families opted for private colleges and universities, but most sent their children to major public institutions. In 1956, a Clarendon County, South Carolina, planter bragged that twelve of the fourteen graduates of the still-segregated Summerton High School of the *Brown* case had left for college. Most went to the University of South Carolina and Clemson University, but he sent his daughter to Sweet Briar.[4]

Across the plantation regions, black schools, and white ones in areas with large numbers of white yeoman farmers or tenants, had split terms adjusted to the three peak labor demands for cotton. White schools that served large numbers of children of planters and small-town businessmen had a traditional academic year that began in late August or early September and ended in late April or May. Schools revealed social snobbery and factionalism that prevailed in a plantation community. A century after the Civil War, the descendants of the poor whites who had filled the common soldier ranks of the Confederate army packed the offensive and defensive lines of football teams. But children of white yeoman and tenant farmers frequently found discrimination in their attempts to enroll in particular courses; to obtain parts in plays; to compete for majorette, cheerleader, and football quarterback; and to be elected most beautiful, most handsome, and most likely to succeed. Local boards of education even went so far as to gerrymander boundaries to exclude areas with large concentrations of poor whites from particular schools.

The Supreme Court ruled that segregated schools were unconstitutional in 1954, but in *Brown II* in 1955, the Court gave directives as to how desegregation was to proceed. The task of ending segregation was placed on public school officials. The officials were to begin a "prompt and reasonable start," and desegregation was to proceed "with all deliberate speed."[5] "Reasonable start" and "all deliberate speed" were intentionally vague to provide a period for white Southerners to accept the idea and for methods of desegregation to be devised and implemented. Unfortunately, some Southern leaders chose to block rather than to implement the Court's order. Throughout the plantation regions, the initial reaction of most whites to the 1954 Supreme Court decision was disbelief that integration would ever actually happen. So outlandish seemed the threat that only one hundred persons showed up at a public meeting held in Indianola, Mississippi, in June 1954 to discuss methods to prevent desegregation of the local public schools.[6]

One ploy of whites to prevent desegregation was to improve black public schools to such an extent that blacks would be satisfied with separate but equal. Public funding for black schools began to increase following the Second World War, but after the 1954 *Brown* decision it rose dramatically.

The pending *Brown* case helped spur even the Mississippi legislature to enact a public education equalization bill in 1953.[7] New school buildings for blacks sprang up across the plantation regions, and in the mornings and afternoons yellow buses filled with black pupils joined those filled with white ones on the highways and the dusty and muddy back roads of the plantation South. In some cases the improvement in facilities for blacks was an exceptional leap. Black pupils literally left decrepit one- and two-room buildings in the spring and entered modern comprehensive twelve-year plants the following fall. In 1959, twelve hundred black students moved into the large new Hancock County Training School at Sparta, Georgia, which replaced eight small dilapidated buildings. With most of the old eroded fields planted to pine trees, the assessed value of Hancock's land had dropped to only $6.50 per acre. The once great cotton county that was the home of David Dixon, who invented the sweep plow and championed scientific farming, was so poor by the 1950s that the new school was built with a $750,000 state appropriation.[8] To whites the mammoth building with its four acres of floor space was the symbol of equality in education, but to blacks the new "training" facility with its emphasis on "practical education" meant not only the persistence of segregation but their continued condemnation to manual labor. New buildings and buses, however, hardly made the dual school systems equal. Black children continued to be taught by undereducated black teachers, most of whom did not have college degrees. Despite the increases in funding and hundreds of new schools for blacks across the South, education was still unequal as well as separate (fig. 9.1). At the close of the 1950s, six years after the *Brown* decision, Mississippi spent $79 on a black child, compared with $140 on a white one.[9]

Should the ploy to give blacks equal schools fail, the provisional plan to prevent desegregation consisted of various legislative and legal devices. By 1956, organized "massive resistance," a term coined by Virginia senator Harry Byrd, scion of a family powerful and legendary since the colonial period, had begun to emerge at the state, sectional, and national levels. According to Numan Bartley, "the southern mood made massive resistance possible; Citizens' Councils provided the working cadres and crusading fervor; and entrenched southern politicians contributed much of the leadership, direction, and strength." The Byrd political machine, together with Mississippi senator James Eastland and the Federation for Constitutional Government, which he helped organize, was especially significant in formulating and promoting the strategy of massive resistance. From Congress in March 1956 emerged the infamous Southern Manifesto, which denounced the *Brown* decisions and was signed by nineteen senators and eighty-two representatives.[10] In 1954, Mississippians ratified an amendment to the state constitution which permitted the legislature to close public schools. The Georgia and Alabama legislatures passed similar bills in 1956 and 1957.[11]

On cool October nights in the late 1950s and early 1960s, parents, grandparents, brothers, sisters, aunts, uncles, and cousins in small insular worlds spread across the plantation South watched teenagers of their school inflict touchdown after touchdown on the boys from a nearby community and were reassured of their invincibility. Although the school band at halftime was squeaky and notes were missed, *America the Beautiful* and *Dixie* played with a dozen large United States and Confederate battle flags flapping briskly in the chilly breeze heightened the sense of invulnerability. The Confederate flag was resurrected after the *Brown* decision as a symbol of defiance. The Georgia legislature went so far in 1956 as to make the Confederate banner part of the state flag. The ensign was that of a short-lived, defeated country, but those in the stadium regarded themselves its hereditary citizens. The audience was reassured by what the owner of the hardware

FIG. 9.1. The waning years of segregated public schools in the plantation South as depicted by the 1958 graduating class of R. L. Norris High School, McDuffie County, Georgia. From among blacks of this generation came the young local leaders for the confrontation phase of the civil rights movement and the new black political leaders who arose during the last quarter of the twentieth century. Courtesy of the Georgia Department of Archives and History

store, who was head of the local Citizens' Council, said at halftime. The state legislature had nullified the Supreme Court's desegregation order and had passed laws to guarantee that a black pupil would never set foot in their school. The euphoric October nights at the public high school football stadiums would go on, unmarred by black teenagers on the field, forever.

A DECADE OF COURT SKIRMISHES

Although the most overt effects of school desegregation occurred in the nation's metropolises, during the first sixteen years of federal efforts to enforce *Brown* a significant emphasis centered on the South's nonmetropolitan plantation regions. After 1970, the primary focus in school desegregation shifted to Southern metropolises, and after 1972 to metropolises of the North and the West. The concept of the effects of segregation on black children and the primary methods used to measure and to achieve desegregation, including racial percentages and busing, originated in court cases from the South's nonmetropolitan plantation regions.[12]

The struggle to desegregate the public schools in the plantation regions extended over almost two decades and was divided into three distinct phases: the court skirmish and delay stage from the summer of 1954 through the summer of 1964; the token integration stage from the autumn of 1964 through 1969; and the stage of massive desegregation from the winter of 1970 into the autumn of 1972 (fig. 9.2). Fueled by inflammatory rhetoric of opportunistic demagogues, who included Arkansas governor Orval Faubus, Mississippi governor Ross Barnett, and Alabama governor George Wallace, the first phase was characterized by tauntful defiance of federal courts and arrogant boasting among the white populace that school desegregation would never occur. One of the most important cases during the court skirmishes was *Briggs v. Elliott*, in which a federal district court in South Carolina rendered a narrow interpretation of *Brown* which held: "The Constitution . . . does not require integration. It merely forbids discrimination."[13]

In May 1964, ten years after the Supreme Court's landmark decision, 9.3 percent of the black pupils in the South were enrolled in desegregated public schools (map 9.1).[14] In the border South, Missouri, and the District of Columbia, 92 percent of the school districts were desegregated, and 54.8 percent of the black pupils attended desegregated schools. The sparring in the federal courts between 1954 and 1964, however, produced little actual desegregation in the plantation regions. Ten years after *Brown*, the public schools of Clarendon County, South Carolina, and Prince Edward County,

Virginia, remained completely segregated. Fewer than 1 percent of the black pupils attended formerly all-white schools in North Carolina, South Carolina, Georgia, Alabama, Louisiana, and Arkansas, and fewer than 3 percent were enrolled in desegregated schools in Virginia, Tennessee, and Florida. Mississippi had no desegregated public schools. In the eleven states that made up the Southern Confederacy, only 34,109, 1.2 percent, of the black pupils attended desegregated schools.[15]

In federal court–mandated desegregation cases in the plantation regions, local school officials subverted the judicial guidelines in such a way that few, if any, black children actually attended white schools. White leaders of Clarksdale, Mississippi, used a variety of measures to prevent or lessen the impact of desegregation. The city's white pupils were withdrawn from the new white consolidated Coahoma County High School and returned to the old city high school, which was reopened.[16] In April 1964, seventeen black parents filed a court suit that sought desegregation of the Clarksdale school system. The federal district court ordered that the first grade be desegregated in September 1964 and the second grade in September 1965 and that nine attendance zones be established. When the 1964–65 year began, no black pupil registered at an all-white school. The attorney for the plaintiffs accused the city's school board of gerrymandering boundaries in such a way that few blacks were included in the attendance zones of

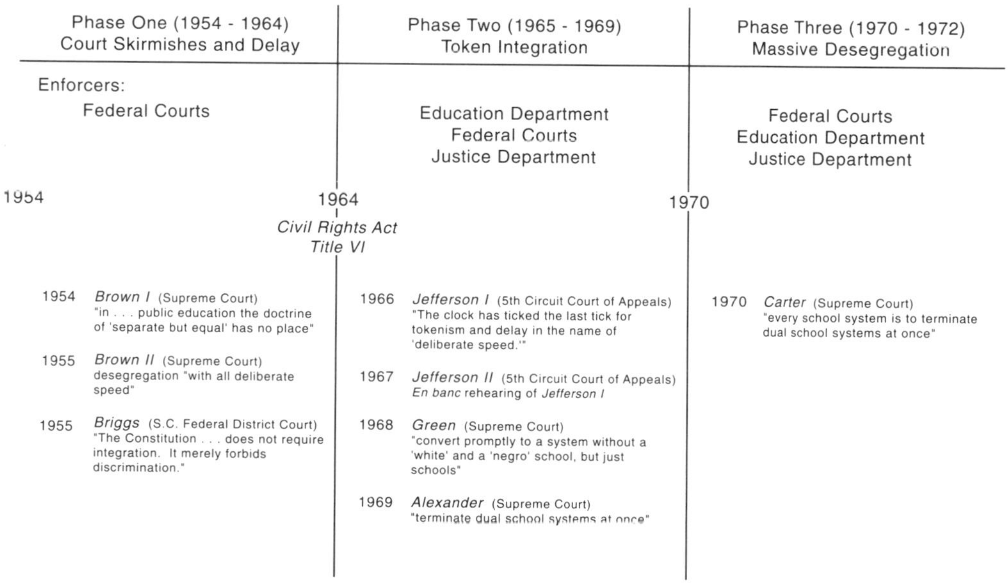

Phase One (1954 - 1964) Court Skirmishes and Delay	Phase Two (1965 - 1969) Token Integration	Phase Three (1970 - 1972) Massive Desegregation
Enforcers: Federal Courts	Education Department Federal Courts Justice Department	Federal Courts Education Department Justice Department
1954	1964 Civil Rights Act Title VI	1970
1954 *Brown I* (Supreme Court) "in . . . public education the doctrine of 'separate but equal' has no place"	1966 *Jefferson I* (5th Circuit Court of Appeals) "The clock has ticked the last tick for tokenism and delay in the name of 'deliberate speed.'"	1970 *Carter* (Supreme Court) "every school system is to terminate dual school systems at once"
1955 *Brown II* (Supreme Court) desegregation "with all deliberate speed"	1967 *Jefferson II* (5th Circuit Court of Appeals) *En banc* rehearing of *Jefferson I*	
1955 *Briggs* (S.C. Federal District Court) "The Constitution . . . does not require integration. It merely forbids discrimination."	1968 *Green* (Supreme Court) "convert promptly to a system without a 'white' and a 'negro' school, but just schools"	
	1969 *Alexander* (Supreme Court) "terminate dual school systems at once"	

FIG. 9.2. The three stages in desegregation of public schools in the southern plantation regions, 1954–72

white schools. Black families who remained in the attendance zones were encouraged to move.[17]

It was more in their role as social centers than educational institutions that public schools were to whites sacred icons that must not be violated by integration of the races. Many whites believed that throwing white and black youths together in schools ultimately would lead to violation of sexual taboo. Martin found that "in the end without a single exception" Southern whites "come around to sex: school desegregation leads to close association, and close association leads to miscegenation, amalgamation, mongrelization."[18] Knowing that whites held such a belief, during the ten years following the *Brown* decision not many blacks in the plantation regions were bold enough to risk transgression of the supreme taboo by attempting to integrate schools, even after the local board of education was under court order to admit them.

TOKEN INTEGRATION

The new civil rights act that Lyndon Johnson signed into law in July 1964 brought two additional federal allies into the desegregation fray, the Office of Education, a branch of the Department of Health, Education, and Welfare, and the Civil Rights Division of the Justice Department (fig. 9.2). The Economic Opportunity Act, which the president signed a few weeks later, added a fourth federal partner, the new Office of Economic Opportunity. The second phase of school desegregation, which lasted from the fall of 1964 through 1969, began with efforts of the Office of Education to enforce Title VI of the 1964 Civil Rights Act. In the bill, largely unnoticed and little debated by Congress, were two sections that affected school desegregation, Title IV and the especially potent Title VI. Title IV authorized the Justice Department to file public school desegregation suits on behalf of persons who were not financially able to undertake them. Title VI prohibited discrimination on the basis of race, color, or national origin in federally financed programs and authorized termination of funding if a recipient failed to comply.[19]

Title VI was expected to have minor significance because most of the South's school districts did not receive much federal money. Moreover, the belief prevailed among Southern politicians and school officials that the small amounts of federal funds which were received would not be terminated to force integration because they supported such essentials for black children as the school lunch program. A few months after Johnson signed the 1964 Civil Rights Act, Congress passed the Elementary and Secondary

Education Act as part of the War on Poverty. This act, which provided large federal grants to school districts under a formula that favored poor districts in the South, made Title VI a potentially potent weapon in the desegregation effort.[20] The Office of Education was drawn into the school desegregation battle with reluctance. No precedent existed for significant intervention by federal officials in local school affairs. During the decade following the *Brown* decision, the agency ignored demands by civil rights organizations that federal funds be used as a desegregation weapon. Even in 1964, the Office of Education interpreted the *Brown* decree very conservatively. At a July 1964 meeting with Mississippi school officials, David Seeley, an assistant for civil rights affairs in the Office of Education, explained his interpretation of *Brown*: "So far as I know the court decision did not require . . . integration, it merely required desegregation. In other words it doesn't require you to force Negro students to go to school with white students. . . . If you have . . . a free choice system . . . and you end up with most of the Negro students going to one school and the white kids going to another, or all of them for that matter, this could, conceivably, be found to be non-discrimination."[21] Seeley's statement, which is essentially a paraphrasing of the *Briggs v. Elliott* decree, is a capstone to the first decade of school desegregation which helps to explain why so little was accomplished in the Deep South.

Once the 1964 Civil Rights Act was signed into law, the Johnson administration pressed for firm enforcement of Title VI. The initial Office of Education Title VI regulations became effective on January 3, 1965. A school system could comply with Title VI in one of three ways: under a final federal court order, by submission of a voluntary desegregation plan, or by affirmation of complete desegregation.[22] Except for a few that were under court order, school systems in the plantation regions could comply with Title VI only by the submission of voluntary desegregation plans. The method used to assign pupils to schools and the pace of desegregation were the two critical features scrutinized in voluntary plans.

At the time the Title VI guidelines were drafted, Burke Marshall, assistant attorney general for civil rights, suggested that the Office of Education should require all school systems to follow the pattern of desegregation ordered by the Fifth Circuit Court of Appeals, which at the time had jurisdiction over much of the plantation South, including Alabama, Georgia, and Mississippi. The court's most recent decisions required integration of the initial grade and then an additional grade each year until all were desegregated. The shocking Selma "Bloody Sunday" confrontation at the Edmund Pettus Bridge in March 1965 occurred in the midst of refining

Title VI guidelines. Selma cast a long shadow. The bridge incident convinced Marshall that white resistance to desegregation had to be dramatically broken. Public schools were the most overt symbols of segregation; the time had come to integrate these cherished icons. The weak Title VI guidelines, which only a few weeks earlier Marshall thought were too aggressive, he now believed were inadequate, and he pushed the Office of Education to strengthen them.[23]

When the final Title VI guidelines were issued in April 1965, they required that for a school system to be in compliance, race could not be a factor in the assignment of pupils or teachers and that no activity or facility, including buses, could be segregated. The most potent part of the guidelines was the timetable for desegregation. At least four grades were to be desegregated in the fall of 1965, and all grades were to be desegregated by the beginning of the 1967–68 year. School systems were given three options for the assignment of students: unitary nonracial attendance areas, free selection of a school by a pupil and his or her parents or guardians (freedom of choice), or a combination of attendance areas and freedom of choice.[24] Because it offered potential for only token integration, freedom of choice was the method of compliance adopted by school systems in plantation regions.

The Office of Economic Opportunity became involved with the education of the nation's children through the Head Start component of the Community Action programs. The close relationship between the civil rights movement and the War on Poverty meant that OEO officials took an aggressive stance and required that Head Start programs be racially integrated. In August 1965 Samuel Yette, the civil rights coordinator for the OEO, boasted that 74.6 percent of the Head Start centers in Alabama were desegregated, while only 6.7 percent of the state's public school districts were. However, the Commission on Civil Rights reported that its study of Head Start in more than forty Southern communities found that "the majority of the children . . . were enrolled in segregated projects located in segregated schools and taught by teachers of their race."[25] Yette announced that in 1966 "freedom of choice" would not be permitted in Head Start centers because the practice had resulted in lack of compliance with desegregation policy. In August 1966 OEO officials terminated twenty-five Head Start programs for failure to desegregate. Some of the terminated programs were sponsored by public school systems that continued to receive other federal funds because they were in compliance with the Office of Education's Title VI guidelines.[26]

A survey of school desegregation during 1965–66 revealed a "slow pace of integration in the Southern and border States" which "in large measure" was "attributable to the manner in which free choice plans" had "operated."[27] An effort was made by the Office of Education to strengthen the freedom-of-choice guidelines for 1966–67, but it remained the principal method of compliance with Title VI. All of the Alabama and Mississippi and 83 percent of the Georgia public school systems adopted freedom-of-choice plans. Between May 1964 and May 1967, the number of black pupils in desegregated public schools increased from 9.3 to 24.4 percent (map 9.1). In the border South, the number rose from 54.8 to 67.8 percent, and in the eleven former Confederate states, from 1.2 to 16.9 percent. Although the accomplishments appear impressive when compared with the previous decade, in the former Confederate states, only 12.5 percent of black pupils attended public schools that were less than 95 percent black.[28] In most states of the plantation South, fewer than 10 percent of the black pupils were enrolled in desegregated schools. In Alabama, Mississippi, and Louisiana, fewer than 5 percent of black pupils attended desegregated schools, and a number of systems had no desegregated schools.

Within the plantation regions, few tangible results were produced by the combined efforts of the federal courts, the Justice Department, the Office of Education, and the Office of Economic Opportunity. The segregation barrier was breached under freedom of choice, but tokenism rather than true desegregation prevailed. Only a handful of county school systems had as many as fifty black pupils enrolled in formerly all-white facilities. Dual public education was still firmly entrenched, despite a mandate by the Office of Education that school districts with little or no desegregation had to have 10 to 15 percent of black pupils enrolled in formerly all-white schools during 1966–67.[29] In retrospect, admission of a few token blacks to white schools under "freedom of choice" was traumatic and momentous. For some Southern whites, segregation was not a matter of degree but all or nothing. To other whites, token desegregation was acceptable as long as it was limited to the "right kind" of blacks. Desegregation, however, must not be permitted to go beyond mere tokenism. The breach in the wall of the segregation fortress had to be guarded to prevent masses of black children from pouring through into the white schools.

Several factors among blacks, in addition to the fear of violation of sexual taboo, contributed to keep freedom of choice from achieving more than token desegregation in the plantation regions. One was a belief that black schools should be preserved. Black teachers and administrators especially thought that abolishment of the dual school system would result in their firing or demotion. Their concerns were not ill founded, for enforcement of Title VI caused 668 black teachers to lose their jobs in 1965–66. Other black teachers opposed desegregation because they believed that many black children could not compete academically with white children. A fatalism ingrained by decades of discrimination dictated that blacks be realistic about their employment potential and recognize that their high schools with emphasis on vocational training prepared students better for the real world than white schools with their accent on scholastic education.[30] Chronic poverty of many black households was another factor that contributed to token integration. In the American South, as in other plantation areas of the world, dress is a mark of caste and class. Although black children

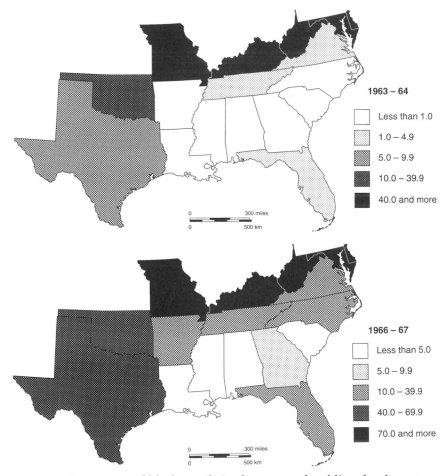

MAP 9.1. Percentage of black pupils in desegregated public schools, 1963–64 and 1966–67 (sources of data: U.S. Commission on Civil Rights 1964, 1967)

frequently showed up on the first day of classes at desegregated white schools overdressed in neatly pressed new Sunday dresses and patent leather shoes and new suits, white shirts, and ties, most black households did not have the financial resources to meet the socially mandated dress codes of white schools. Poor blacks also could not afford to purchase the plethora of workbooks, supplemental textbooks, science equipment, athletic uniforms, and other supplies that the public schools of the plantation South did not furnish but parents of white children were expected to buy as part of their subsidies to the underfunded educational system. Black parents often were overwhelmed by teachers' constant requests for money, which they sometimes interpreted as forms of discrimination and harassment.[31]

Actions of whites were the primary reasons freedom-of-choice plans resulted in only token desegregation. One tactic of school boards was to channel new federal funds and programs into the black schools rather than into the white ones. When free lunches began to be provided to pupils under Title I of the 1965 Elementary and Secondary Act, some districts such as North Panola and South Panola County, Mississippi, made them available only in black schools. Economic coercion, a weapon effectively employed by whites against blacks who attempted to register to vote, was also used to prevent blacks from enrolling in white schools under freedom of choice. Incidents of economic reprisals were scattered across the plantation regions. Tenant farmers were evicted, and teachers, school bus drivers, maids, and janitors were fired if they enrolled their children in the previously all-white schools.[32] In addition, physical intimidation, which ranged from foolish threats among children to cruel acts of violence against families, was occasionally directed against blacks.

The most effective method whereby whites negated freedom of choice was for school boards simply to ignore or impede implementation of Title VI desegregation plans filed with the Office of Education. A study of desegregation in 1966–67 found that "the great majority of school districts in Alabama, Georgia, Louisiana, Mississippi, and South Carolina failed to meet the standards of the guidelines," but "only a small fraction" of the districts had been "subjected to enforcement action." In the majority of the districts investigated, school officials "conformed their procedures closely to the technical requirements of . . . [Office of Education] guidelines . . . , but little actual desegregation was achieved."[33] Freedom of choice broke the segregation barrier in the plantation regions, but without further federal initiative it merely insured that only a few black children would sit among whites in classrooms.

THE CLOCK HAS TICKED THE LAST TICK

School officials in the plantation regions should have clearly understood that token desegregation was not the primary objective of the Office of Education in the enforcement of Title VI. As early as March 1965, an Office of Education official publicly announced, "Desegregation plans based on freedom of choice are . . . no more than transitional devices that ultimately will give way to unitary zoning." In January 1967 Harold Howe, commissioner of education, informed local school officials that the "most important requirement in remaining eligible for Federal assistance [was] to make further progress in 1967–68 in eliminating the dual school system."[34]

By 1965, the Fifth Circuit Court of Appeals in New Orleans, whose district stretched from Georgia through Texas, had spent a decade hearing appeals on desegregation and voting rights cases. The court's load was staggering, made even more burdensome by the same cases that kept reappearing. A major problem that the Fifth Circuit Court confronted was the lack of a uniform set of standards that could be employed in all school desegregation cases. When the Office of Education issued its Title VI school desegregation guidelines, the court seized them as the standards it so desperately needed. The Fifth Circuit Court ruled that the Title VI guidelines were legal and deferred to the agency, which the justices believed was better prepared than the courts to evaluate desegregation plans of various school systems. Judge John Minor Wisdom also attacked the 1955 *Briggs v. Elliott* decree, stating that it could not be used to prevent black pupils from transferring to white schools.[35]

Almost as soon as the Fifth Circuit Court deferred to the Office of Education as the primary enforcer of school desegregation, the federal courts began to reassume the primary role (fig. 9.2). The new initiative came in decisions made in four critical cases between December 1966 and January 1970, the most significant ones since *Brown I* and *Brown II*. The first set of decisions was from the Fifth Circuit Court. In the *United States v. Jefferson County [Mississippi] Board of Education*, which consolidated seven school desegregation suits from Alabama, Louisiana, and Mississippi, the court departed from judicial precedent in both the speed and the method of school desegregation. In the *Jefferson* decision Judge Wisdom pronounced an end to the period of "all deliberate speed." "Now after twelve years of snail's pace progress toward school desegregation, courts are entering a new era. . . . The clock has ticked the last tick for tokenism and delay in the name of 'deliberate speed.' " The decree attacked "apartheid by dual zoning" as the "central vice in a formerly de jure segregated school system." Dual zones were to be converted into a single system, and faculties, as well as student bodies, were to be integrated. Office of Education Title VI guidelines were mandated throughout the Fifth Circuit district. Use of percentages by the Office of Education to measure desegregation was upheld.[36]

Following the lower court's lead, in May 1968, thirteen years after *Brown II*, the Supreme Court in *Green v. School Board of New Kent County [Virginia]* acted to change the guiding principle of desegregation. Like two of the five cases that made up *Brown*, the *Green* case was from the rural South. New Kent County, located between Williamsburg and Richmond, was a geographical microcosm that exposed the basic spatial components of the complexities of school desegregation in plantation regions. Had the Court intentionally sought a simple geographical model that succinctly portrayed the fundamental issues of "freedom of choice," the situation to which "all deliberate speed" had led, a better example than New Kent could not have been found. The county had no cities or even a large town. Approximately half of the county's population was white and half was black. The settlement pattern imposed by the plantation system in an earlier age had now come to haunt the planters' descendants, for the dwellings of whites and blacks were intermingled across the rural landscape in New South fashion. The Court specifically noted, "There is no residential segregation in the county; persons of both races reside throughout." Although the Court failed to grasp the subtle residential segregation, the oversight was negated by New Kent County's having only two schools, New Kent and Watkins, which had a combined enrollment of 1,290 pupils, 550 whites and 740 blacks. There were no attendance zones; both schools served the entire county. Bus routes for the two corresponded; black and white children were bused past one another to racially distinct schools.[37]

In August 1965 the New Kent County School Board adopted a freedom-of-choice desegregation plan in response to a court suit and in order to remain eligible for federal aid under Title VI. Three years later, only 115 black pupils (15%) attended formerly all-white New Kent School, and no whites had enrolled at the all-black Watkins School. To the Court, the central question was "whether . . . a 'freedom-of-choice' plan . . . constitute[d] adequate compliance with the Board's responsibility 'to achieve a system of determining admission to public schools on a non-racial basis.'" It found that the "pattern of separate 'white' and 'Negro' schools in New Kent County" was "precisely the pattern of segregation to which *Brown I* and *Brown II* were particularly addressed, and which *Brown I* declared unconstitutionally denied Negro school children equal protection of the laws." The Court decreed: "The New Kent School Board's 'freedom-of-choice' plan cannot be accepted as a sufficient step to 'effectuate a transition' to a unitary system. In three years of operation not a single white child has chosen to attend Watkins school and although 115 Negro children enrolled in New Kent school in 1967 . . . 85% of the Negro children in the system still attend the all-Negro Watkins school. In other words, the school system remains a dual system." The Court stated that the county school board was "to convert promptly to a system without a 'white' and a 'Negro' school, but just schools."[38] Most significant, the Court accepted percentages as the method of determining whether or not a dual system was desegregated and dismantled. Following the *Green* decision, the federal courts employed racial percentages to measure whether or not a desegregation plan actually worked.[39]

Alexander v. Holmes County Board of Education and *Carter v. West Feliciana Parish School Board* were the third and fourth critical court cases that finally abolished the traditional segregated school systems of the Deep South. The Supreme Court's rulings in the two cases activated the doctoral shift of *Jefferson* and *Green*. They brought the freedom of choice phase (1964–69) to a sudden end and initiated the massive school desegregation phase (1970–72) (fig. 9.2). By the fall of 1968, 20 percent of the black students in the eleven former Confederate states were enrolled in formerly all-white schools, and federal funds had been terminated to 130 Southern school districts that refused to desegregate.[40] With the election of Richard Nixon as president that fall, the tenacious efforts by the Johnson administration to desegregate the South's school systems came to a close. As part of his Southern campaign strategy, Nixon attacked the Office of Education's Title VI desegregation guidelines and endorsed freedom-of-choice plans that were not ruses for continued segregation. Under the Nixon administration the primary tactic was to shift further the enforcement of desegregation from the Office of Education and the Justice Department to the federal courts. During 1969, the pace of desegregation began to wain across the Deep South as school districts sought to prevent further integration.[41]

Thirty-three Mississippi school districts were among those that had not desegregated schools and sought further delay. Most were rural and small-town districts and included Amite, Holmes, Pike, and Neshoba Counties, places that had become notorious during the civil rights struggle. The Fifth Circuit Court ordered the thirty-three districts to implement desegregation plans by August 25, 1969. To retain the assistance of powerful Mississippi senator John Stennis in the passage of a defense bill, the Nixon administration intervened in the *Alexander* case. Robert H. Finch, secretary of health, education, and welfare, wrote the Fifth Circuit Court and asked for a delay in the implementation of desegregation. The court approved a three-month delay. NAACP attorneys appealed the delay to the Supreme Court, which reversed the Fifth Circuit Court.[42] In remanding the case to the Fifth Circuit Court, the Supreme Court held, "Continued operation of segregated schools under a standard of allowing 'all deliberate speed' for desegregation is no longer constitutionally permissible. . . . The obligation of every school district is to terminate dual school systems at once and to operate now and hereafter only unitary schools."[43]

The Fifth Circuit Court did not interpret the Supreme Court's order to mean sudden desegregation, for this would involve relocation of thousands of children in the middle of the 1969–70 year. The judges planned immediate desegregation in two stages. Faculty, staff, athletic activities, and bus routes were to be integrated by February 1, 1970, but student bodies did not have to be desegregated until the start of the 1970–71 school year.[44] The Fifth Circuit Court's plan was quickly reversed by the Supreme Court. In *Carter v. West Feliciana Parish School Board*, a case similar to *Alexander*, the Court ruled on January 14 that "insofar as the Court of Appeals had authorized deferral of student desegregation beyond February 1, 1970, it had misconstrued the holding in Alexander v Holmes County Board of Education . . . as to the obligation of every school district to terminate dual school systems at once and to operate now and hereafter only unitary schools."[45]

When the Supreme Court issued its ruling in the *Alexander* case in November 1969, 3,000 of the approximately 4,000 school districts in the South were desegregated. Of the remaining districts, 500 were in the process of desegregation to comply with Title VI, 390 were under court order to desegregate, and 120 had had federal funds terminated for failure to desegregate. "Southerntown" was among the recalcitrant districts. Because of failure to accept even token desegregation, federal funds to the Indianola, Mississippi, Separate School District were terminated in June 1966, and two months later the system was placed under court order.[46] For Indianola and other defiant rural and small municipal school systems of the plantation regions, the Court's *Alexander* and *Carter* rulings meant that the decade and a half of postponement was over. The walls of the segregation fortress were demolished; massive desegregation was imminent. Immediately after the *Carter* ruling, the Fifth Circuit Court, together with the Fourth Circuit Court, whose territory stretched across Maryland, Virginia, and the Carolinas, issued prompt desegregation orders for all cases.[47]

In the face of decades of warnings by Southern demagogues of the dire consequences of throwing black and white children together, open white resistance collapsed, and massive desegregation came to the schools of the plantation regions relatively peacefully. Representative comments by white and black children were, "They're not as dumb as lots of us figured," and "They're not as mean as lots of us thought." Such cursory comments, however, did not imply whites' sudden acceptance of blacks as equals or that public schools had no racial problems. Integration produced few difficulties among younger children, but from the fifth or sixth grade through high school, a mood of suspicion and hostility often existed.[48]

THE RISE OF PRIVATE SCHOOLS

The sudden collapse of white resistance to school desegregation did not mean that all whites were reconciled to integration. The principal way that whites subverted desegregation was by fleeing from public schools to private ones. By the time massive school desegregation came in 1970, part of the white opposition to desegregation had collapsed because the most vehement segregationists had abandoned the public schools. In the fall of 1970, a number of segregated private schools existed across the plantation regions. Although a few of these schools opened in the 1950s and early 1960s, most were chartered after 1964.

During the long struggle to prevent desegregation, the laws passed by legislatures of Deep South states included ones to facilitate the creation of segregated private schools. Among them were provisions for state tuition grants, tax exemptions, publicly purchased textbooks and supplies, and the lease and sale of publicly owned school buildings to private nonprofit corporations. The Mississippi legislature passed a bill in 1964 which made $185 annual tuition grants available to students enrolled in private schools. Although the amount of the grant might seem small, in the context of the time, it was more than some public school systems spent per white child.[49] To many whites in the plantation regions, the reality of school desegregation loomed only after passage of the 1964 Civil Rights Act and enforcement of Title VI. The Citizens' Councils, an early advocate of private schools as a means of maintaining racial segregation, did not actually create their first institution until ten years after the *Brown* decision. In 1966, Mississippi had only thirty-five private segregationist schools and Alabama only thirteen. The first private segregationist school in nonmetropolitan counties of the lower Georgia Piedmont opened in Hancock County in 1966. John Hancock Academy, housed in the creaky Gothic building that was Sparta's white public school from 1890 until 1959, initially had 180 students. A few years later the academy moved to a new sheet-steel warehouselike structure, located just south of the town (fig. 9.3, map 9.2).[50] At the time of the Supreme Court's order in the *Carter* case in January 1970, there were approximately four hundred segregationist private schools in the South with a total enrollment of between 350,000 and 500,000.[51] The order for massive desegregation produced hasty formation of a new group of private segregationist schools and caused whites in the plantation regions to flee the public schools in droves. In some instances the effects of the Court's order was a cataclysm. A former teacher recalled that when the Tunica County, Mississippi, school was ordered to integrate in the middle of the 1969–70 year, "it was like Pompeii." The white children "just walked out and left everything, books and all."[52] The refuge was Tunica Institute of Learning, a private school chartered in 1966 and built less than a mile from the public one.

The percentage of whites who left the public schools varied across the South. Munford, who studied white flight from the thirty-three Mississippi systems of the *Alexander* case, found that the percentage of a public school district's population which was black was the most important determinant of the degree of white exodus. Systems whose populations were "overwhelmingly-white," only three of the thirty-three, experienced little exodus in white students. Systems with black populations in "a middle range" retained at least half of their white students, but "black-belt" systems, those in which the population was predominantly black, lost almost all of the whites.[53] Because most public school districts across the plantation regions were predominantly black, massive desegregation resulted in the exodus of significant numbers of white pupils. What happened in Holmes County, Mississippi, against whose board of education Beatrice Alexander filed her suit in the landmark desegregation case, is illustrative. Token desegregation came to Holmes County in the autumn of 1965. The integration of four grades a year under a court-ordered freedom-of-choice plan resulted in only a few blacks enrolling in the former all-white schools. A few blacks, however, were enough to cause almost all white parents with children in integrated grades to remove them from the public system. Three private schools were quickly created, and between the autumn of 1964 and that of 1966, white enrollment in the Holmes County public school system dropped from 1,500 (20%) to 1,000 (14%), while the black enrollment remained approximately 6,000. By the fall of 1966, most of the white students in the Loess Plains section of the county had returned to the integrated schools, but in the Delta section, where the ratio of blacks to whites was substantially greater, the white boycott persisted. With abolishment of the dual public education system and massive integration after the Su-

FIG. 9.3. The Modern South Landscape. John Hancock Academy in Hancock County, Georgia, in 1987. Compare the dress of the children with that of 1941 (fig. 5.10). Charles S. Aiken

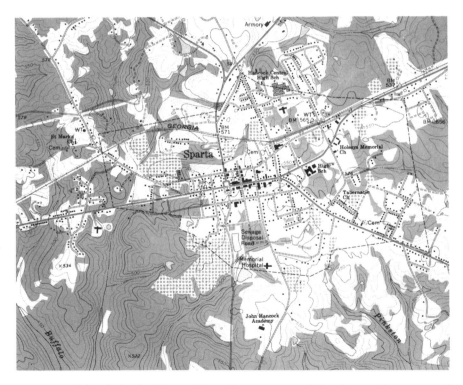

MAP 9.2. Schools in the Sparta, Georgia, area, 1974. Schools were desegregated just prior to collection of data for the map. The high school in the center of Sparta was the former all-white county school, and Hancock Central High School on the municipality's northern boundary was the former county school for blacks. John Hancock Academy is a private school that was created in 1966 during the token integration phase of school desegregation (U.S. Geological Survey, Sparta, Georgia, Quadrangle, 1:24,000, 1972).

preme Court's *Alexander* ruling, almost all white parents removed their children from the public schools. Two decades later, the Holmes County public school system was still essentially all black; only 2 of the 4,344 pupils were white.[54]

The private schools that arose in the plantation regions were denounced for their segregation, public subsidy, and poor academic quality. Many were housed in old or makeshift buildings and lacked adequate textbooks, equipment, and faculty. Although a number of teachers left the public schools for the private ones, the new schools had to employ some non-degreed teachers. Most private schools paid lower salaries than public ones and frequently did not offer fringe benefits such as retirement programs and health insurance. Initially, the new private schools had no accreditation agencies and operated without specified educational standards.[55]

State laws that provided subsidies to private segregated schools were declared illegal in a series of federal court decisions. The federal courts also barred local boards of education from supporting private schools through the donation of buses, textbooks, and supplies and the lease and sale of buildings. In 1971, segregated private schools lost their tax-exempt status with the Internal Revenue Service.[56] The shaky financial condition and the other problems faced by a number of the private schools made it appear that they were on the verge of closure from the time they opened. "There are signs throughout the South that the magnetism drawing hundreds of white parents to 'segregation academies' is substantially less than permanent," wrote a journalist in 1970.[57] Loss of public subsidies, inept leadership, and the inability to overcome marginal financing contributed to the failure of a number of the private schools. Hillcrest Academy in eastern Tate County, Mississippi, opened in 1970 in the old Thyatira Consolidated School, whose buildings were sold after it fell victim to reconsolidation. Hillcrest closed in 1978, only eight years after it had opened with much boasting, fanfare, and 148 students.[58]

At the state level, the number of whites who left the public schools was never as large as media reports led one to believe. In 1990, the percentage of children attending private schools in Minnesota (9.6%) was greater than in Mississippi (8.7%); the percentage in Pennsylvania (16.1) more than in Alabama (7.8); and the percentage in Connecticut (11.9) larger than in Georgia (7.8). In addition, the belief that most of the private schools in the South are operated by Bible-thumping Protestant fundamentalists is a myth. In 1976, only 3,248 of the children in private schools in Alabama and 1,427 of those in Mississippi attended schools affiliated with Baptist churches, while 15,639 and 10,335 pupils in the two states were enrolled in Roman Catholic schools.[59]

Although a number of the private schools failed, many survived and by 1980 had became permanent parts of communities across the plantation regions. Hillcrest Academy closed, but the larger and more soundly funded Magnolia Heights Academy in nearby Senatobia survived by absorbing students from Hillcrest and other surrounding private schools that foundered. Private schools toned down their racial rhetoric, began touting quality education, and became an accepted "alternative" to public education. Some of the older, well-established schools even began enrolling small numbers of black pupils as a symbol of their enlightenment. The private schools in the plantation regions survive, in part, because of the perception by white parents that integrated public schools are in a state of decline whereas the private ones offer quality education. The principal reason that most private schools persist in the plantation regions is still the one that initiated them. The percentage of the population in a school district which is black is the primary factor that governs the percentage of white children enrolled in the public schools. As in the 1960s, at the close of the twentieth century, the private schools are primarily refuges from public schools that are predominantly black. Despite its advertisement of "quality education in a wholesome environment," Autauga Academy in Prattville, just north of the Alabama Black Belt, flaunts a logo with the image of a Confederate officer, the founding date 1969, and the motto "We Dare Defend Our Rights."[60]

THE NEW GEOGRAPHY OF SCHOOL SEGREGATION

By the last decade of the twentieth century, school desegregation in the plantation regions was an accomplished fact for more than two decades, and the second generation of black and white children were enrolled. However, a new type of segregated educational structure had replaced the old dual public school systems. Across the plantation regions, more whites attended integrated public schools than private ones, but significant spatial variation existed in the percentage of school-age whites enrolled in public schools (map 9.3). Like large metropolitan school systems, nonmetropolitan ones in the plantation regions which appear desegregated at the system level have significant degrees of segregation at the local school level. In part, the variations in the percentages of white and black pupils enrolled in different public schools reflect the historical demographic spatial patterns, but they also are the product of new racial dynamics produced by integration and a refuge strategy of whites.

Two decades after massive desegregation, the degree of integration in county public school systems of the plantation regions ranged from ones that were essentially all black (90% or more) to ones in which blacks constituted fewer than 30 percent of the pupils. The Alabama Black Belt illustrates the school racial situation in plantation areas with relatively large black populations. In 1990, the school-age population of the Black Belt counties, exclusive of Montgomery, was 50,733, 74.7 percent of which was black. Fifty-two percent of the 12,852 school-age whites did not attend public schools, while only 2.8 percent of the 37,881 school-age blacks were not enrolled.[61] Most of the whites not enrolled in public schools attended private ones. Many of the private schools were small, with fewer than 300 students, but combined, they educated approximately half of the white children of the region.

A more spatially complex type of refuge for white children of the Black Belt is public schools in which whites are the racial majority or a predominant minority. Such schools are at least 40 percent or more white. Sixty-eight of the Black Belt's ninety-five public schools in 1990–91 were more than 90 percent black (fig. 9.4). Only twelve of the ninety-five had enrollments that were 60 to 90 percent black, but in seventeen schools blacks constituted fewer than 60 percent of the pupils. The schools that were 40

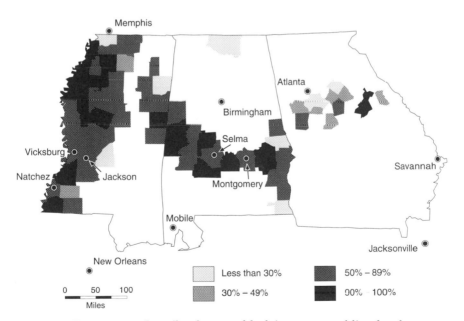

MAP 9.3. Percentage of pupils who were black in county public school systems, 1989–90 (sources of data: Alabama Department of Education 1990; Mississippi Department of Education 1989; *Atlanta Journal/Atlanta Constitution*)

percent or more white were located primarily in two types of places, the margins of the historical plantation areas, and municipalities with their own school systems. A four-county cross section of the eastern Black Belt illustrates the spatial situation of schools that were 40 percent or more white (maps 9.4, 9.5). The black soil prairie zone, where most of the plantations were concentrated, is a narrow strip across central Alabama. All Black Belt counties are partly in the black soil zone and partly in the adjoining hill lands, which historically were dominated by white yeoman farmers. In 1940, 50 to more than 90 percent of the population of the black soil zone was black, but across the adjoining hill lands the black population decreased from 50 percent to less than 10. The black population is still concentrated in the historical plantation zone, and the fringe areas are still majority white. In the old plantation area most of the schools are 90 percent or more black. Ten of the seventeen with large percentages of white students are in the plantation fringe areas, including the Moundville sec-

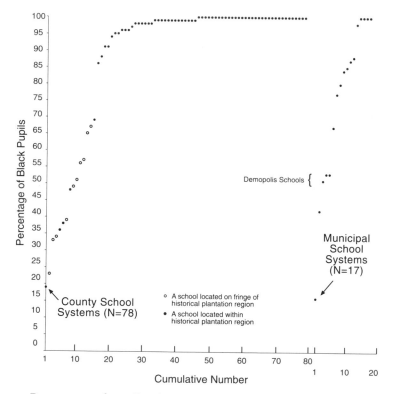

FIG. 9.4. Percentage of pupils who were black in Alabama Black Belt public schools in September 1990. Montgomery County is not included (source of data: Alabama Department of Education, 1990).

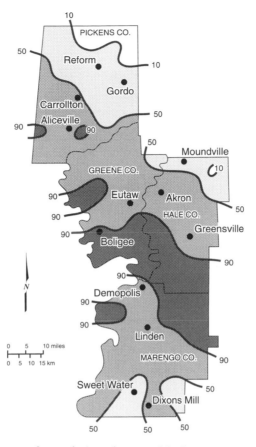

MAP 9.4. Percentage of population that was black in western Alabama Black Belt, 1940 (source of data: U.S. Bureau of the Census 1993)

tion of Hale County, the infamous setting of James Agee and Walker Evans' classic study of rural white poverty.[62]

A second type of place in which white pupils are a majority or only a slight minority is a municipality with its own school system where whites have been able to control the size of the black population. Only three Black Belt municipalities—Selma, Demopolis, and Linden—have school systems. Like certain other municipalities scattered across the plantation South, Demopolis is a type of white refuge in the age of integration. In 1990–91 the town's school system had 585 pupils, 53 percent of whom were black.[63] The white population of Demopolis grew 3 percent between 1970 and 1990, while the black population declined 3. The desire of whites to keep the black pupils in certain public schools a minority, or not a threatening majority, is among factors that prevent consolidation of small community schools with limited curricula into large comprehensive ones. In 1990–91, Hale County, Alabama, had five community high schools, all of which had fewer than 500 pupils. Three of the schools were all black; whites were slight majorities in the one at Moundville and the former all-white high school in Greenville.[64]

The geography of school desegregation alludes to recent local population redistribution and changes in settlement patterns in which race is an overriding concern. The demise and mechanization of plantation agriculture and new racial dynamics initiated by the termination of segregation by law, restoration of voting rights to blacks, and new federal social and economic initiatives introduced by the War on Poverty produced a new political and settlement geography.

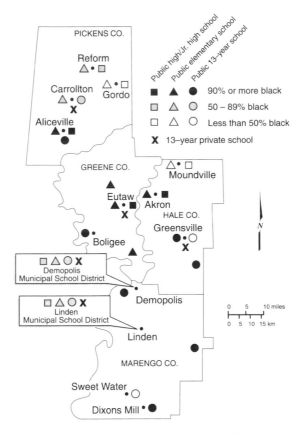

MAP 9.5. Percentage of pupils who were black in public schools in the western Alabama Black Belt, September 1990 (source of data: Alabama Department of Education, 1990)

THE

COTTON
PLANTATION
REGIONS

IN THE MODERN SOUTH

III

CHAPTER TEN

The Right to Vote—
An Illusive Black Power

꽃 꽃 꽃

*If you don't have power, you're begging. It's time to get power
that every other group has. We've got to remove them and make sure
they are gone. We're going to take over and get black sheriffs
and black tax assessors.*

Stokely Carmichael, 1966

*I have learned what suffering means. . . . I think I can understand
something of the pain that black people have come to endure. I know
I contributed to that pain, and I can only ask your forgiveness.*

George Wallace, 1979

THE DAWNING OF A NEW POLITICAL EPOCH

The passage of the 1965 Voting Rights Act began a new epoch in the
relationships between blacks and whites in the South's plantation regions.
The 1964 Civil Rights Act was, perhaps, more dramatic in immediate effects, but the Voting Rights Act was just as significant in long-range consequences. Other than central cities of the nation's metropolises, the local
impact of blacks' ballots is greatest in the plantation regions where the
residual black population is relatively large. The Voting Rights Act ended
the long struggle by blacks in the plantation regions to regain their political
rights, but it initiated a new era in which they sought to capitalize on the
restored right to vote while whites attempted to manipulate, dilute, and
nullify the black franchise. Two controversies erupted immediately after
Lyndon Johnson signed the new bill into law. One concerned the Justice
Department's method of enforcing the legislation; the other was a dispute
among blacks over the idea of black power, what the concept meant and
how the ballot should be employed.

The Justice Department's preference for mediation rather than confrontation to persuade county officials to register blacks continued after the

passage of the Voting Rights Act. Although the right of the attorney general to send federal examiners into counties to register blacks was a powerful new weapon, Nicholas Katzenbach chose to use it sparingly (map 7.2). Civil rights leaders believed that Lyndon Johnson and the attorney general failed to enforce the Voting Rights Act vigorously because of the influence of powerful Southern congressmen and senators. Senator James Eastland, who chaired the Judiciary Committee, publicly boasted to his white constituents that he "objected to the sending of every Federal registrar into Mississippi."[1] Examiners were not sent to Sunflower County, Eastland's home, until April 1967. In the summer of 1966, only 17 percent of the voting-age blacks in Sunflower County were registered, compared with 53 percent in adjoining Leflore County, to which examiners were sent immediately after Johnson signed the voting rights bill. Martin Luther King Jr. charged that there was "great disappointment with the federal government and its timidity in implementing the civil rights laws. . . . What was minimally required . . . was the appointment of hundreds of registrars and thousands of federal marshals to inhibit Southern terror."[2]

In part, the failure of all eligible blacks to register to vote was due to factors other than what they confronted at county registrars' offices, since traditional barriers, including literacy tests and poll taxes, were made illegal by the Voting Rights Act. Recognizing this, the Johnson administration encouraged civil rights organizations to employ voter registration campaigns to mobilize the black electorate. Nicholas Katzenbach emphasized to civil rights leaders, "Counties which have seen extensive Negro voter registration, whether by local officials or by federal examiners, are counties in which registration campaigns have been conducted."[3] Because SNCC and CORE had become increasingly radical and lost much of their support, the voter registration campaigns were supported primarily by the NAACP, the SCLC, the Southern Regional Council, and the National Urban League. The Southern Regional Council revived the Voter Education Project, which originally promoted black voter registration in Mississippi during the COFO's efforts from 1962 into 1964. The new greatly expanded project, which was legally separated from the Southern Regional Council, grew to provide various types of assistance to local political groups and to black candidates. Vernon Jordan, a former NAACP official, was chosen to direct the new Voter Education Project. The initial strategy was to focus on areas where federal examiners were sent and to work with local black leaders and civic organizations.[4]

Although registration of black voters was not as rapid as civil rights leaders would have preferred, significant progress was made during the two

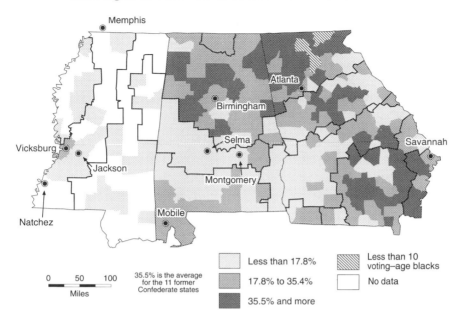

MAP 10.1. Percentage of voting-age nonwhites registered prior to passage of the 1965 Voting Rights Act (source of data: U.S. Commission on Civil Rights 1968b)

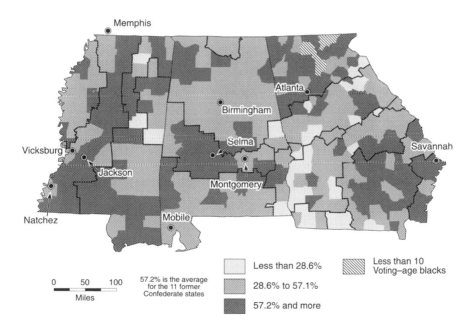

MAP 10.2. Percentage of voting-age nonwhites registered two years after passage of the 1965 Voting Rights Act (source of data: U.S. Commission on Civil Rights 1968b)

years after the Voting Rights Act. Between August 1965 and August 1967 the nonwhite voting-age population that was registered in the eleven former Confederate states increased from 36 to 57 percent. In Alabama, Georgia, Louisiana, Mississippi, and South Carolina, states with counties to which federal examiners were sent, the increase was from 12 to 62 percent. More than half of the new nonwhite voters in the five states were enrolled by the federal agents.[5] The changes in the geography of black voter registration were profound (maps 10.1, 10.2). The patterns of the percentages of non-whites registered just before and two years after passage of the Voting Rights Act are essentially inverse. Before enactment of the law, the plantation regions had the lowest percentages of registered voting-age nonwhites in Deep South states. Two years later, large areas of the plantation regions had the highest percentages. The Alabama Black Belt and the Mississippi Loess Plains are especially striking in relative increases of nonwhite voters. The lag of the Yazoo Delta, a region where the plantation system remained highly viable, lends credence to charges by civil rights leaders that intimidation by whites, especially by planters, prevented larger numbers of blacks from registering to vote. The percentages of nonwhites who were registered in August 1967 also reveal that the strategy by white leaders in counties such as Panola, Mississippi, to comply with federal law worked better in controlling black political participation than the tactics employed by leaders in counties like Dallas, Alabama, who resisted it to a bitter end. More than 70 percent of the voting-age nonwhites were registered in Dallas County, to which federal examiners were sent, compared with 53 percent registered in Panola County, which evaded Justice Department enforcers.[6]

BLACK POWER

On a sultry evening during the Meredith march in June 1966, Stokely Carmichael stood before a euphoric, chanting crowd in Greenwood, Mississippi, and shrieked, "The only way we gonna stop them white men from whuppin' us is to take over. We been saying freedom for six years and we ain't got nothin'. What we gonna start saying now is Black Power!" The crowd responded in unison "BLACK POWER!" and the cheer "BLACK POWER!" "BLACK POWER!" went on and on into the Delta night. Cleveland Sellers thought that this moment, not Lyndon Johnson's signing the 1965 Voting Rights Act into law ten months earlier or the Supreme Court's *Brown* ruling a dozen years before, was the pivotal point in the struggle for black rights in twentieth-century America.[7]

A controversy quickly erupted over exactly what black power meant.

According to Sellers, "Black Consciousness" not only "signaled the end of the use of the word *Negro* by SNCC's members," but it permitted them to relate their struggle "to the one being waged by Third World revolutionaries in Africa, Asia, and Latin America." "It helped us understand the imperialistic aspects of domestic racism. It helped us understand that the problems of this nation's oppressed minorities will not be solved without revolution."[8] To some whites, especially to particular politicians, federal officials, and military officers and to many residents of torched Northern and Western metropolises, black power implied, at best, a race war and, at worst, a revolution to overthrow the government. North American whites' phobia of this type of black power was at least as old as William Byrd's 1736 apprehension over impending slave revolts that would "tinge our Rivers . . . with blood."

Taken literally, the militant rhetoric of Carmichael, Sellers, Malcolm X, and other civil rights leaders, including even Martin Luther King Jr., who increasingly spoke in more confrontational terms and took a stand against the United States' involvement in the Vietnam War, gave even some moderate whites cause for concern. Malcolm X articulately shouted such statements as "White America is doomed! Death and devastating destruction hang at this very moment in the skies over America."[9] What began as a simple effort of blacks to gain fundamental rights guaranteed to Americans by the Constitution now seemed to some whites a crusade that had gone out of control and sought to destroy law and order. The Federal Bureau of Investigation under the directorship of J. Edgar Hoover and segments of United States Army intelligence certainly took black power seriously. The FBI had long conducted surveillance of King, Ralph David Abernathy, John Lewis, Robert Moses, Stokely Carmichael, James Farmer, and other black civil rights leaders and their organizations. The intelligence branch of the United States Army also began a file on Martin Luther King Jr. as early as 1947 after he was spotted with other Morehouse College students leaving a meeting of Dorothy Lilley's Intercollegiate Council, a suspected Communist organization. FBI agents assumed King was a Communist when he addressed the twenty-fifth annual meeting of the Highlander Folk School at Monteagle, Tennessee, in 1957, and thereafter they persistently spied on him.[10]

In the summer of 1967, the FBI began a "new Counterintelligence Program" that was designed to "neutralize" the "tremendous increase in black nationalist activity." The goals of the project were "to prevent the coalition of militant black nationalist groups, . . . the rise of a leader who might unify and electrify," the increase of "respectability" of the groups, and the "growth of these groups among America's youth." The fear that "vociferous

Stokely Carmichael, perennial exponent of black extremism" and "foremost exponent of the black power movement," might become the messiah subsided only after he left the United States for Africa in December 1968.[11] At the time of King's assassination in Memphis in April 1968, both army intelligence and the FBI were greatly concerned over the SCLC'S impending Poor People's Campaign. Planned to take advantage of the growing antiwar sentiment, which an army general later acknowledged was a good idea tactically, the campaign was to begin at Marks, Mississippi, in the Yazoo Delta and to climax in Washington, D.C.[12]

The impending Poor People's Campaign promised that 1968 would be as bad as, if not worse than, the previous years. Major riots involving thousands of persons had erupted in the nation's cities for five consecutive summers, and in the fall of 1967, two hundred thousand antiwar demonstrators, using civil rights leaders' strategy, marched on the Pentagon. The scale and organization of the Pentagon march caused some top military leaders to become even more anxious over the expanding protest against the United States' involvement in Vietnam. Leaders of the FBI's Counterintelligence Program proposed to sabotage the SCLC's fund-raising for the Poor People's Campaign by providing "cooperative news media" with a bogus story about King's "embarrassing position of having too much money."[13] Perhaps nothing revealed the army's paranoia over the growing racial unrest and its link to the antiwar movement more than one of its Special Forces units. To deal with subversive groups of blacks, army intelligence had secretly recruited, trained, and armed a Ku Klux Klan klavern at Cullman, Alabama, which it designated the Twentieth Special Forces Group.[14]

The concept of black power found its strongest support and created the most controversy in the nation's metropolises. But the idea emerged in the rural plantation South, conceived by SNCC workers in the Alabama Black Belt and unleashed by SNCC's new chairman, Stokely Carmichael, in the Yazoo Delta. That the cry "BLACK POWER" signaled, to some whites, the beginning of the long-prophesied bloody race war proved baseless and was less important than what the idea meant to blacks. According to Van Deburg, one of the most important traits of black power was that it was multifaceted, viewed and interpreted in different ways, "but it was essentially cultural." Black power "was a revolt in and of culture that was manifested in a variety of forms and intensities. In the course of this revolt, the existence of a semipermeable wall separating Euro- and Afro-American cultural expression was revealed."[15] In essence, black power helped African Americans, especially younger ones, redefine themselves and their places in the United States and its history. Although certain elements of what Van

Deburg defined as the black power era, such as clothing and hairstyles, were fads, others, including intellectual interest in the history and culture of African Americans, endured. For many African Americans in the plantation regions, black power helped to create a newly found pride and self-respect. The simple expression gave them hope that they had in their hands the capacity to effect change and implied they could help accomplish change through the ballot box. *Black* quickly superseded *Negro* and *colored* as the preferred racial term, and more than a quarter century after it was introduced, it was still favored over the later *Afro-American* and *African American*.

Fundamentally, black power is more a political concept than a cultural one. Most cultural implications of black power soon washed across America. Remaining was the core of the idea, the power to cause change through the ballot box. However, the new political power that blacks in the plantation regions thought they held quickly proved to be illusory. The struggle to regain the right to vote had ended, but a new conflict, the battle to use the vote effectively, had just begun.

An essential spatial dimension underlies black power. Neither of the words that make up the expression overtly implies geography, but only in political spatial units where voting-age blacks are a majority can they dominate elections, and only in ones where they constitute at least 20 percent of the electorate are blacks a serious minority. In 1960, blacks accounted for only 11 percent of the United States population, and they had long ceased to be a majority population in any Southern state. Even in Georgia, Alabama, and Mississippi only 25, 26, and 36 percent of the voting-age population in 1960 was black, and blacks were a majority voting-age population in only eighty-five counties of the eleven states that made up the Southern Confederacy.[16] By 1990, continuing population redistribution had produced further relative declines in the strength of voting-age blacks at the state and county level in the South. Blacks constituted 25 percent or more of the voting-age population in only four states: Mississippi (32%), Louisiana (28%), South Carolina (27%), and Georgia (25%). The black voting-age population had declined to 23 percent in Alabama, 20 percent in North Carolina, and 18 percent in Virginia.[17] The number of counties in which blacks were the dominant political force also had declined.

Blacks can be a political majority in two types of places. One type is governmental units that exist primarily for administrative purposes. Most such places are counties and municipalities, and residents readily identify with them. Boundaries are relatively overt, and changes usually require a formal public process. A second type of place where blacks can be a political majority is ones that exist essentially for electoral purposes. Such places include congressional districts, state senate and house districts, and districts for election of representatives to county and municipal governing bodies, including school boards. Unless electoral districts coincide with administrative places or communities, inhabitants do not identify very strongly with them or pay much attention to their boundaries and changes in them. In the South in 1960, blacks were a majority population in a

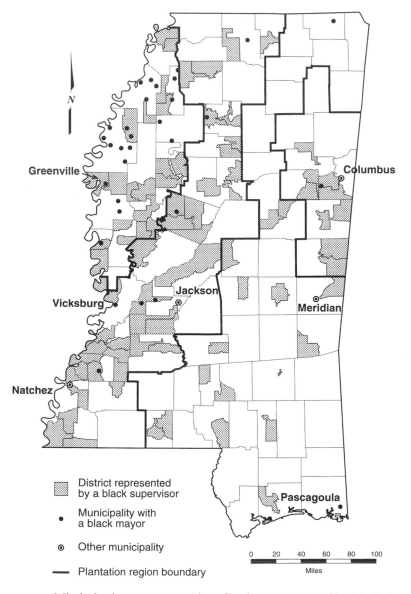

MAP 10.3. Mississippi county supervisor districts represented by blacks in January 1993 (source of data: Joint Center for Political and Economic Studies 1994)

number of counties, municipalities, and electorial districts across the old plantation crescent from Maryland into Texas. However, what might appear to be large spatial concentrations of black population and black elected officials often are split among different administrative and electorial units, as illustrated by the distribution of black county supervisors (commissioners) in Mississippi (map 10.3). Although concentrations of black supervisors existed in the Black Belt, the Natchez district, the Yazoo Delta, and the northern Loess Plains, black officials were not a majority on administrative boards in most counties in the areas.

EFFORTS TO NEGATE THE VOTING RIGHTS ACT

Because for many whites the civil rights era had moved to the acquiescence phase and because in several Southern states relatively large numbers of blacks voted prior to passage of the Voting Rights Act, there was no organized massive defiance similar to that for school desegregation. Across much of the plantation South, six decades of out-migration had greatly diluted the black population in many counties and minor civil divisions where blacks once had been an overwhelming majority. In addition, an age structure in which children made up a large part of the population and the political apathy of the blacks in counties where they were a slight majority lessened their political threat in certain areas of the plantation regions. As of 1965, little school integration had been accomplished in the plantation regions beyond mere tokenism, and whites were confident that the Voting Rights Act could be rendered as meaningless as the *Brown* decision. The Commission on Civil Rights reported "resistance to change in varying degrees in the Deep South States of Mississippi, Alabama, Louisiana, Georgia, and South Carolina and isolated incidents in other Southern States."[18]

After 1965, the politics of race in the plantation South moved to a more sophisticated level. Several methods were employed by whites to nullify and dilute the black vote and especially to prevent the election of blacks to public office. The methods were not new, nor were they entirely Southern. The business and professional upper class of American cities had long employed an effective group of techniques to dilute and control the political power of ethnic and racial minorities.[19] The methods included gerrymandering and consolidating election districts, making consolidated districts multimember, and altering the selection process for certain administrative offices, such as school superintendent, from elected to appointed. However, Congress and the federal courts extended the life and broadened the scope of the Voting Rights Act to help checkmate efforts to negate ballots of blacks. The struggle continued as the twentieth century drew to a close, three decades after passage of the original legislation.

Mississippi white leaders moved swiftly to counteract the 1965 Voting Rights Act. The urgency of turning back blacks who were about to pour through the political gates was shown in a special session of the Mississippi legislature in 1966. Thirteen bills designed to lessen the effects of the Voting Rights Act were adopted. Court decisions and the new federal laws, however, had already begun to change Mississippi. Defiant racial rhetoric was missing from speeches, debates, and media interviews, and the racial motives that underlay the bills affecting elections were indirectly implied but not openly discussed. The shooting of James Meredith beside U.S. Highway 51 on his "March against Fear" occurred during the special session. The Mississippi House of Representatives passed a special resolution that condemned the ambush by an "outsider" as a "most deplorable, unfortunate and criminal act." Less than four years had passed since Meredith's admission to Ole Miss was marked by the infamous bloody riot, but the resolution specifically noted that he was a Mississippian by birth and the first Negro graduate of the state university.[20]

Except for Mississippi, resistance to blacks' voting was not well organized at the state level, and efforts to negate or render the black vote less effective were largely through various measures at county and municipal levels. Six weeks after passage of the Voting Rights Act, the Alabama legislature enacted a reapportionment bill. A three-judge federal district court held the plan to reapportion the state senate was constitutional. But reapportionment of the state house was not valid because predominantly black counties were deliberately grouped with counties that were predominantly white "for the sole purpose of preventing the election of Negroes to House membership."[21]

The Mississippi legislature tampered with electoral districts more than with administrative ones. The boundaries of congressional districts, state senate and house districts, and county supervisor districts were altered in ways to dilute the black vote. Legislation was also passed which allowed local administrative units to switch from district to at-large elections and to change the method of selection of school superintendents from elected to appointed.[22] An effort to gerrymander congressional districts which began in 1962 did not end until 1982 and two decades of court litigation. Historically, the Second and the Third Mississippi Congressional Districts had the largest black populations, but by 1960 only the Third, which was essentially the Yazoo Delta, had a population that was more than 50 percent black (map 10.4). Because of population decline, Mississippi lost a congressional seat following the 1960 census. The state legislature made the adjustment from six to five representatives by combining the Second District, the northern Loess Plains, with the Third, the Delta. The strategy was to pit Congressman Frank Smith, a racial moderate who represented the Delta, against Congressman Jamie Whitten, a more militant segregationist and an obstructionist, who served the Second. Although the new Second District was more than 50 percent black, so few blacks voted that they posed no serious threat to Whitten, who easily won the 1962 election.[23] Registration of large numbers of blacks following passage of the 1965 Voting Rights Act resulted in the redefinition of Mississippi's congressional districts by the legislature in 1966. The Delta with its large black population was split among the First, Second, and Third Congressional Districts, only one of which had a slight majority in black population (map 10.4). In 1966, the Mississippi legislature also passed a bill that authorized elections of county boards of supervisors to be changed from districts to at-large.[24]

Allen v. State Board of Elections, a case from Mississippi which eventually reached the Supreme Court, sought the application of the Voting Rights Act to four changes in Mississippi's election laws, including at-large elections for boards of supervisors and appointment, rather than election, of superintendents of education. The changes were denied.[25] Following the *Allen* decision, approximately half of the counties in Mississippi hired a private company to devise new redistricting plans based on data from the 1970 census. Traditionally, in Mississippi supervisors not only made up a county's governing body but also were responsible for road maintenance in their districts. To equalize road mileage, as well as population, the company plotted districts that split municipalities as well as rural areas. To dilute black voting power, major black communities of municipalities and in the countryside were divided among two or more of a county's five supervisor districts.[26]

The microscale redistribution of population within plantation counties since 1960 complicates identification of deliberate attempts by white leaders to dilute the black vote. The First Supervisor District in Tate County, Mississippi, historically included a major plantation area and had a population that was majority black (map 10.5, table 10.1). The district's whites were yeoman farmers and rural nonfarm blue-collar workers concentrated in the communities of Crockett and Strayhorn. As blacks were severed from plantation agriculture, few remained in the district because the area had few black landowners. In addition, most planters would not sell blacks homesites as a final act of traditional paternalism. The black population of the First District plummeted. In 1960, the First District was 56.1 percent

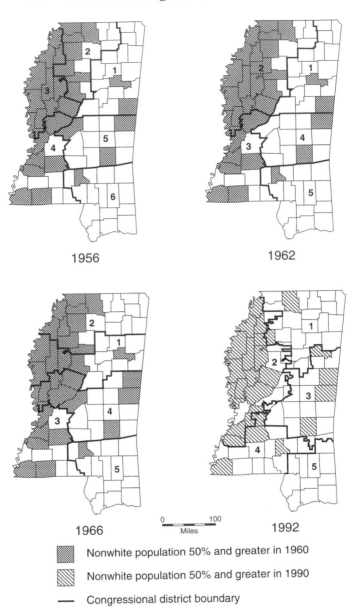

MAP 10.4. Mississippi congressional districts, 1956–82. Districts in 1956. Districts in 1962 as redrawn by the Mississippi legislature after loss of one representative following the 1960 census. Redistricting plan adopted by the legislature in 1966 following passage of the 1965 Voting Rights Act. Redistricting plan adopted in 1982 under a federal court order to enforce Voting Rights Act (U.S. Commission on Civil Rights 1966; Mississippi Election Commission, 1992)

TABLE 10.1 Population of Tate County, Mississippi, Supervisor Districts, 1960 and 1990

District	Total Population 1960	Black Population 1960	Percentage Black 1960	Total Population 1990	Black Population 1990	Percentage Black 1990
1	2,428	1,361	56.1	4,151	521	12.6
2	2,470	1,739	70.4	4,055	1,608	39.7
3	3,787	2,271	60.0	4,224	1,693	40.1
4	6,503	3,437	52.9	4,836	2,400	49.6
5	2,950	1,631	55.3	4,166	1,195	28.7
Total	18,138	10,439	57.6	21,432	7,417	34.6

Source: U.S. Bureau of the Census 1963a, 1991.

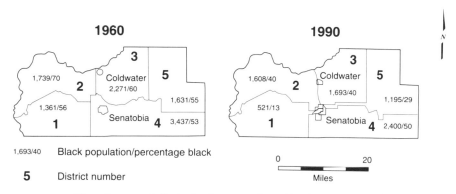

MAP 10.5. Tate County, Mississippi, supervisor districts, 1960 and 1990 (U.S. Bureau of the Census 1963a, 1991)

black, but in 1990 blacks constituted only 12.6 percent of the population in the redefined district, which extended into Senatobia. The Senatobia part of the district had 1,824 whites and only 267 blacks. Exclusive of the Senatobia addition, the 1990 white population was 1,792 and the black only 254. Without boundary changes, the black population of the First District would have declined from 56 to approximately 12 percent. It is also evident that white leaders of Tate County did not attempt to create majority-black supervisor districts. The Fourth District, a plantation area in which several large rural black hamlets emerged, could easily have been made majority black by slight boundary shifts in Senatobia or the countryside.

EXPANSION OF THE VOTING RIGHTS ACT

To counter efforts at the local and state levels to neutralize the Voting Rights Act, the federal courts and Congress expanded and extended the legislation. The 1965 act, which was to expire after five years, was extended in 1970 for five more and in 1975 for an additional seven. In 1982, Congress modified the act and extended it for twenty-five years. Among the alterations to the original bill were the inclusion of minority groups other than blacks and the strengthening of and attempt to clarify Section 2 of Title I. The potent Section 5 of Title I, which requires that any changes in voting qualification or procedure be approved either by the attorney general or a declaratory judgement from the Federal District Court for the District of Columbia, was retained.[27]

The role of the federal courts in support and expansion of the Voting Rights Act began shortly after the passage of the original bill in 1965. In *South Carolina v. Katzenbach* (1966), the Supreme Court refused South Carolina's request to void the act because it permitted federal intrusion into state matters. Three years later in *Allen v. State Board of Education* (1969), the Court indicated that Section 5 would not be limited to just registration and voting but would be directed at any "standard, practice, or procedure." Black vote dilution was a primary consideration in the decision. In *Perkins v. Matthews* (1971), the Court ruled that annexation of territory by a municipality and the relocation of polling places had to be precleared under Section 5.[28] Two years later, the Court held in *Georgia v. United States* and *White v. Regester* (1973) that redistricting plans and multimember voting districts were subject to Section 5 scrutiny. That same year, the Fifth Circuit Court in *Zimmer v. McKeithen* ruled that at-large elections for school board members diluted the votes of blacks.[29] That the *Georgia*, *White*, and *Zimmer* rulings came eight years after passage of the Voting Rights Act delayed the election of significant numbers of blacks. Only after the redistricting following the 1980 census did the full impact of the 1973 decisions begin to be realized.

In *Mobile v. Bolden* (1980), the Supreme Court handed down a decision that civil rights activists thought was a retreat from a decade and a half of aggressive enforcement of the Voting Rights Act. The Court ruled that a voting rights suit based on the Fourteenth Amendment had to demonstrate that a law or practice was created with intent of discrimination. As a consequence of the *Bolden* decision, Section 2 of the Voting Rights Act was revised in 1982 in such a way that purposeful discrimination does not have to be proved. A violation can be established by "the totality of circumstances," including the "extent to which members of a protected class have been elected to office in the State or political subdivision."[30]

Thornburg v. Gingles (1986) was another significant voting rights decision by the Supreme Court. Following the 1980 census, the North Carolina legislature redistricted the state house and senate. Blacks initiated a suit that held that five multimember districts violated the Voting Rights Act. A unanimous Supreme Court upheld the federal district court's ruling for the plaintiffs. A significant aspect of *Gingles* is that the Court defined a three-part test for determining whether or not vote dilution exists: "[T]he minority group must be . . . sufficiently large and geographically compact to constitute a majority in single-member district. . . . [T]he minority group must be . . . politically cohesive," and "the white majority votes sufficiently as a block to enable it . . . to defeat the minority's preferred candidate."[31] Under the *Gingles* decision, blacks sought to create at the local and state scales electoral districts in which they were a majority. The geographical approach was to devise a contiguous district in which black voting-age population was a majority regardless of the shape.

By 1988 the federal district courts had begun to question electoral districts that had what were perceived to be peculiar geographical shapes that implied gerrymandering. In *Shaw v. Reno*, writing the majority opinion, Justice O'Connor warned in strong language that "reapportionment is one area where appearances do matter." "[A plan] that included in one district individuals who belong to the same race, but who are otherwise widely separated by geographical and political boundaries and who may have little in common with one another but the color of their skin bears an uncomfortable resemblance to political apartheid."[32] The *Shaw* decision opened the gates for numerous court suits challenging new majority-black and majority-Hispanic congressional, state, county, and municipal districts created to comply with the Voting Rights Act.

Correct geographical shape in Western culture is usually understood to mean a compact area with a configuration similar to a square, a circle, or a rectangle that is not too elongated. This perception of spatial correctness is what was defined for the *Gingles* test as "geographic compact." Bizarrely configured districts can be accepted as correctly shaped if the boundaries follow geographical features, usually interpreted as physical features such as a river or an escarpment or an overtly visible cultural feature such as a major road. Extensions from the main part of an area, an excessive number of sharp corners along an uneven boundary, an area that is excessively elongated, and several large areas connected by narrow corridors are regarded as odd in shape. In advanced Western societies, many historical notions of the correct shape of areas defined for political reasons have been rendered obsolete by innovations in communication and transportation and by changes in settlement patterns. However, the fear of the fictional gerrymander from a more geographically simplistic past constrains even some of the best American judges.

PHOENIX-LIKE HE WILL RISE UP AND COME AGAIN

Black power in the plantation regions evolved through five stages: regaining the right to vote, registration to vote, effective use of the vote, overcoming efforts by whites to nullify and dilute black votes, and election of blacks to office. The passage of the 1965 Voting Rights Act concluded the first stage, and by 1966 the second was well under way. Local elections in 1966 demonstrated that the third phase had begun. Whites who were overt symbols of oppression of blacks began to lose their elected offices, and whites who hoped to keep their offices in places with large black populations began to appeal to the black electorate. Black votes in 1966 defeated Dallas County, Alabama, sheriff James Clark, who was replaced by the more moderate Wilson Baker.[33] Arrogant white secretaries in tax assessor and chancery clerk offices began to deal with blacks more tolerantly, and a few blacks even began to appear behind the counters. Unpaved roads through rural black enclaves began to be improved, and some municipalities began to improve the infrastructure of black residential areas. Streets were paved and water and sewage lines were installed, usually with federal grants from programs that were instigated during the War on Poverty. Certain municipalities and regional housing authorities began to seek federal funds to improve housing. By the mid-1980s mayors of such places as Sunflower and Senatobia, Mississippi, could brag that almost all substandard housing had been eliminated. Racist political demagoguery largely disappeared from the plantation regions. Even some of the most vehement segregationists became political survivors in the new age. Not only did George Wallace renounce his past and publicly apologize for his treatment of blacks, but his populist causes were fashionable, and he carried a critical part of the black vote in repeated successful campaigns for governor of Alabama.

The elections of 1966 also demonstrated that the fourth and fifth phases of black power were illusive and that a new struggle lay ahead to overcome the impediments whites began using to negate and dilute the black vote and to keep blacks from being elected to public office. Of the fifty-one black candidates for public office in Alabama in 1966, twenty-five, some running on black-party tickets, survived the primaries, but only four were elected to office. Three of the four, including the sheriff and the tax collector, were elected in Macon County, which had not only a majority-black electorate but also a long history of political activism led by Charles Gomillion and other well-educated blacks affiliated with Tuskegee University. Lucius Amerson was the first black elected sheriff of a Black Belt county since Reconstruction.[34] The 1966 elections also illustrated to perceptive black leaders that attempts to win through black power political organizations were largely futile. There were insufficient bases of black voters beyond the county scale, and some blacks, as well as whites, were repulsed by the militancy and the rhetoric. The Black Panther Party (officially the Lowndes County Freedom Organization in Alabama) created by Stokely Carmichael failed to have one of its candidates elected even though Lowndes County was majority black. Carmichael's "Ignorant, smelly, with our noses running, we're going to take political power"[35] rhetoric was good media material, but it also was repulsive to many blacks. The Mississippi Freedom Democratic Party suffered a similar defeat in its attempt in 1967 to demonstrate black power in Sunflower County, Mississippi.[36] Although black independent parties enjoyed a few victories, most black leaders realized that their political success at the local and national levels lay with the establishment, which for them meant not the party of Abraham Lincoln, Ulysses Grant, Dwight Eisenhower, and Richard Nixon but the party of Franklin Roosevelt, Harry Truman, John Kennedy, and Lyndon Johnson.

The black leadership, encompassing potential leaders, in the plantation regions during the late 1960s was hardly a homogeneous or unified group. The traditional leadership, which included preachers, schoolteachers, businessmen, and larger landowners, began to be superseded by a new group. The civil rights movement infused three new types of leaders into the plantation South. Some SNCC, SCLC, and CORE workers remained to become local leaders and also activated a group of young indigenous blacks by challenging them to become involved in the movement. A number of rural blacks had startling first encounters with civil rights workers similar to that of Unita Blackwell—"they were pointing the finger at me!" One such leader, who was among the first blacks elected to the Alabama legislature, was characterized in 1965 by the president of a small-town bank as a person he "could have put in . . . jail on several occasions for crooked financial deals." An OEO investigator, however, found the person to be "an obvious leader" whom "other Negroes look[ed] to . . . with . . . awe because he [was] not afraid of whites." A third type of black leader arose from the local bureaucracy of the War on Poverty and its related federal programs.

Though beneficiaries of the civil rights movement, members of this third group of new leaders usually were not overt participants in it. They were relatively well educated and, though sometimes outspoken, communicated with moderate whites. Competent in administration, these new leaders often began their rise in token minority positions. Some of them remained within the federal and state bureaucracies, while others made transitions to politics and to private companies. Bennie Gooden, the deputy director of Coahoma Opportunities, soon left the agency and formed a company to manage the HUD and FmHA housing complexes built by nonprofit organizations in the lower Mississippi Valley as part of the War on Poverty. By 1990 Gooden's Southland Management operated more than sixty housing projects in Arkansas, Mississippi, and Tennessee which were occupied almost entirely by blacks. Gooden lived in a large new house in Clarksdale and was considered to be the wealthiest black in the city and one of the most influential in the Delta.[37]

The number of black elected officials in the states of the plantation South began to increase following passage of the Voting Rights Act. Through court suits, the spatial redefinition of electoral districts, and local population redistribution, blacks became a majority voting-age population in various places across the plantation South. By the spring of 1973, blacks held 1,179 elective offices in the eleven former Confederate states. They held 149 in Alabama, 104 in Georgia, and 152 in Mississippi. The number in the Confederate states grew to 4,924 in 1993. With 751, Mississippi had the largest number of elected black officials among the states. Alabama had 699 and Georgia 545.[38] Most blacks hold offices in counties and small municipalities. Although some scholars dismiss the gains by blacks in local offices as insignificant, at the grassroots level black officials make an important difference in access to federal, state, county, and municipal programs, services, and jobs. They also contribute to black pride and to improvement in the treatment of blacks by local whites. The black political caucuses in the Alabama, Georgia, and Mississippi legislatures became quite effective in bargaining on issues important to blacks.

In November 1968 Hancock, Georgia, was the first plantation county to pass into the control of blacks since Reconstruction. The victory was hardly the sensational "takeover" that the front page of the *Atlanta Constitution* proclaimed, for blacks won only five of eighteen elected offices. However, they won two of the three county commission seats and the offices of probate judge and clerk of the superior court.[39] Once in political control, blacks did not let go. Significant changes came to Hancock County, especially in access of blacks to federal antipoverty programs and in local government's response to the needs of blacks. In July 1969 Greene County, Alabama, became the second county to come under the control of black officials. Blacks were elected to four of the six Greene County commission seats under the auspices of the National Democratic Party of Alabama, a black independent organization that was active in several areas of the state. Macon County, Alabama, and Holmes County, Mississippi, also soon passed into the hands of blacks. However, another two decades went by before the most overt symbol of white resistance, Dallas County, Alabama, came under the control of blacks.[40]

Unfortunately, the efforts by whites to dilute blacks' voting power and to prevent their election to office have increased racial tension and split certain communities.[41] Although certain rhetoric and gestures upon taking public office are understandable in light of past discrimination, some black leaders also have contributed to racial tension. Following assumption of office in Greene County, Alabama, in August 1969, participants in the jubilant freedom march broke into the all-white swimming pool. During the celebration, SCLC president Ralph David Abernathy reportedly stood on the diving board and "invited everyone on the outside to come in" because the "pool is integrated now." Unita Blackwell hardly helped the cause for Mayersville, Mississippi, with local whites by a statement that appeared in a widely circulated international journal: "I am the law. I am over the slave owners that use to be over me. I am their mayor. I'm the judge."[42]

Initially, blacks placed emphasis on election of county commissioners, tax assessors, law enforcement officials, and state senators and representatives. Increased success caused them to seek both lower and higher offices. The election of L. Douglas Wilder as governor of Virginia in 1989 illustrates just how far Southern politics had advanced since the 1960s. However, in the 1990s political battles between blacks and whites continued and included ones fought over control of municipal and county school boards, appointment and election of black judges, and creation of majority-black congressional districts.

The efforts to create majority-black congressional districts illustrate the problems that blacks confront in defining places in which they are the majority electorate. No black served in Congress between the time George W. White, Republican from North Carolina, left office in 1901 and the election of Oscar DePrist, Republican from Illinois, in 1928. In his sad farewell address as the last black congressman, White foretold the political rebirth of Southern blacks, who "Phoenix-like" would "rise up and come again." At the time of the 1954 *Brown* decision, William L. Dawson and Adam Clayton Powell Jr., Democrats from Illinois and New York, were the only blacks in Congress. The number of black representatives began to increase after the passage of the 1965 Voting Rights Act. In 1972, Barbara C. Jordan, Democrat from Texas, and Andrew Young, Democrat from Georgia, became the first blacks elected to Congress from the South since disfranchisement.[43]

Congressional redistricting following the 1990 census brought increased demands from blacks for more majority-black districts to comply with the Voting Rights Act and the *Baker* and *Gingles* Supreme Court decisions. Majority-black congressional districts created in response to federal court orders and Department of Justice mandates using the power of Section 5 of the Voting Rights Act contributed to the election of a significant group of blacks to Congress in 1992, 1994, and 1996. Ironically, the Republican Party aided in the effort to create majority-black congressional districts. Most members of America's largest minority group remain loyal Democrats, for with the 1968 presidential campaign of Richard Nixon, Republicans overtly began using racial issues in their appeal for Southern white votes. Republican strategists learned not just the overt but the subtle geography of the new racial politics. Creation of majority-black congressional districts helps black Democrats win office, but it also aids Republican candidates. Removal of blacks from surrounding congressional districts to form a minority one dilutes the Democratic vote. The Democratic National Party estimated that five congressional seats were lost to Republicans in 1992 and ten in 1994 as a result of the creation of majority-black districts.[44] The number of blacks in the House of Representatives increased from twenty-four in January 1989 to thirty-eight in January 1993. Eighteen of the thirty-eight were from the South, and sixteen were from the eleven former Confederate states. Most of the districts were inner-city metropolitan ones. Seven were in the Southern plantation regions (map 10.6).

In 1996 all but one of the black members of Congress were reelected. By July 1997, nine of the seventeen majority-black districts in the South had been held unconstitutional by the federal courts. The fear prevailed among civil rights advocates that the number of blacks in Congress might decline, especially when unconstitutional majority-black districts are eliminated following the census of 2,000.[45]

Three of Georgia's congressional districts were represented by blacks in 1993. The Fifth District, which included part of Atlanta's inner city, was the one from which Andrew Young was elected in 1972. Young was replaced by John Lewis, who rapidly rose within the Democratic leadership. Two additional majority-black congressional districts were created by the Georgia legislature following the 1990 census through pressure brought by the Jus-

tice Department (map 10.6). The Second District centered on the Inner
Coastal Plain plantation region and included Albany. The Eleventh District
extended from the edge of the Atlanta metropolitan area across the lower
Piedmont, the Inner Coastal Plain, and the Sea Island plantation regions
into the edge of metropolitan Savannah. Although the federal district court
ruled that the Eleventh was gerrymandered and had to be redrawn and five
justices of the Supreme Court concurred, the district was spatially logical in
light of the plantation past. When the Georgia legislature failed to agree on a
plan to redefine the state's congressional districts, the federal district court
redrew them. Two of Georgia's three majority-black districts were eliminated. Only the Atlanta district represented by John Lewis remained.[46] In
November 1992, Earl Hilliard became the first black elected to Congress
from Alabama in 105 years. Hilliard was elected from a majority-black
federal district court–mandated district that included the western Black
Belt and counties on the southern edge. The effort by blacks to obtain a
second majority-black congressional district in the eastern Black Belt was
denied by both the federal district court and the Supreme Court.[47]

Whites' efforts to negate or dilute the black vote and expansion of the
Voting Rights Act by Congress and federal courts led to a preclearance
quagmire in the Civil Rights Division of the Justice Department. Because
the division does not have a staff to conduct field investigations, it cannot

MAP 10.6. Congressional districts with majority-black voting-age populations
in January 1993 (Rector 1995)

research each submission, much less find violations that are never reported. Lacking a field staff to investigate and enforce the Voting Rights Act, the division relies heavily on "expert" witnesses to research, interpret, and explain individual cases of vote dilution and other methods used to render the black vote less effective. The need for expert witnesses has resulted in a host of cottage industries operated primarily by political scientists, planners, and sociologists, who use census data, computers, and off-the-shelf sets of analytical procedures. The fragmentation of cases among various courts and the investigative efforts by numerous persons, some of whom are not proficient in geography or in law, led to unfortunate results.

More than thirty years after passage of the 1965 Voting Rights Act, Congress and the federal courts appeared not to have a spatial or historical grasp of the exact state of minority voting, its impediments, and the results. Despite expansion of the Voting Rights Act by Congress and the federal courts, unlike school desegregation, no critical sweeping breakthrough that effectively removed all new barriers, especially forms of vote dilution, were made in the three decades following passage of the 1965 act. This caused judicial confusion over the creation of majority-minority congressional and other types of electoral districts which prevailed in the late 1990s, as the Supreme Court attempted to distinguish between permissible districts defined by minority community of interest and impermissible ones that were racially gerrymandered. Rulings by a divided Court in *Miller v. Johnson* (1995), *Bush v. Vera* (1996), *Shaw v. Hunt* (1996), and *Abrams v. Johnson* (1997) declared reapportionment schemes in Georgia, Texas, and North Carolina, which created majority-minority congressional districts, were too narrowly tailored with respect to race or ethnicity. *Abrams* also questioned the role of the Justice Department in pressuring state legislatures to create majority-minority districts.[48] These decisions by the Court appeared to be a retreat from more than a half century of efforts to give blacks an adequate political voice at the local and national levels.

CHAPTER ELEVEN

New Settlement Patterns

❧ ❧ ❧

This [roadside demonstration] is a boil that has
come to a head, indicating a widespread condition
in the cotton region. It is probably that nothing less than
a great resettlement campaign, involving both
housing and land, could materially improve
living conditions among these folk.

Report on the Missouri sharecroppers demonstration, 1939

We've become a much more segregated society now than
we were 30 years ago. Then it was segregated on the
basis of race. But the white and black folks
worked in the same field and they rode in the same
trucks and they toted the same watermelon.
. . . They knew each other.

Jimmy Carter, 1992

The decline and mechanization of the Southern cotton plantation was accompanied not only by significant migration of blacks from the plantation regions but also by redistribution of the black population that remained. Urban and rural nonfarm black populations increased while the farm population declined. Even in the Yazoo Delta and other regions where plantation agriculture remained viable, the black rural farm population decreased while the rural nonfarm and urban populations increased absolutely as well as relatively. The population figures imply much more than they reveal. Severing of blacks from the plantation was accompanied by profound changes in the settlement patterns. The new settlement patterns have not been fully studied, but rural hamlets, underbounded municipalities, and a new type of black municipal ghetto are among the spatial components.

RURAL HAMLETS

The New South settlement pattern of "scattered cabins," which in 1950 David Cohn saw beginning to "give way to small villages as in European agriculture," soon became a dominant geographical trend. By the 1990s, much of the rural black population in the countryside of the plantation South was concentrated in hamlets. Although some of the hamlets house employees on neoplantations, most are home to a rural nonfarm population that has little or no relationship to the old plantation landholdings that surround them.

Nucleation of the black population occurred in the countryside as households built dwellings close together on small lots, creating hamlets, groups of five or more dwellings, none of which is more than two-tenths of a mile from another.[1] This new pattern of rural settlement exists whether the plantation is viable, as in the Yazoo Delta, or extinct, as on the lower Georgia Piedmont. Two basic types of hamlets developed, unplanned assemblages of dwellings and planned subdivisions. Unplanned assemblages frequently evolve on farms owned by blacks, in some instances reinforcing and redefining traditional black enclaves. The first stage in their development usually is associated with an extended family through the laying out of lots as dwelling sites for family members. Parents who own a farm or small tract give children a plot on which to construct a house or locate a mobile home. In black enclaves that historically were composed of several farms, construction of two or three dwellings by children on each landholding can create a sizable hamlet. Large hamlets usually evolve beyond the extended family stage and enter a second one in which lots are sold to nonrelatives. Frequently, a black church that was the focus of the black farm enclave also serves as the nucleus of the new black hamlet. Larger rural hamlets usually have one or more stores and other businesses that include juke joints and motor vehicle repair shops. The role of farms owned by blacks in the development of hamlets is illustrated by eastern Tate County, Mississippi (maps 5.1, 11.1). Most of the hamlets in 1990 were enclaves of black landowners in 1940. Independent Tyro and New Hope were still in the extended family stage, but Freedonia had evolved into a large hamlet in which lots were sold to nonrelatives.

Although some unplanned assemblages are groups of well-kept, modern houses resembling suburban subdivisions, others contain a hodgepodge of dwelling types (figs. 11.1–11.3, map 11.1). Within a single unplanned hamlet may be relocated tenant shacks, mobile homes, shell frame houses completed by owners, and large brick-veneer houses with attached garages.

Houses usually are close to the road, and beside or behind the main structure may be one or two dwellings occupied by family members. Rural plantation counties, most of which had no building codes until the 1970s, were forced to enact laws to manage the newly emerging rural hamlets of blacks and whites. Most counties require a lot of 1 or 1.5 acres for each dwelling to accommodate the drainage field of a septic tank for sewage disposal. Rural hamlets with three or more dwellings per acre evidence either lack of adequate county building codes or failure to enforce them.

Planned rural subdivisions are hamlets in which lots are platted and the development officially registered at the county courthouse. Although profit from the sale of land often is not the primary factor in the development of unplanned assemblages of dwellings, it is fundamental to subdivisions. Subdivisions that are intended solely for blacks have been created across the plantation regions by white and black developers. Developers usually

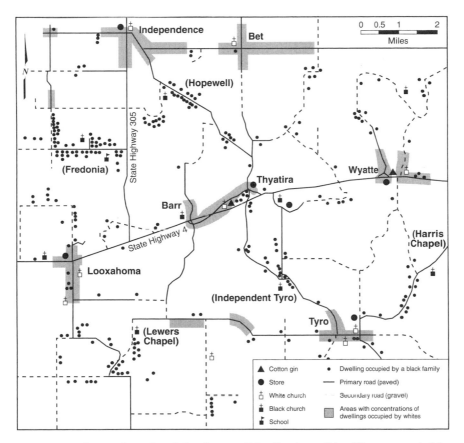

MAP 11.1. The Modern South landscape. Distribution of dwellings occupied by blacks in eastern Tate County, Mississippi, 1990. Charles S. Aiken

FIG. 11.1. Transition from New South to Modern South landscape. Hopewell, eastern Tate County, Mississippi, in 1980. The new shell house replaced the abandoned dwelling, which was razed. The older dwelling was built as a tenant house prior to the Farm Security Administration's purchase of the plantation in 1940 and subdivision of it into several small tracts that were sold to black farmers under the agency's tenant-purchase program. Charles S. Aiken

FIG. 11.2. The Modern South landscape. Three houses of an eight-dwelling extended-family black rural hamlet in Tunica County, Mississippi, in 1985. The occupants commute 30 miles to employment in Memphis, Tennessee. The houses were financed by the Farmers Home Administration. Charles S. Aiken

are local entrepreneurs who perceive the market potential. The War on Poverty infused into the plantation regions large sums to replace and renovate housing. A special type of planned rural subdivision is one that is sponsored by the Farmers Home Administration or the Department of Housing and Urban Development (fig. 11.4). Although the FmHA was originally created to aid farmers, its role in rural nonfarm housing escalated after 1961 when the agency's restriction to farm housing was removed by Congress and the number of nonfarm dwellings financed significantly surpassed the number of farm.[2] Because a large percentage of the nation's low-income nonmetropolitan households are in the South, the FmHA's housing programs have had the greatest relative importance in the Southern states. During the 1980s, Mississippi frequently received the greatest annual expenditures among the states. The New South furnish merchant stores and plantation commissaries are gone, but the rural store survives in a Modern South form to serve largely nonfarm black and white populations (fig. 11.5).

Northern Tunica County, Mississippi, is representative of the new rural settlement landscape. In 1960, almost all of the area was cropland, and cotton was still the most important crop (map 11.2). Plantations dominated the area. Kirby, Abby-Leatherman, Bowdre (Owen Farm), and Holbert

FIG. 11.3. The Modern South landscape. An extended-family rural black hamlet between Thyatira and Hopewell in eastern Tate County, Mississippi, in 1985. Charles S. Aiken

FIG. 11.4. The Modern South landscape. A street in a Farmers Home Administration–sponsored subdivision in Jonestown, a black ghetto in Coahoma County, Mississippi. The houses were built in the 1980s. Charles S. Aiken

FIG. 11.5. The Modern South landscape. Patrick's Store at Chulahoma, Marshall County, Mississippi, in 1993. The store is representative of the type of rural store found across the plantation regions in the late twentieth century. Although primarily a grocery that functions as a convenience market, the store stocks feed and other basic farm supplies and has a laundry. The furnish merchant tradition continues in the large credit business that such a store usually has, with bills paid on a weekly or monthly basis. Charles S. Aiken

were among the largest and were in various stages of transformation from the New South to Modern South plantation. In 1960, as in 1860, more than 80 percent of the area's population was black. The only concentration of dwellings occupied by whites consisted of the houses of planters, merchants, and farm managers along the main street of Robinsonville, an unincorporated village. Except for several small, family-owned farms, blacks in the area lived on plantations or in Robinsonville. Most of the dwellings were tenant houses in various stages of repair. The houses in the fields, including those on Bowdre Plantation, were second-generation, post–Civil War ones. The dwellings removed from the fields and rebuilt along roads, together with a few new ones constructed among them during the first phase of plantation reorganization, constituted the third generation of post–Civil War housing. As a consequence of the decline of tenancy, Tunica County's black population decreased 30 percent between 1940 and 1960, declining from 19,335 to 13,342. Some shacks stood vacant, falling into ruin. A number of the occupied dwellings housed families of machinery

MAP 11.2. A portion of northern Tunica County, Mississippi, in 1960 (U.S. Geological Survey, 1960, Horseshoe Lake Quadrangle, Ark.-Miss.-Tenn. Quadrangle, 1:62,500)

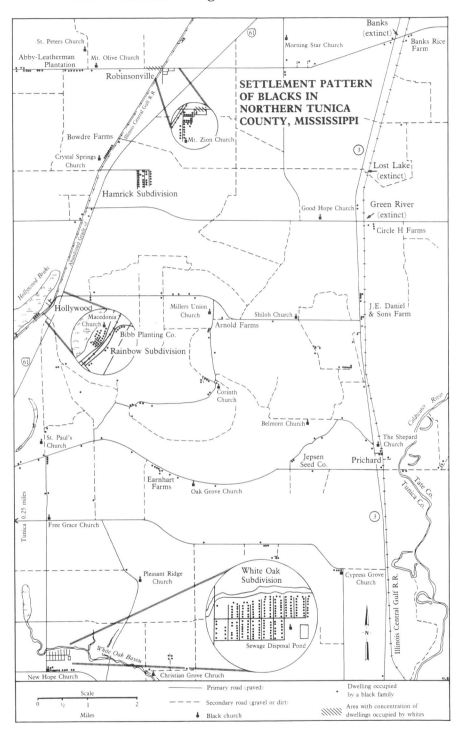

**SETTLEMENT PATTERN
OF BLACKS IN
NORTHERN TUNICA
COUNTY, MISSISSIPPI**

MAP 11.3. The Modern South plantation landscape. Settlement pattern of blacks in northern Tunica County, Mississippi, in 1985 (Aiken 1985). Courtesy of the American Geographical Society

operators, who were paid weekly wages, and day laborers, who barely subsisted on what they intermittently earned weeding and harvesting cotton. Churches and schools of blacks were prolific in the area. Most of the small one- and two-room public schools were closed when the large new consolidated Rosa Fort school for blacks was built in the 1950s just outside the town of Tunica as part of Mississippi's effort to make black schools equal to white ones.

Twenty-five years later, by 1985, the landscape of northern Tunica County was fundamentally different. Mechanization of agriculture had been completed for two decades, and the end of de jure segregation, the Voting Rights Act, and the War on Poverty had made an impact on the landscape (map 11.3). Not only did the new settlement pattern depict the Modern South neoplantation, but it conspicuously portrayed the severance of blacks from the plantation. Although northern Tunica County was settled before the Civil War, no overt evidences of the Old South plantation remained. Catastrophic floods, the worst in 1927, intermittently erased the land. The once bustling river town Commerce, which was taken by the Mississippi more than a century ago, would be nothing but a name if Tunica County's approval of legalized gambling had not reestablished the place in the early 1990s as a site for river-barge casinos. Even the relics and fossils of the New South era were rapidly vanishing. The giant mule barn on Abby-Leatherman Plantation, once one of the largest and tallest structures in Tunica County, was razed in the early 1980s. A few abandoned or converted commissaries and furnish merchant stores survived on some plantations and in Robinsonville and Hollywood, but even these structures were disappearing. The most prominent residuals of the New South plantation were a smattering of former tenant shacks and the churches and cemeteries of blacks. A few of the houses were still occupied, but most were empty, awaiting demolition. Churches, the only unyielding New South–era objects on some plantations, sat forlorn amidst soybean, cotton, and rice fields (fig. 11.6).

Almost all Tunica County's 6,148 blacks in 1990 lived in nuclei of dwellings. Fewer than 10 percent still resided on farms and plantations. Rows of dwellings for wage employees on neoplantations were on Bowdre Farms, Abby-Leatherman Plantation, and Earnhart Farms and in Hollywood for employees of the Bibb planting company. Most of the blacks in northern Tunica County resided in the unincorporated villages of Robinsonville and Hollywood and in hamlets that developed after 1964 just beyond the boundaries of plantations. Hollywood became virtually extinct with the demise of the New South plantation, but the creation of Rainbow Subdivision for blacks by a white entrepreneur in 1965 redefined the place. Several hamlets of blacks developed in the countryside. The small hamlet at Corinth Church is that of an extended family; another extended-family hamlet is 2 miles west of the church. Hamrick is a subdivision that was created in 1970 on a quarter section owned by a black family. As Rainbow Subdivision, Hamrick evolved into an assortment of dwellings and contrasts sharply with the surrounding uninhabited expanses of fields. White Oak, an FmHA-sponsored subdivision occupied by blacks, is the largest and most prominent hamlet. Constructed in the late 1970s, this 180-house development, 3 miles from the town of Tunica, is isolated among cotton, soybean, and rice fields. In 1990, more than 900 persons, one-seventh of Tunica County's black population, lived in White Oak.[3]

The late-twentieth-century landscape and economy of the lower Georgia Piedmont differ substantially from those of Tunica County, Mississippi. The area surrounding the site of the Barrow Sylls Fork Plantation is dominated by pine forest (map 11.4). Pastureland is infrequent, cropland exceptional, and the plantation extinct. Scattered through the forests are relics and fossils of both the Old and the New South plantations, which survive largely because, rather than remaining viable, the plantation and crop agriculture perished. Ironically, this area from which the plantation has vanished appears more stereotypical Deep South than Tunica County, where the plantation survives and the Modern South landscape of industrial agriculture resembles California's Great Valley. Decaying and restored antebellum big

FIG. 11.6. The vanishing New South landscape. Morning Star Church, northern Tunica County, Mississippi, in 1985. Charles S. Aiken

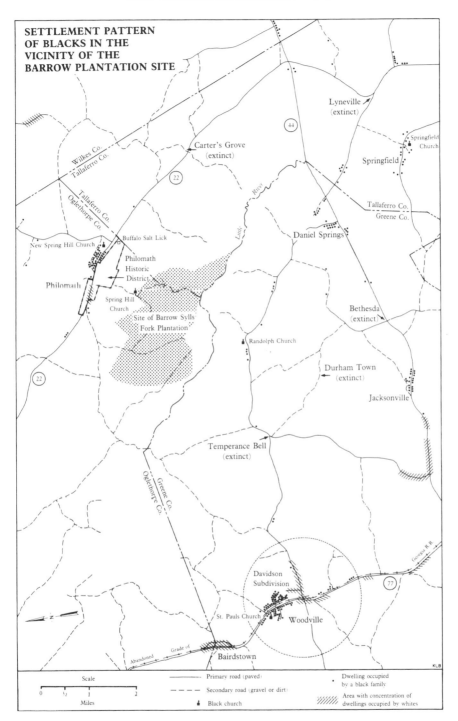

MAP 11.4. The Modern South landscape. Settlement pattern of blacks in the vicinity of the site of the Barrow family's Sylls Fork Plantation in 1985. Timber products companies now own many of the former plantation properties, including Sylls Fork (Aiken 1985). Courtesy of the American Geographical Society

houses, forsaken by the slave quarters, barns, and other buildings that once surrounded them, are the most prominent remnants of the Old South era. Philomath and Woodville have concentrations of such dwellings. Slave burial grounds and artifacts at the sites of slave quarters are the principal remnants of the first period of black settlement. The landscape residuals from the New South plantation are more numerous. Relic store buildings of furnish merchants are in Philomath, Woodville, and the countryside. Remnants of tenant farmer shacks rot away in the woods. The few that are still occupied house the poorest black households in the area. As in Tunica County, the most significant New South residuals are the churches that were organized by former slaves shortly after freedom. A mile from the highway in the pine forest, a small group of elderly blacks still meet occasionally at Spring Hill Baptist Church, which was rebuilt in 1937 on the site given to the congregation by David Barrow in 1870 (fig. 11.7).

More prominent than the landscape residuals of the Old and New South eras is the new pattern of black settlement which emerged after 1964. Although the plantation is extinct, its cultural traditions and landscape influences survive. The settlement pattern of blacks differs little from that of northern Tunica County. Most blacks live in five unplanned hamlets and in Woodville, which has a small FmHA subdivision. The hamlets of Daniel Springs and Springfield began on the farms of blacks who sold lots to relatives. The large hamlet at Philomath is the creation of an elderly planter

FIG. 11.7. The vanishing New South landscape. In 1985, Spring Hill Church still occupied the site given to the congregation by David C. Barrow shortly after the Civil War. The original building was replaced by a new one in 1937. The cemetery behind the church includes the graves of former Barrow slaves and members of tenant families on Sylls Fork. Charles S. Aiken

FIG. 11.8. The Modern South landscape. New Spring Hill Church is the focus of a large black hamlet of more than thirty dwellings which developed after 1967 on the eastern margin of the old planter village of Philomath. The site of Sylls Fork Plantation, the area's most famous historic feature, and the new black hamlet, the most significant modern link to the past, are prominently excluded from the Philomath historic district. Charles S. Aiken

and his wife who in 1967 began selling inexpensive 1-acre lots to their former tenant farmers as the final act of the old paternalism.[4] The hamlet, which contains more than thirty dwellings that range from neat brick-veneer houses to mobile homes to shacks, developed around New Spring Hill Baptist Church, which was built in 1977 to replace the 1937 building. The original location had become almost inaccessible by the old road that crossed Sylls Fork Plantation. The present church building is an imposing air-conditioned brick-veneer structure with Sunday school rooms and a large auditorium (fig. 11.8). Among the deacon names on the cornerstone are Pope and Maxey, names of tenants on Sylls Fork in 1881 who had been Barrow slaves.[5] The new edifice evidences socioeconomic improvement in the condition of many rural blacks and confirms their commitment to their historical home.

UNDERBOUNDED MUNICIPALITIES:
SOUTHERNTOWN REVISITED

Whereas in metropolitan areas the majority of the black population is confined to inner cities, for numerous small municipalities across the plan-

FIG. 11.9. The vanishing New South landscape. Housing along infamous Sugar Ditch in the old black residential area of Tunica, Mississippi, in 1985. These dwellings were razed and the occupants relocated to an FmHA apartment complex and a line of new mobile homes on the abandoned railroad grade outside the Tunica corporate limits. Charles S. Aiken

FIG. 11.10. Dwellings at the edge of Tunica North Census Designated Place, Mississippi, in 1987. These dwellings, which were constructed circa 1945–64 by blacks displaced from plantation agriculture, are representative of the third generation of housing occupied by blacks after the Civil War. Charles S. Aiken

tation regions, large numbers of blacks live in communities just beyond the corporate boundaries.[6] The reenfranchisement of blacks created a new spatial dimension to local politics in which annexation of black residential areas is opposed by whites who fear loss of political control of municipalities. An ironic geographical situation has developed in some plantation counties where blacks are a political majority, and whites have attempted to retain political rule in municipal enclaves where they are a plurality of voters. Describing Eutaw, Alabama, after blacks assumed political control of Greene County, a white patrician jestfully told a journalist, "We are proud of our Greek revival and our white survival."[7]

Frequently cities seek annexation of territory only to be opposed by suburbanites.[8] For many small municipalities in the plantation regions a reverse situation has developed, for blacks in the suburban fringes seek annexation only to be resisted by white-controlled governments. Though some towns and cities have readily annexed black residential areas, others either have refused to annex any new territory or have selectively expanded the corporate boundaries to exclude blacks. That municipal underbounding involving discrimination against blacks became a problem in the wake of the Voting Rights Act is evidenced by federal court cases seeking annexation of black residential areas.[9]

In the Yazoo Delta, municipal underbounding for racial reasons is a regional problem.[10] As blacks surged toward the municipalities of the Yazoo Delta, the residential areas grew in stages. Three distinct types of black residential areas can be identified. First is the traditional black residential area (map 5.2, fig. 11.9). Older dwellings in this area belong to the second generation of black housing following the Civil War. In some municipalities, the traditional black residential area, even in the 1990s, differed little from that of Indianola described by Powdermaker and Dollard more than half a century earlier. Shacks still dominated. The number of dwellings and population density were actually greater than in the 1930s. A second type of residential area is evidenced in the dwellings constructed between 1935 and 1965 to harbor the blacks who moved to the municipalities during the great exodus initiated by the federal crop reduction programs and mechanization of cotton production. In part the growth in housing occurred by infilling vacant tracts and by crowding additional dwellings onto the small lots in traditional black residential areas. Development also was by outward expansion, with the black residential areas frequently growing beyond the political boundaries of municipalities (fig. 11.10). New black residential areas also evolved just beyond the political boundaries but spatially isolated from the older municipal black sections.

The substantial number of dwellings beyond municipal borders constructed between 1935 and 1965 are the result of the failure of municipal leaders to create new black residential territories within corporate boundaries to meet the increased demand for housing. Initially, unresponsiveness of the white leadership was largely due to the decline of traditional interdependence of whites and blacks. With passage of the New South plantation economy and alteration of municipal businesses that required large amounts of cheap black labor, whites no longer needed to make provisions for increasing numbers of blacks in towns. Even with out-migration, too many blacks remained for the available jobs. Alteration of the age structure to a population with large numbers of the elderly and the young contributed to some white leaders increasingly viewing blacks as an economic liability.

Most of the dwellings in the second type of black residential areas are similar to traditional ones. Municipal dwellings constructed and relocated between 1935 and 1965 belong to the third generation of post–Civil War housing. Many are shacks, including old tenant houses that were either given away or sold for a few dollars by planters who removed them from fields as mechanization of cotton production proceeded. Deteriorated mobile homes are occasionally jammed beside and behind some of the houses. The young Northern urbanites who descended into the Yazoo Delta during the 1964 Freedom Summer saw the black municipal residential areas near the end of the third era of housing. Their letters to parents and friends are filled with graphic descriptions. Of Itta Bena a civil rights worker wrote: "The Negro neighborhood is literally 'on the other side of the railroad tracks.' . . . [It] hasn't got a single paved street. . . . It's all dirt and gravel roads. The houses vary from really beat-up shacks to fairly good looking cottages. The beat-up places predominate. There are lots of smelly outhouses and many of the houses have no inside water."[11]

The War on Poverty created a third area of municipal housing for blacks. Despite the initial apathy of many Delta white leaders, a combination of greed, altruism, and, especially, the fear that civil rights agencies and blacks would assume local control of the new federal programs caused whites to take interest. In 1960, 48 percent of the housing units in the Yazoo Delta were deteriorating or dilapidated.[12] Between 1968 and 1981 more than two hundred million dollars was spent on housing in the region under the auspices of the FmHA and HUD. The new housing was constructed primarily within and on the fringes of municipalities and was part of the fourth generation of post–Civil War housing for blacks (figs. 11.4, 11.11). The impact on the quality of housing was profound. Nearly half of the

37,398 housing units occupied by blacks in the Yazoo Delta in 1980 were built after 1960.[13] A number of blacks were able to make quantum leaps from tenant shacks without plumbing to new dwellings with air-conditioning and other modern conveniences. A destitute day laborer with several children who received national publicity when Robert Kennedy visited her two-room shack in the Yazoo Delta in 1967 moved thirteen years later into a new housing development.[14]

The populations of the fifty-seven Delta municipalities in 1980 ranged from Glendora with 220 persons to Greenville with 40,613.[15] For thirty-six of the fifty-seven, the fringe population was one-third or more the size of the municipal; twenty-two had fringe populations that were 50 percent or more the size of those within the corporate limits. Ten had a fringe population greater than the municipal. Although concentration of a significant black population on the fringe of municipalities evolved through the continual construction of dwellings in traditional black residential areas just beyond the corporate limits, a major factor in the growth of the fringe populations after 1964 was the federal housing programs.

Indianola, Dollard's "Southerntown," and Tunica and Belzoni are illustrative of the municipal underbounding that had developed by 1980.[16] For all three municipalities the proportion of the white population was

FIG. 11.11. The landscape of the War on Poverty. South Gate, a Department of Housing and Urban Development Section 23 leased public housing project just beyond the southern boundary of Indianola, Mississippi, two years prior to its federal court–ordered annexation in 1989 under the provisions of the Voting Rights Act. Charles S. Aiken

larger within the corporate limits than without. The black fringe population of Tunica was larger than the entire municipal population. After 1940, the small Tunica Colored Subdivision grew and became the Census Designated Place of North Tunica (map 11.5). Although the ratio of fringe population to municipal was smaller for Indianola than for Tunica, three thousand blacks were concentrated on the southern fringe of the city. Belzoni, site of Marion Post Wolcott's infamous movie theater photograph that so graphically and symbolically portrayed segregation (fig. 5.5), depicted the new type of residential segregation in the age of integration. The fringe population of Belzoni was almost equal to that within the corporate limits, but whereas blacks constituted 54 percent of the municipal population, they were 79 percent of that in the fringe. Large black residential areas were immediately west of the corporate limits. The Crescent Theater survived federal mandated social changes and was still in business in the late 1980s. Seating was desegregated, blacks were hired, and the balcony was converted into a small, second-screen theater. The facility was patronized almost entirely by black teenagers and young adults.

The role of federally financed housing in the concentration of blacks on the fringes of the three municipalities was highly significant. Thirty-four percent of the housing units on the fringe of Tunica and 64 percent on the fringe of Belzoni were constructed between January 1960 and March 1980. For the fringe of Indianola, the figure was an astounding 98 percent. Most of the housing was sponsored by the FmHA, but part of it was funded by HUD. The new housing included subdivisions of single family dwellings for home ownership built by the FmHA under Section 502 and HUD under Section 235 of the Federal Housing Acts. In addition, HUD Section 8 (new construction) and FmHA Section 515 multifamily rental complexes were scattered through the fringes. Also, large HUD Section 23 privately owned public housing projects leased to the South Delta Housing Authority were in the fringes of Indianola and Belzoni (fig. 11.11).

Most of the housing units occupied by blacks in North Tunica are not new, nor are they federally sponsored. New federally financed dwellings are, however, scattered among the old shacks, and in 1975 the 29-house Park North Subdivision was built under the auspices of the FmHA. In 1986 a 40-unit multifamily FmHA 515 rental project was constructed in a former cotton field just beyond the Tunica corporate limits south of a collection of shacks known as Sunrise, and another 36-unit 515 apartment complex was completed next to Park North Subdivision in 1991. Other black households that remained in the worst of Tunica's housing along Sugar Ditch, a place that became infamous in the 1980s when the living conditions along the

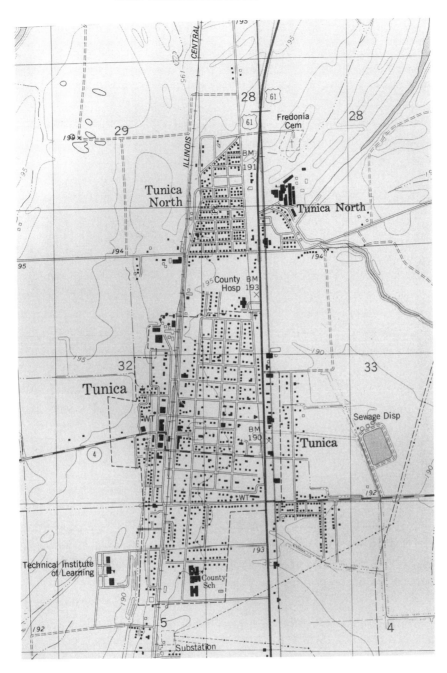

MAP 11.5. Tunica and Tunica North Census Designated Place, Mississippi, 1981. Although Tunica and Tunica North are functionally one place, the boundary of Tunica (dashed line) encloses only the area that contains most of the businesses and white population. In 1990, Tunica was 73 percent white and Tunica North 96 percent black (U.S. Geological Survey, 1982, Tunica, Miss.–Ark. Quadrangle, 1:24,000).

open sewer received national media attention, were relocated to a 36-unit HUD Section 8 project and 17 FmHA-sponsored mobile homes placed in a tight line along the abandoned railroad grade north of the municipal limits.[17] The western fringe of Belzoni includes a hodgepodge of black-occupied federally sponsored housing built since the commencement of the War on Poverty. In addition to FmHA-sponsored subdivisions, the area contains two Section 23 public housing projects leased to the South Delta Housing Authority and two large HUD Section 8 apartment complexes. The largest subdivision is Westgate, a 136-house development that was begun in 1971.[18]

Nowhere in the Yazoo Delta is the impact of the federal housing programs more evident than on the southern fringe of Indianola. During the two decades following the initiation of the War on Poverty, large numbers of shacks in the black residential area of Dollard's Southerntown were razed and the occupants relocated. Subdivisions of small, new brick-veneer houses, including those of a 264-dwelling Section 23 leased public housing project, sprawl across the landscape, interspersed with FmHA Section 515 apartment complexes (fig. 11.11). Although built largely under Republican administrations, symbolically the south side of Indianola is Lyndon Johnson's Great Society. The names of the subdivisions—Grove Park, South Gate, and Green Acres—are those of suburban America. But economically and socially the area is not the suburban stereotype. The population is all black, and in 1980, 40 percent of the inhabitants were poor, 11 percent of the labor force was unemployed, and the majority of the housing was subsidized.

Several factors contribute to the location of much of the new, federally sponsored housing in the Yazoo Delta and other plantation areas on the fringes rather than within municipalities. In some instances, lack of space, especially for large housing projects and subdivisions, prevented their construction within corporate boundaries. The major factor, however, is fear by white-controlled municipal governments that increases in housing for blacks within corporate limits will dilute white voting strength. Testimony given in a lawsuit in which the mayor of Lula, Mississippi, tried to block construction of an FmHA apartment complex reveals this phobia. A prominent white businessman and planter who was willing to sell land on which the project was to be built testified that the mayor said to him: "Bill, what in the world are you trying [to do] to Lula? . . . You are absolutely going to ruin the town and destroy the voter ratio, the Blacks are going to take over. My wife and I have spent considerable money remodeling and refurbishing our house, we will be forced to sell it at a sacrifice and move to Clarksdale."[19]

Acceptance of housing for blacks in the fringe rather than within the corporate limits is a compromise. Even dwellings built just beyond the corporate boundaries require a degree of approval by municipal governments, for water and sewer lines must be extended to them. Persons who work to improve housing of blacks in the plantation regions accept the fringe locations because they realize insistence that projects be located within municipal boundaries usually results in new dwellings not being constructed. For merchants, who are among the political leaders, a principal motivation for acceptance of the new housing in the fringe is that increases in the local black population mean growth of retail sales without growth of black political strength. Almost all of the retail businesses of the municipalities with large fringe populations are within corporate limits, and retail sales within the municipalities are substantially greater than the size of the municipal populations indicate.[20]

The resistance of whites to annexation of black residential areas has resulted in court suits under the Voting Rights Act to force annexation and efforts to incorporate the fringe as a separate municipality.[21] In 1984, after several years of litigation, Indianola was ordered by the federal district court to annex the southern fringe by January 1, 1989. Although in 1985 a similar case in Alabama was reaffirmed by the Supreme Court, no sweeping order mandates annexation, and each annexation case must work its way through the courts.[22]

A NEW TYPE OF BLACK GHETTO

The new racial dynamics in the plantation regions resulted not only in whites of some municipalities seeking to maintain political control by refusing to annex black residential areas but also in the concentration of blacks in other municipalities. As blacks in the plantation regions became increasingly urbanized, changes occurred in the racial percentages of municipal populations. Restoration of the franchise to blacks and desegregation of public facilities, including massive desegregation of public schools in 1970, caused whites to retreat from municipalities in which blacks were perceived a threatening majority, especially if they began to make significant gains in election to municipal offices. At the opposite extreme from municipalities that underbound part of their black populations are historical and newly incorporated all-black towns and municipalities in which the white population has drastically declined relative to the black population. Analysis of population redistribution in the Yazoo Delta also revealed the development of *ghetto towns*, places with relatively large black populations which have traits common to black ghettos in the nation's metropolises.[23]

Although in the Yazoo Delta the movement of blacks from plantations to local municipalities began during the 1930s with cotton acreage reductions under the Agricultural Adjustment Act and the adoption of tractors, the principal surge to the region's towns and cities occurred after 1950. In 1950, the black population surpassed the white in slightly more than half of the Yazoo Delta's fifty-two incorporated places. The black population was 75 percent or greater in only three, two of which were the historic all-black towns Mound Bayou and Winstonville. By 1980, blacks were in the majority in all but eleven of the fifty-eight incorporated places, and they accounted for 75 percent or more of the population in twenty municipalities. Although the proportion of the population which is black might seem high in 1950 as well as near the twentieth century's close, within the local context the alterations in the racial composition of municipalities are profound. That 75 percent or more of a municipality's population is black is significant because this is the figure used to define the core of the metropolitan black ghetto.[24]

The concept of "black ghetto," or "Negro ghetto," has several geographical meanings in the United States, but the principal use is in reference to black population concentrations in American metropolises.[25] In the plantation South the term *ghetto* has rarely been used to refer to the residential territories to which blacks have been confined in towns and small cities. But the territories are ghettos, and the labels that have been given to them, including *quarter*, also have meant places to which blacks are restricted. Recent redistribution of the black population has created a new type of black municipal ghetto in the Yazoo Delta and other plantation regions. As the black ghettos of Chicago and other large metropolises have become cities in themselves,[26] so certain small municipalities in the plantation regions have become ghettos in themselves. By 1980, in the Yazoo Delta the black population was 75 percent or more in the two historic "all-black" towns, four newly incorporated places, and twelve municipalities in which the number of whites declined as the number of blacks increased. These ghetto towns mimic large metropolitan ghettos in particular characteristics. Like metropolitan black ghettos, ghetto towns are places in which a poor minority population is concentrated and segregated. Redistribution of blacks in the Yazoo Delta has resulted not only in concentration of blacks in particular municipalities but also in increased segregation among the region's municipalities.[27]

The increase in segregation occurred, in part, through incorporation of new towns with predominantly black populations. Municipal government allows citizens of a densely populated place to deal more effectively with problems and to qualify for state and federal funds. In the Yazoo Delta, four municipalities—Falcon, Renova, Metcalf, and Mayersville—were incorporated during the 1970s, and one town, Coahoma, was incorporated in the 1980s. However, the increase in segregation among the Yazoo Delta's municipalities occurred primarily because, in certain towns, growth in the black population was accompanied by decline in the white. The changes in racial composition, in turn, produced significant alterations in the geography of municipalities. Small municipalities in plantation regions were traditionally characterized by the three major spatial components: the white residential area, the black residential area, and the business district (map 5.2). Since 1965, two additional components have emerged in ghetto towns that have lost white population and grown in black: a zone of encroachment by blacks into the white residential area and an area of new federally sponsored housing.[28] Shelby, an 87 percent black municipality of 2,806 in the heart of the Yazoo Delta, illustrates the five spatial components (map 11.6). Although Shelby's total population increased 11 percent between 1980 and 1990, its white population decreased 36 percent.[29]

The first spatial component, the core of the white residential area, is shrinking as blacks move into the margins of the traditional white area. The core is the part of the area which historically contained the dwellings of the white elite. Some of the dwellings are large, quaint two-story ones, but others are modern ranch-style houses built during the two decades following the Second World War (fig. 11.12). The core usually contains the white churches and may contain the building that was the white school during the era of segregation. The school building may be abandoned or may have been converted to other uses, including a private academy for whites. Population density in the core of the white residential area is low, averaging fewer than two persons per dwelling in many of the blocks. The population is composed primarily of middle-aged and elderly persons. In 1990, 44 percent of the whites in Shelby were sixty-five and older. Because the area is one of white flight, the number of houses available for sale to whites is greater than demand. Several of the houses are vacant, and a few dwellings may even have begun to fall into genteel ruin, but they usually are not overtly for sale, with signs in the yards and advertisements in newspapers (fig. 11.13). Houses that must be sold sometimes are bought by residents of the core to prevent the possibility of blacks occupying the dwellings. Vacant lots with concrete house foundations and sidewalks evidence places where dwellings have been razed to protect the core from incursion by blacks (fig. 11.14).

The old black residential area is the second component of black ghetto towns (map 11.6). The area contains many substandard dwellings, some of which date from the establishment of the town and others that were built during the 1940s, 1950s, and 1960s when the black population began to surge in municipalities. Some dwellings, however, may have been renovated with federal funds. Major improvements have been made in the infrastructure of the old black residential area. Federal grants and loans have been used to expand and rebuild the infrastructure, including installation of water and sewage systems and paving of streets. Empty areas and new parks evidence the use of federal funds to raze the worst of the old housing and relieve residential crowding. Population density is high in the old black residential area. In Shelby in 1990, it averaged more than four persons per dwelling in some of the blocks. The area contains the churches of blacks and may include the former school for blacks from the era of segregation. Though the school may still be virtually all black, since desegregation it has been the town's only public school. It may have been enlarged one or more times, an indication of the large proportion of the black population who are children.

The third component of municipalities in which the white population is decreasing while black is increasing is one of the new elements that spatially

FIG. 11.12. The house of an affluent white family in the core of the white residential area of Shelby, Mississippi, in 1987. Charles S. Aiken

FIG. 11.13. The aftermath of the civil rights movement. Vacant house in a state of deterioration at the edge of the white residential core of Friars Point, Mississippi, in 1987. Initially, evidence of vacancy is subtle, but after several years of neglect dwellings begin to deteriorate. Charles S. Aiken

FIG. 11.14. The aftermath of the civil rights movement. House being razed in the white residential area of Tchula, Holmes County, Mississippi, in 1986. The house, which was vacant, was torn down to prevent the possibility of blacks occupying it in the shrinking white residential core. Charles S. Aiken

distinguish them (map 11.6). Whereas the traditional pattern of munici-palities in the plantation regions was one in which the boundary between the black and white areas was sharply defined, a zone of encroachment of blacks into the old white residential area has developed. The zone is similar to that found at the fringe of black ghettos in metropolises.[30] In 1980, the index of dissimilarity was 82.6 for Shelby, which had a well-developed zone

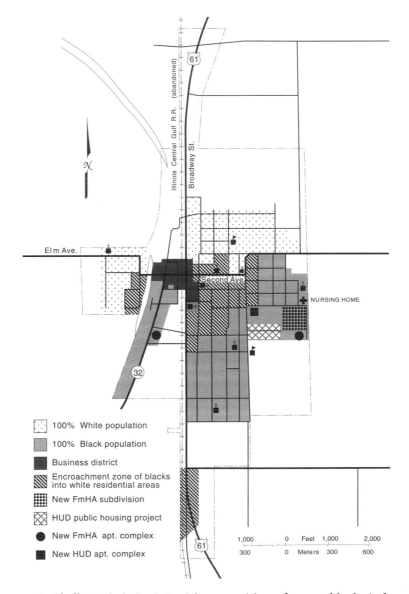

MAP 11.6. Shelby, Mississippi. Racial composition of census blocks is for 1980. FmHA- and HUD-sponsored housing is for 1988 (Aiken 1990). Courtesy of the Association of American Geographers

of encroachment of blacks into the white residential area. Although it might seem that the movement of blacks into white residential areas has been produced by federal fair-housing legislation, the principal factor is an economic one. Urban housing models assume that older dwellings filter down to a less affluent population. However, this is not the case in small Southern municipalities where race historically has been all consuming. Housing either does not filter to blacks, or the process is controlled and retarded. Because there is essentially no market among whites for dwellings in municipalities with significant black populations, buying houses to prevent blacks from doing so and leaving them vacant eventually becomes an economic burden. Finally a stage is reached when houses on the fringe of the white residential area are sold to blacks. "Block busting" tactics used by metropolitan realtors, however, in the nonmetropolitan South are constrained as much by social and economic controls as by legal measures that realtors confront in nonmetropolitan areas.

If significant numbers of whites begin to leave a municipality, panic flight by whites occurs similar to that which sometimes happens in metropolitan neighborhoods in racial transition. "For Sale" signs even appear in yards of houses in the core of the white residential areas. The exodus of whites has reached a critical stage, and it seems to the ones remaining that the town is lost to blacks. This stage was reached in Shelby in the mid-1980s. One of the principal factors that produce panic among whites is blacks using political power that was restored by the 1965 Voting Rights Act to elect black municipal officials. By the mid-1980s, Shelby had a black mayor and a black alderman.

The fourth component of ghetto towns is also one not traditionally found in municipalities of plantation regions. Growth in population occurs in part by movement of blacks into new federal housing.[31] The new housing usually is built close together and as additions to the old black residential areas. In Shelby, the federal housing in 1990 consisted of a twenty-four-unit HUD public housing project and forty-unit Section 8 apartment complex, together with an FmHA section 501 subdivision and two Section 515 apartment complexes (fig. 11.15).

The fifth component of the ghetto towns is the business district. New housing and growth in population normally contribute to viable retail business, but in municipalities where the white population is rapidly decreasing while the black is increasing, the opposite happens. The business districts are in decline, and retail trade has both deteriorated and changed. In general, businesses in the nation's small towns, especially those with populations less than twenty-five hundred, have declined during recent

FIG. 11.15. The landscape of the War on Poverty. Farmers Home Administration Section 515 apartment complex in Shelby, Mississippi, in 1987. Charles S. Aiken

decades.[32] Factors that produced the declines nationally, including decreases in the rural population and competition from larger municipalities, also have been important in the plantation South. In the plantation regions, however, the role of racial changes in the municipal population is also a factor in the decline and alteration of the retail structure.[33] Shelby and other municipalities that lost significant numbers of whites have visual evidences of retail decline, including vacant store buildings and buildings with lower-order businesses than previously. The retail decline in Shelby began during the 1970s and accelerated during the 1980s. In 1970, Shelby had a strong, diversified business structure of 110 firms that included specialty shops, building materials companies, new automobile dealers, and professional services. Between 1980 and 1990 actual dollars of retail sales declined 24 percent, and by 1990 the number of firms had decreased to 63. Both the Ford and the Chevrolet dealerships closed (figs. 11.16, 11.17). The greatest relative declines were in "miscellaneous retail," the category that includes specialty shops, and in "miscellaneous services," the one that incorporates physicians, lawyers, and accountants.[34]

Increase in the black population and decrease in the white is accompanied not only by a decline in retail trade but also by its restructuring from one partly oriented to relatively affluent whites to one that almost exclusively serves low-income blacks. In Friars Point, an old Mississippi River port town of 1,334 which in 1990 was 90 percent black, the alteration of the business district is complete (map 11.7). The role of the federal

FIG. 11.16. The aftermath of the civil rights movement. A variety store in Shelby, Mississippi, in the process of going out of business, in 1986. Charles S. Aiken

FIG. 11.17. The aftermath of the civil rights movement. A closed motor vehicle dealership in Rolling Fork, Mississippi, in 1986. Charles S. Aiken

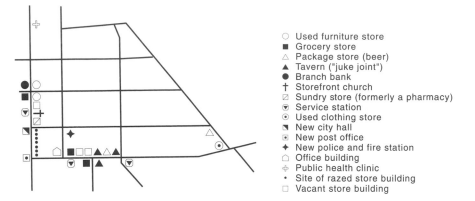

MAP 11.7. Business district of Friars Point, Mississippi, 1988 (Aiken 1990). Courtesy of the Association of American Geographers

government in the transition is evidenced by the vacant space on the town's main street where store buildings that formed the heart of the business district were razed with HUD grants and in the new federally funded city hall and fire and police station. The used furniture and clothing stores, the small independent groceries, the juke joints, and the storefront church are among the types of establishments found in low-income metropolitan areas dominated by blacks.[35] A small branch bank survives in Friars Point only because of the infusion of Social Security, AFDC, and other types of transfer payments. Though not as numerous as in the Yazoo Delta, black ghetto towns are distributed throughout the plantation regions.

Ghetto towns are but extreme versions of particular geographical characteristics that are depicted more subtly in municipalities where the relative black population is less than 75 percent but where serious racial discord exists. Selma, Alabama; Albany, Georgia; and Greenwood, Mississippi, are places where white resistance and racial unrest continued long after the cities disappeared from national television and newspapers. A result is the exodus of whites. Selma's white population dropped from 51 percent in 1960 to 42 percent in 1990. The white population of Albany declined from 64 to 45 percent and that of Greenwood from 49 to 41 percent between 1960 and 1990. In all three cities, the white population decreases were absolute as well as relative. Almost a third of Selma's whites left between 1970 and 1990.

BEYOND THE PLANTATION

Within the span of a little more than a century, the local distribution and the settlement pattern of blacks in the plantation regions of the American

South dramatically changed twice. Blacks were nucleated in slave quarters during the Old South era, dispersed across the countryside during the New South period, and in the modern South are again nucleated. The blacks who inhabit municipalities, their fringes, and rural hamlets often live within view of the sites of slave quarters where their ancestors dwelled. Although there are superficial spatial parallels between the modern rural hamlets and Old South slave quarters, there also are profound geographical differences. With each of the two major changes in settlement pattern, blacks in the plantation regions made further progress in their prolonged journey toward freedom. As the dispersed settlement of the New South plantation was an expression of a severely circumscribed freedom, so the new nucleated pattern may be interpreted as an expression of a newly augmented, but in certain respects a still restricted, freedom.

For numerous blacks, local population redistribution has meant exchange of shacks isolated among fields and forests for modern dwellings. However, the rural hamlets, municipal fringes, and ghetto towns in which rural blacks now live are not parts of the plantation system but lie beyond the margins of the old landholdings. This same marginal pattern of settlement exists whether the plantation is extinct, as it is on the lower Georgia Piedmont, or whether it survives in modern spatial form, as in the Yazoo Delta. Literally, and symbolically, most blacks in the plantation regions of the American South are now beyond the plantation.

As blacks have undergone recent microscale population redistribution, they have left behind the agrarian objects of white domination and control. Missing from their new settlements are the big house, the furnish store, and the old plantation paternalism. With each year, remembrance of these objects recedes further into history. At the close of the twentieth century more than three-fourths of the blacks in the plantation regions are too young to have known the agricultural system. The New South era of tenant farming is just as obscure to them as the Old South era of slavery.

In the process of settlement change, new types of important, but obscure, black enclaves have been created. Some of the new places inhabited by blacks, such as Woodville, Daniel Springs, and Hollywood, are built on the ruins of a past that persists only in names on the landscape. Though Philomath is protected by a historical district that overtly excludes the new hamlet of blacks but encloses a house touted as the site of the last meeting of cabinet members of the Confederate States government and what may be the Great Buffalo Salt Lick described by William Bartram, the place is more synonymous with the present than the past and is known locally for its large black population.[36] Other new black hamlets, such as Hamrick, are unnamed, while the names of extinct places like Carter's Grove and Banks, which were important in the bygone eras of plantation agriculture, continue to be printed on maps. Whether or not the new places where blacks reside are officially named, they are important local concentrations of black political strength.

Even municipal black ghettos are obscure places. They are concealed by spatial distribution and by environment. Small corporate nuclei of poor blacks spread over a large agricultural region such as the Yazoo Delta hardly have the geographical impact, visually or statistically, of black ghettos in metropolises. Environment further obscures the Delta's black ghettos. In the lush farmlands, even rural slums such as Falcon and Mayersville lose part of their shocking conspicuousness and become merely quaint places on the American rural landscape. Towns such as Shelby, Friars Point, and Sunflower, which are greatly ameliorated by the federal impress initiated by the War on Poverty, appear nothing like stereotypical metropolitan black ghettos. Despite a population that is 83 percent black and a poverty rate of 55 percent, at first glance Sunflower, with its new subdivisions and apartment complexes, is reminiscent of a prosperous agricultural town or a metropolitan suburban community in an idyllic agrarian setting. Superficially, it seems actually to be the amenity village that is advertised on a large, highway sign—"Sunflower, Country Living on the Sunflower River."

Although the new settlement pattern represents for blacks a new freedom, it also displays the problems they confront at the end of the twentieth century. Lost are the historical, paternalistic rights of blacks to hunt and fish on surrounding land once farmed by them and their ancestors. A type of enclosure movement, similar in some ways to that which occurred in Europe more than a century earlier, commenced in the plantation regions as the New South era waned and the Modern South era emerged. New dwellings in rural hamlets, ghetto towns, and just beyond the boundaries of municipalities have helped to ameliorate certain problems of blacks, but they also have contributed to the creation of new ones. From a public policy perspective, the locations of many of the new dwellings of blacks illustrate the consequences of federal programs introduced and administered without adequate geographical perspective and policy. Although federal programs alone did not create rural hamlets, ghetto towns, and underbounded municipalities, they are important to their sustenance and growth. A few federal officials were wise enough to foresee long-range repercussions of unguided federal programs, especially the housing programs, in plantation regions designed, in part, to decrease migration of blacks to Northern and Western metropolises. In 1967, Robert E. Levine, deputy director of the Office of Economic Opportunity, advised Sargent Shriver: "The economic basis [of the South's plantation counties] simply will not support the current population. . . . Neither will possible industrial development. . . . Given modern industry, which is not labor intensive, industrialization projects for such areas . . . can use up an awful lot of money, create very few jobs and of those jobs they do create, reserve even fewer for the rural poor. . . . Housing programs for the poor in these areas where there are no jobs and will be none are the height of cruelty."[37] However, Levine's and a few other similar warnings concerning location decisions of federally sponsored housing in nonmetropolitan areas went unheeded.

Despite the problems, the new settlement patterns depict for a segment of the black population an ameliorated living standard and a new level of affluence which promises hope to the others who are now beyond the plantation. Historically, discrimination and poverty have been the overriding problems of blacks in the plantation regions. In the past, blacks had no potential to influence and change the difficulties they faced. The civil rights movement and the War on Poverty, federal court decisions, and congressional legislation took them additional steps toward control over their destinies. But adequate solutions to the economic problems that confront many of the blacks who still live in the old plantation regions as the twentieth century draws to a close remain elusive.

CHAPTER TWELVE

Quest for a Nonagrarian Economy

❧ ❧ ❧

Considering the dimensions . . . nationally, . . . it would seem
that substantial alleviation of rural poverty is not within the reach
of the [Economic Opportunity] act. Attainment of this objective
would require broadening the economic base through investment
of capital in labor-intensive industrial and commercial activities.
To do this for all rural areas where poverty is acute, might be
beyond the reasonable expectation of national economic growth
in the foreseeable future.

Comptroller general of the United States, 1969

In many respects Mississippi, especially the Delta, is akin to
an underdeveloped country. You've got a lot of Third World
characteristics in this one little place. Where do we look for change?
Twenty years ago if you were on one side of the table you
pointed the finger at Whitey. Who do you point the finger at
when you're the government? Yesterday I was the activist,
today I'm the elected official.

Larry Farmer

DECADENCE OF THE PLANTATION

Arthur Raper was one of America's preeminent plantation scholars at the middle of the twentieth century. After thorough studies of the cotton plantation in Greene and Macon Counties, Georgia, Raper came to the conclusion that the agricultural system was "a millstone about the neck of the civilization which it cradled." The plantation was a "preface to peasantry," for it prepared "the land and the man for the emergence of a peasant rather than for the appearance of the traditional independent American farmer." The system led to "economic and cultural dependency," which planters deliberately created to achieve "satisfactory workers."[1] At the end of the study, which was published in the midst of the Great Depression, Raper issued a bleak prophecy for those trapped in the agricultural system:

> For the rank and file of plantation workers there is little hope either in the rejuvenation of the present Black Belt plantation or in its collapse. The cotton plantation can be revitalized only by mechanizing it, which, without a change in the prevailing philosophy, will result in the employment of fewer people without materially raising the level of the majority of those still employed. On the other hand, as the plantation system disintegrates it disinherits its owners and enforces many of its workers to emigrate without materially improving the lot of those who remain behind. Dependent and fatalistic people, exhausted soil, and crippled institutions are the natural remnants of the collapsing plantation economy.[2]

In spite of his early praise for the plantation, an older U. B. Phillips had come to a similar dismal conclusion several decades earlier. Phillips contributed to the conception of an Old South, but his perspective from secure academic jobs at the University of Wisconsin, University of Michigan, and Yale University increasingly was that of a refugee from the New South. Phillips concluded that the plantation system "concentrated wealth . . . within the hands of a single economic class and within certain distinctive geographical areas."[3] A diversified economy and a large middle class could not develop in an area dominated by plantations, for the agricultural system strangled other types of economic development. Although a plantation economy offered employment to large numbers of low-skilled workers, it provided few opportunities for the ambitious. The ultimate result was that "the plantation system was in most cases not only the beginning of development, but its end as well." Stated succinctly, "the system led normally to nothing else."[4]

At the close of the twentieth century, large parts of the landscapes of the South's plantation regions prove Raper and Phillips accurate prophets. Pine forests and sedge fields are conspicuous features across vast areas of the older plantation regions. Although the alluvial Mississippi Valley and the Inner Coastal Plain remain among America's most intensely farmed and productive agricultural regions, so automated is modern plantation agriculture that some lonely machines operate for hours without any care by humans. The landscapes of the plantation regions at the end of the twentieth century are aesthetically superior to the mature New South land-

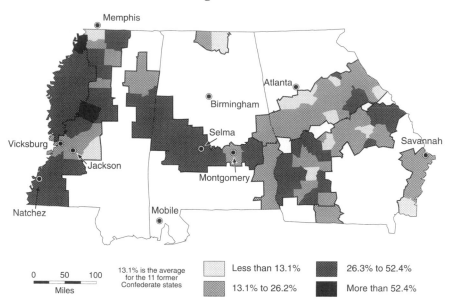

MAP 12.1. Percentage of persons who were poor, 1989 (source of data: U.S. Bureau of the Census 1993)

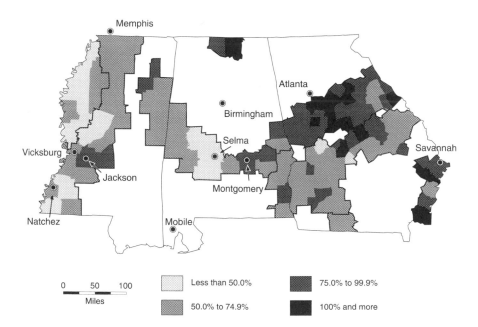

MAP 12.2. Median income of black households as a percentage of the national median income of black households, 1989 (source of data: U.S. Bureau of the Census 1993)

scape at the time of the depression. From a pedagogical perspective, the contemporary landscapes, superficially, are less interesting than those of the New South. Sharecropper shacks amidst scraped, swept yards bereft of any vegetation where wretchedly clothed children with symptoms of hookworm and rickets played were more striking than the modest, ill-kept frame and brick-veneer houses and mobile homes with few external signs of life which one now sees. But the modern landscapes of the old plantation regions are deceitful. Despite the socioeconomic improvement of segments of the descendants of the blacks and whites who provided the labor for plantations, the landscapes conceal the types of places that Harrington called "assembly points for the poor and the almost-poor."[5] Across the plantation regions, rural hamlets of blacks, ghetto towns, and the fringes of certain municipalities are among the assembly points.

The obscure blacks and whites in the plantation regions at the close of the twentieth century are not the down-and-out tenant farmers of half a century ago. In an absolute sense, life in the plantation regions is profoundly better than it was at the time of the depression, or the time of the civil rights movement, and a larger and more affluent black middle class has arisen. However, in a relative sense large numbers of persons in the plantation regions still live in pathetic conditions. By almost any measure of socioeconomic well-being, many inhabitants of the plantation regions are still at the bottom of the ladder. Large parts of the plantation regions are among the areas of the United States with the highest relative poverty, and the poverty is persistent decade after decade (map 12.1).

Scholars have long written of two Souths: the Upland South, the Lowland South; the Old South, the New South; the rural South, the urban South. As the twentieth century drew to a close, a new, fourth concept of two Souths emerged, a South that is doing well economically and a South that is left behind. Figuratively, there is a Sunbelt South, and there are "shadows" in the Sunbelt. Together with the inner cities of metropolises and parts of southern Appalachia, many counties that make up the old plantation crescent are parts of the South that is left behind.[6] Across the South's plantation regions, black household incomes are low, unemployment rates are high, and transfer payments are the leading source of personal income (maps 12.2–12.4). Median incomes of black households in a number of counties are less than 75 percent of the national median income and in other parts of the crescent are less than 50 percent. Lack of strong regional and local economies are also reflected in poor health care, high infant mortality rates, badly underfunded public schools, low educational attainment, and excessive school dropout rates.

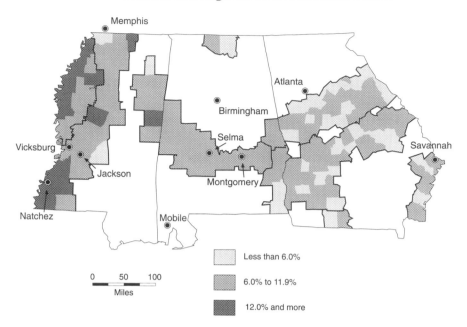

MAP 12.3. Percentage of labor force unemployed, 1990 (source of data: U.S. Bureau of the Census 1993)

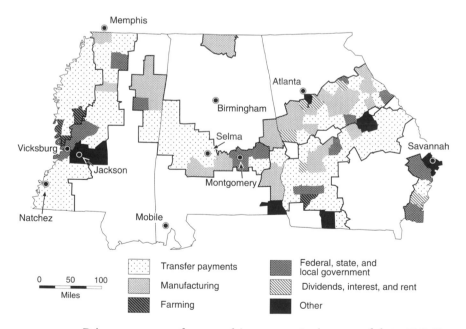

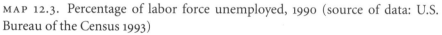

MAP 12.4. Primary source of personal income, 1989 (source of data: U.S. Department of Commerce, Bureau of Economic Analysis, 1991)

WHY NOT KEEP THEM DOWN ON THE FARM?

From the Reconstruction period until 1920, the number of "nonwhite" farmers in the United States and the South continually increased. In 1920, 97 percent of the nation's 950,000 nonwhite farmers were in the South. The number of nonwhite farmers declined to 719,000 in 1940 and to only 23,000 in 1987. Although by 1920, 24 percent of the nonwhite farmers in the South had achieved the former-slave dream of owning a farm, 76 percent were tenants, nearly half of whom were lowly sharecroppers. Most of the farms of nonwhite owners were relatively small, averaging only 77 acres in 1920. The precarious situation of black farmers compared with white ones is revealed by the peaking of the number of nonwhite farmers fifteen years before the number of white ones climaxed in 1935.[7] The peaking of the acreage held by nonwhite farm owners in 1910 is sometimes interpreted as evidence of the loss of land by blacks to unscrupulous whites who took it from them. An alternate interpretation is that in the exodus of blacks from the rural South, the more affluent and ambitious landowners and their children, perceiving greater opportunity and better conditions in cities, especially those in the North, were among the first to leave. Black tenant farmer households lagged behind those of landowners at the commencement of the Great Migration and appear to have been "pushed" from the land by the boll weevil disaster and the collapsing plantation system more than "pulled" by the opening of urban jobs to blacks.

The decline in the numbers of black farmers and land in farms controlled by blacks is also often mistakenly equated with the loss of land ownership by blacks.[8] In part, this idea is based on misinterpretation of census figures, which reveal land farmed by blacks rather than land owned by them. Forced loss of land by blacks is a problem in certain areas of the South, including the Sea Island plantation region, where tourism and retirement have inflated land values, and in the rural-urban fringes of Southern metropolises, where the encroaching city has engulfed rural black enclaves. However, most black families retain title to tracts that are no longer farmed by them. The decline in black farmers primarily was a result of the demise of the New South plantation with its large numbers of tenants and of alterations in the American economic system which increasingly placed small owner-operated and rented farms at a competitive disadvantage. In decrying the decline of black farmers, critics should remember that most were tenants who lived in the depths of poverty. The majority of blacks who left the rural South in the Great Migration did not nostalgically yearn for the good old days on the plantation. With each succeeding generation of urban-born, knowledge of farming and ties to the rural South waned. It should also be remembered that a substantial part of the land owned by blacks consisted of relatively small tracts that had physical and locational drawbacks. If whites rejected ownership of such land during the New South era, they hardly desired to take it away from blacks as agriculture declined and economies of scale, which accompanied mechanization, had to be achieved through the assembly of large tracts. By the mid-twentieth century, the former-slave dream of owning 40 acres and a mule had become a mythical historical icon that had no genuine contemporary economic or social relevance. At the end of the twentieth century, small owner-operated farms can hardly give American blacks, or non-blacks, independence and economic security, much less material and social affluence.

The displacement of agricultural farm workers in the South by mechanization following the Second World War was part of the commencement of a restructuring of the American economy, which by the 1960s had begun to affect other sectors. The decline and relocation of blue-collar metropolitan industries and the increase in white-collar jobs that demanded well-educated workers contributed to the failure of many among the growing numbers of blacks in inner cities to find employment, especially after 1960.[9] Beginning in the summer of 1963, metropolises of the North and West began to be rocked by racial turmoil that continued throughout the remainder of the decade. Bloody and costly riots erupted in the black ghettos of New York, Chicago, Detroit, Los Angeles, and other major cities. The riots, which corresponded with the last phase of the modernization of the South's plantations, initiated a second era of public concern over agricultural mechanization and exodus from the land. Migration of blacks from the rural South frequently was cited by journalists, mayors, and other officials as causes of the riots and other urban ills.[10] If migration could readily be accepted as the principal cause of the urban racial problems, assertion that displacement of Southern tenant farmers by mechanization was the primary cause of the mass migration of blacks to the nation's metropolises was a simple next step. Stated succinctly, "the black migrant to Newark was driven out of the South by the mechanical cotton picker."[11] Illustrative of pronouncements supporting this view is one by former secretary of agriculture Orville Freeman, who in 1970 wrote: "Displacement of large numbers of rural people by mechanization is more responsible for the big city problems which resulted in the burning of cities in the United

States in the 1960s than any other factor. . . . I have no hesitation in confirming this analysis."[12]

Placing the blame for the riots on black migrants was a continuation of urban America's propensity to make scapegoats of the most recent immigrant groups. By the 1920s, blacks had begun to supersede southern and eastern Europeans as the culprits who were thought to be the basis of urban social and economic problems. Earlier, Irish, Italians, and Slovaks, and later, Vietnamese, Haitians, and Mexicans, were among the urban immigrants who endured the same judgment as "Negro refugees." Such a view treats black migration as an exceptional type of emigration. Collapse and modernization of plantation agriculture, segregation and discrimination, and increased urban employment opportunities were major causes of black migration from the rural South to the nation's metropolises during the twentieth century. Only when viewed within the narrow context of the nation's urban racial problems, rather than in the scope of national and world geography and history, do American blacks' loss of traditional employment, persecution, and perception of a better, distant economic opportunity seem unique.[13] That migration of Southern blacks was the primary cause of the urban riots in the sixties was questioned and shown to be a popular but naive explanation.[14]

The agrarian myth assumes racist overtones when extended to blacks. One form of the myth holds that blacks are inherently peasants, a simple rural people who neither desire urban life nor are culturally or socially suited to it. A black yeomanry did not develop because blacks were compelled to migrate from the rural South to American metropolises. They were driven off plantations by machines and denied opportunity to become small landowners. Their forced migration created a host of social, economic, and political problems that ruined the nation's great cities. In the midst of the 1960 urban riots, from the New York headquarters the executive secretary of the National Sharecroppers Fund wrote of the bucolic life poor blacks could enjoy if they remained in the plantation South:

> The recent racial riot in the Watts ghetto of Los Angeles was different only in virulence . . . from those that have been occurring in many Northern cities over the last two summers. . . . The fact simply stated is that long overcrowded Northern slums have been gutted with Negro refugees from the rural South. . . .
>
> There is ample evidence that many of those who are being pushed off the land to the unwelcoming city slums would prefer to remain. . . . In the rural areas there is clean air, living space, more wholesome

family life, and a closeness of community which is lacking in the cynicism and despair which greet the farm poor when they give up and migrate to the city slums. . . .

Immediately needed is the development of imaginative programs to make use of the total resources of government to reverse the out-migration of the uneducated, jobless poor from the rural South to the city slums. . . .

This could well lead to a renaissance of our most impoverished rural areas. . . . It might result in a return migration of many city slum dwellers who now face a bleak future.[15]

The idea that small farms, alternative agriculture, and handicrafts are appropriate types of economic enterprises for the poor, especially for Southern plantation blacks and highland whites, has been subscribed to in various forms and degrees. At one extreme are those who view farming and handicraft projects as temporary instruments whereby the rural middle-aged and elderly can partially ameliorate their chronic poverty. At the other extreme are those who believe that small semisubsistence farms and cooperatives are long-term economic solutions to poverty and America's racial problems.[16] This is especially so for projects that make blacks small yeoman farmers. For more than a century, as a part of the agrarian myth, the idea has prevailed that throngs of small-scale, semisubsistence farmers, including persons who cannot be smoothly integrated into modern urban economies, can contentedly exist amidst American affluence and profitably operate in the face of economies of large-scale commercial agriculture. Because farming talents are assumed to be simple innate skills, agriculture is the profession of last resort which can and should absorb all who cannot fit elsewhere in the modern American economy.[17] To find their niche, small farmers need either to emphasize unique plants or animals or to employ uncommon methods in the production of conventional crops and livestock. To survive the competition of large-scale farmers and agribusiness companies, yeomen must band together in cooperatives.

The solution to poverty among blacks, as among whites, has not lain in the creation of small farms. The agrarian myth influenced the 1964 Economic Opportunity Act, including the provision for poor farmers to organize cooperatives. A program to make the rural poor into small landholding farmers was part of the original legislation that did not survive congressional alterations. When Congress amended the act in 1967, it included the statement that the OEO "should not . . . encourage the poor to migrate to urban areas because . . . [they] tended to further congest the overcrowded slums and ghettos." The OEO was "to provide employment and other opportunities to the poor living in rural areas to enable them to remain."[18]

The OEO's agricultural cooperative program harked back to the cooperative projects of the Farm Security Administration and its predecessor, the Resettlement Administration, more than a quarter century earlier. Although the FSA projects made landowners of a small number of the South's legions of black and white tenant farmers, by the 1960s most of the cooperatives were defunct, and the majority of the small landowners, though they might still live on their properties, no longer farmed because only a meager arduous livelihood could be eked from agriculture. The all-black Gee's Bend Homestead Association and its affiliated cooperative in Wilcox County, Alabama, faced financial difficulty from the time of their creation in 1938. The estimated profits from the one hundred 80-acre farms and the cooperative's store, cotton gin, and gristmill were barely enough under ideal circumstances to meet annual debt payments. After payment of all farm and family expenses, the projected net annual farm income was only $5.50 per family. By the early 1950s, the Gee's Bend Cooperative Association was experiencing annual operating losses of more than $5,000 annually, which led to the demise of the organization. Although the black landowners of the FSA's Mileston Farms Cooperative tenant resettlement project were critical to the civil rights movement in Holmes County, Mississippi, in the 1990s, the county was still among the poorest in the nation. By the 1970s, Gee's Bend, Tallahatchie Farms, and Mileston were places with unplanned hamlets in which rural nonfarm populations lived in dwellings that depicted various levels of affluence. Many, however, were in deteriorated houses and neglected mobile homes.[19]

Among the handicraft cooperatives that had a relatively lengthy period of success was the Freedom Quilting Bee, which operated in Wilcox and other eastern Alabama Black Belt counties during the late 1960s and 1970s.[20] The Southwest Alabama Farmers Cooperative Association, which served ten counties and was headquartered at Selma, was a failure. Its purpose was to enable small farmers to market cucumbers, okra, field peas, and other vegetables. Not only did the cooperative have to contend with opposition from the state and local county governments, but it also suffered from inefficient management and unrealistic goals. Salaries and other operating costs were estimated to be $473,000 in 1970, but only $302,000 was to be realized from the sale of vegetables. A federal grant was expected to pay the deficit.[21]

FACTORIES AMONG THE PINE FORESTS
AND COTTON FIELDS

Because modern agriculture offers few opportunities and because few young persons view farming positively, in recent years most leaders who have worked for economic development and creation of jobs in the plantation regions have looked to manufacturing. Ironically, some of the world's oldest factories and some of the United States' least industrialized counties are found in the plantation South. Although nonmetropolitan industrialization in the South is often considered a post–Second World War phenomenon, the erection of factories in rustic locations began on the lower Piedmont shortly after the diffusion of the Industrial Revolution from Great Britain to the United States. The narrow, shallow creeks and rivers of the Piedmont with their numerous shoals and consistent annual flow offered inexpensive factory sites to entrepreneurs. The water power potential of the lower Piedmont was comparable to that of New England. The horsepower at Cedar Shoals on the Yellow River in Newton County, Georgia, east of Atlanta, not an exceptional site, was approximately that of the Merrimack River at Lowell, Massachusetts. Bolton, the first textile mill in the South, was constructed in 1811 on Upton Creek in Wilkes County, Georgia, at a site the ill-fated partnership of Eli Whitney and Phineas Miller had purchased for a toll cotton gin.[22] By the beginning of the Civil War, textile mills and their villages were scattered along rivers and creeks at several locations on the Piedmont.

Manufacturing, primarily textiles, continued to grow on the Piedmont after the Civil War, and factory towns and villages were built across the region from the 1870s into the depression. Mills and their communities sometimes were at isolated locations in the countryside, but usually they were built just beyond the borders of existing towns and cities. Although local entrepreneurship was responsible for much of the immediate post–Civil War frenzy to industrialize, after 1890 New England textile companies increasingly built branch plants and eventually relocated to the Piedmont. In 1890, New England had 10,934,000 cotton spindles and the South only 1,570,000. By 1925, the South with 17,292,000 spindles had surpassed New England, with 15,975,000. Clarkdale, constructed in the countryside west of Atlanta in 1932 by Coats and Clark, a Scottish thread company,[23] was the last New South–era textile company town built on the lower Georgia Piedmont. Because the New South rage to industrialize was primarily an Appalachian Piedmont and Ridge and Valley phenomenon, the Black Belt,

Loess Plains, alluvial Mississippi Valley, and other plantation regions largely escaped the quest for factories.

The industrial creed accepted in plantation regions was codified by William Gregg, the most overt promoter of Southern manufacturing during the Old South era, in a series of widely circulated essays originally published in the 1840s. Although Gregg thought that black slaves were the most efficient industrial workers, he believed that the primary source of operatives for textile mills should be "the thousands of poor, ignorant, degraded white people" who lived "in comparative nakedness and starvation."[24] Gregg's ideas were extended after the Civil War by Daniel Tompkins, Henry Pinckney Hammett, and other New South strategists. The primary modifications were spatial segregation of workplaces and vertical segregation of jobs. For a century, from the 1870s into 1960s, the relationship of blacks to manufacturing in the plantation regions was governed by the idea that factories were primarily for whites while plantations were for blacks. Stated succinctly by Broadus Mitchell, "mill life" was "the only avenue open to . . . poor whites . . . to provide an escape from competition with the blacks" in the fields.[25]

By 1900, the textile mill community was one of the most conspicuous signatures of racial order imposed by segregation on the New South landscape. According to Cell, the textile mill's "rigidly segregationist labor and residential systems were the closest possible approximations to the new order of race relations that simultaneously was being created in law and political action across the South."[26] Like plantations, textile towns and villages were small, self-contained isolated worlds. Porterdale, a mill community built at Cedar Shoals just outside Covington as part of the New South industrialization frenzy, reached its heyday during the Second World War when 2,500 operatives worked in the Bibb Manufacturing Company's three rope and twine mills. Only 5 percent of the town's 3,116 inhabitants in 1940 were black. In 1940, blacks accounted for only 2.1 percent of the employees in the United States textile mill industry and only 4.5 percent in 1960.[27]

Wilbur Cash, and others who followed him, interpret the Southern textile mill community as copied from the Southern plantation. According to Cash, during the New South era, "progress was being accomplished so completely within the framework of the past that the *plantation* remained the single great basic social and economic pattern of the South—as much in industry as on the land." "For when we sound the matter, that is exactly what the Southern factory almost invariably was: a plantation, essentially indistinguishable in organization from the familiar plantation of the cotton fields. . . . The institution, in short, had literally been brought bodily over."[28] Contrary to Cash's belief, the Southern plantation and factory community are similar, not because the plantation was the pattern for the other, but because both originated from the same European concepts and copied features of the same spatial model, the feudal estate, for the employment of capital and labor to the large-scale production of agricultural commodities and manufactured goods. Except for modification for the South's biracial population, the ideas of the Southern industrialists as to the creation and operation of factories were hardly original. Inclusive of the paternalism, villages for workers, and castle- and fortresslike architecture of mill buildings, they were borrowed from New England industrialists who had copied them from the British fathers of the factory system.

The rule that called for segregation of workplaces was justified by planters and industrialists as a means to prevent racial strife through elimination of competition between blacks and whites for the same jobs and for protection of white women, who, because of the sex taboo, should not work alongside black men. What few factory jobs were open to blacks in plantation regions were vertically segregated by task and wage. Even in rudimentary factories such as steam-powered cotton ginneries, the skilled jobs of bookkeeper, engineer, ginner (gin stand operator), and pressman were usually filled by whites, while the unpleasant, low-skilled tasks of fireman, suction-pipe operator, and bale handler often were held by blacks.

Although only a few blacks found jobs in Lowland South factories, large numbers of poor whites were increasingly drawn into the plantation system, where they sometimes competed with blacks as tenants. By the time the number of sharecroppers peaked in 1930, approximately half were blacks and half whites (table 2.1).[29] The competition for plantation jobs developed partly because across much of the plantation South planters successfully opposed factories and other alternative sources of employment which might drive up the price of agricultural labor. A 1941 study of Wilcox County, Alabama, found that "the old families" had "discouraged all efforts to bring manufacturing to the county seat."[30] Planters in Mississippi, Louisiana, and Arkansas were especially successful in thwarting industrialization during the New South era. In 1934, the planter-dominated Natchez Chamber of Commerce persuaded the owners of one of the sawmills to reduce the wages of blacks to seventy-five cents a day because the high pay of more than one dollar threatened the caste system. When Armstrong Tire and Rubber in 1937 announced plans to build a factory in Natchez, company officials assured local and state leaders that the plant would employ only a few blacks in the lowly jobs of porters and those mixing carbon black.[31] Even on the lower Georgia Piedmont, which by the depression was among the most industrialized areas of the South, some counties, including Hancock and Oglethorpe, had essentially no manufacturing because large landowners were apathetic toward, if not opponents of, factories. Many planters also opposed manufacturing because factories brought an infusion of alien owners and managers who threatened their economic, social, and political dominance. Not only were the newcomers usually better educated than members of the indigenous white power structure, but they had economic, political, and social connections beyond the small provincial worlds of planters. Few planters rose above the status of member of the local bank board or, if he contributed enough to a governor's political campaign, honorary state "colonel." In agreement with an earlier similar observation by Powdermaker, Tony Dunbar thought that many of the planter lords who ruled the Yazoo Delta at the time of the civil rights movement were barely middle class in the national socioeconomic context.[32]

Most of the manufacturing that developed across the Alabama Black Belt, the Loess Plains, and the alluvial Mississippi Valley during the New South era consisted of primary industries related to agriculture and forestry. Dollard found the "small industrial section" of Indianola in the 1930s "devoted to ginning cotton and processing cotton seed." In the early 1930s, saw and planning mills were the only manufacturing establishments in Natchez and Adams County, Mississippi. The largest employed 190 whites and 510 blacks. In 1948, Rubin discovered that two cotton gins, a gristmill, an ice plant, a small veneer mill, and a window frame workshop were all that constituted Camden, Alabama's manufacturing.[33]

Planters who opposed factories found allies in a group of conservative agrarian intellectuals, who lamented alterations in the economic and social structures of the South, which they blamed on increased industrialization and urbanization. Among the attacks on manufacturing was the infamous *I'll Take My Stand*, which was written by twelve Southerners, the core of whom were associated with Vanderbilt University in Nashville.[34] The twelve's perception of the South was a mythical one that overlooked the severe poverty, ill health, and racial segregation of large numbers of the realm's inhabitants. Like many agrarians before and after them, the twelve were charlatans who did not really live the simple rural life that they espoused. Nashville, even in the 1930s, was hardly a rustic place, for it had the innovations and luxuries that had revolutionized American urban life but were not found far beyond the limits of cities. The group included Stark Young and Andrew Lytle, both of whom were of the plantation tradition. Young wrote *So Red the Rose*, *Heaven Trees*, and other works that romanticized the Old South. He lived in New York City but spent a few weeks each summer on his cousin Caroline McGehee's plantation in Tate County, Mississippi, which is the setting of *Heaven Trees*. Even though Lytle romanticized the yeoman farmer in *I'll Take My Stand*, he was a member of an Alabama planter family and was educated in expensive private schools and universities. According to Harry Crews, one of his students, he and Lytle came from different Souths. "The Crewes were a 'white trash, no-teeth, tobacco-chewing family'; the Lytles were landed gentry who sent their son to school in France."[35]

New South industrialists who built factories outside the plantation regions in the Upland South were not as constrained by the segregation dogma, and some liberally employed blacks in factories. The entrepreneurs of Alabama's iron and steel industry, which was established in Birmingham, Bessemer, Gadsden, and other places north of the Black Belt, quickly learned subtle methods to keep wages of skilled whites and unskilled blacks low by adroit use of race phobia. Large numbers of blacks were recruited to the New South city of Birmingham, which was chartered in 1871, and were given low-wage industrial jobs that paid more than they could earn on Black Belt plantations. In 1910, 75 percent of Birmingham's iron and steel workers were black. Vertical job segregation that confined black workers to the bottom tiers, combined with vaunted racial superiority, kept skilled white labor docile.[36] Little wonder that half a century later, Birmingham's seething racial stew needed only a little additional heat from Martin Luther King Jr. and the SCLC's "Project C" to overflow so violently.[37]

Increasingly, during the New South era, leadership in plantation counties became concentrated in large towns and small cities and was less focused on plantations, hamlets, and villages in the countryside. When out-migration in response to the decline in and the mechanization of agriculture threatened the livelihood of town and city businessmen and their children even began leaving because of lack of jobs, some local leaders looked to manufacturing as the panacea for their economic problems. They thought that factory jobs would give more stability to town and small-city economies and help them even to grow through new employment opportunities for white yeoman and tenant farmers and their wives and children. By continuing the traditional plantation region practice of restricting employment in factories largely to whites, businesses could be preserved while surplus black labor continued to migrate.

In the depths of the depression, political leaders of Mississippi began to view manufacturing in a new light. Persons such as Mississippi governor

Hugh White saw the limitations of future employment in lumbering and even in agriculture with the commencement of mechanization. White initiated the Balance Agriculture with Industry program in 1936, which depended on government subsidy to recruit manufacturers to the state.[38] The efforts to bring factories, which began modestly during the depression, blossomed into major campaigns in Southern states following the Second World War. In 1949, Camden, Alabama's newly organized Chamber of Commerce recruited a branch plant of a shirt company which employed seventy-five white women in minimum wage sewing jobs. With the decline of cotton and the shift to cattle, Wilcox County planters and businessmen thought that the new factory would help to reverse the exodus of whites.[39] To attract manufacturing, plantation counties joined older industrialized ones of the South in touting cheap labor, low taxes, and antiunion laws. Added incentives included buildings constructed with public bonds, state-funded training of workers, and other economic lures.

In 1977, the Fantus Company, which advises manufacturers on factory location decisions, listed "pro-business" state and local governments, a large and willing labor force that believed in the "work ethic," industrial development bonds, a mild climate, and relatively inexpensive land and electric power as major industrial attractions of Southern states.[40] Antiunion right-to-work laws were vital parts of the pro-business stance. The head of the Greenville, Mississippi, Industrial Foundation candidly stated in 1978 that the organization opposed unions and that a unionized factory would not be welcome. In Mississippi, Georgia, and Alabama, 7.6, 11.7, and 14.3 percent of the workers were unionized in 1989, compared with 23.8 percent nationally.[41] Regressive taxation laws, which placed relatively large tax burdens on low- and middle-income households and relatively light taxes on high-income ones and on corporations, were also major parts of the pro-business stance.

What the plantation regions snared in the Southern recruitment effort were labor-intensive, low-wage types of manufacturing which filtered from metropolises seeking cheaper production costs and nuisance types that were forced to locations where little or no opposition was offered to their pollution and other hazards. Low-wage garment plants increasingly joined their textile mill cousins in the South. Food processing, furniture, chemicals, and electrical appliances also were heavily represented among the types of manufacturing which were located in the plantation regions by the last decade of the twentieth century. Proximity to raw material and lack of local adversaries to their water and atmosphere pollution helped to attract pulp and paper mills. The nauseating sulfur odor of paper mills became a smell of the South. Reforestation in the lower Piedmont, the Alabama Black Belt, and other areas that underwent significant agricultural decline helped in recruitment of pulp and paper mills to the plantation regions. The Hammermill Paper Company's factory at Selma, Alabama, was constructed in the mid-1960s without any odor control and with minimal waste-water treatment facilities. The pro-business Alabama legislature had passed no law regulating air pollution, and the state's waste-water law was so weak that it was essentially ineffective.[42]

In addition to the opposition planters pose to factories, officials of certain manufacturing companies are not drawn to counties with large black populations. In 1969, Lonsdale reported that a large black population was among the "barriers" to rural industrialization in the South. The avoidance of such counties goes beyond race. The plantation's legacy is a labor force that was poorly educated in underfunded schools, together with poverty and "general unattractiveness." Although manufacturing companies and state officials do not publicly confirm it, for certain industries a black population of 30 to 35 percent within a county is the threshold for exclusion from consideration for factory location.[43] Not only do race and socioeconomic characteristics enter into industrial redlining of counties, but manufacturing executives believe that blacks are easier and more likely to unionize than whites. The efforts to recruit factories resulted in an uneven pattern of personal income from manufacturing which had developed by 1990 (map 12.4).

The roles that local recruitment of and opposition to factories and racial mix of the labor force played in the distribution of manufacturing by the last decade of the twentieth century is illustrated by the state of extremes—Mississippi. The white yeoman "hill counties" in the northeastern part of the state, together with the northern counties of the Mississippi Black Belt, had significantly more industrial jobs and higher levels of personal income than the Delta and parts of the Loess Plains. The decline of agricultural economies relative to urban and rural economies in which manufacturing and services are paramount is revealed by the value of real and personal property in Mississippi. The per-square-mile wealth is now greater in many of the planter-disdained hill counties, such as Union, than in some of the Delta counties, such as Sunflower.[44]

A NEW EMPLOYMENT STRUCTURE

The civil rights movement, together with the fair-labor and affirmative action laws it spawned, opened more manufacturing jobs to blacks. In addition, by the 1970s blacks who remained in the plantation regions were among the nation's last pools of cheap indigenous labor. Prior to the passage of the 1964 Civil Rights Act, few laws existed for effectively dealing with discrimination in employment. Title VII of the legislation initiated active federal intervention in employment discrimination. It created the Equal Employment Opportunities Commission as an agency under the president and provided for private and federal court suits to end discrimination. All private employers and labor unions with twenty-five or more employees were under the act. In 1972, Congress modified Title VII by giving the EEOC enforcement powers, extending coverage to state and local agencies, and lowering the threshold to firms or agencies with fifteen or more employees.[45] Decisions in federal court cases against Crown Zellerbach Paper Company in Louisiana in 1969, Duke Power Company in North Carolina in 1971, and other corporations that discriminated began to motivate businesses and government agencies to hire more blacks and to give them access to jobs other than menial ones. The federal initiative in fair hiring gave companies that wished to tap the pools of cheap black labor in the plantation regions a reason to transgress the tradition that certain types of jobs were for whites only. Controversies over affirmative action and court cases involving racial discrimination in hiring and promotion practices, however, continue as the twentieth century draws to a close. In the 1990s suits that charged discrimination in hiring and promotion were brought against companies that ranged from international corporate giants such as Texaco to regional fast-food restaurant chains.

Between 1960 and 1990, a profound change occurred in the structure of the employed black labor force in the plantation regions (maps 12.5, 12.6). In 1960, the employment structure was still that of the New South era. Agriculture and personal services were the primary types of employment. By 1990, the structure had been altered toward that of the Modern South. Manufacturing had replaced agriculture, and professional services had superseded personal services as the main occupations. In the Yazoo Delta, the Alabama Black Belt, and scattered areas of the other plantation regions, blacks constituted more than half of the manufacturing labor force (map 12.7). In 1990 on the lower Georgia Piedmont, 32 percent of employed blacks worked in manufacturing and 19 percent in professional services. Only 2.3 percent worked in agriculture and 2 percent in household services. In the Yazoo Delta, where farming was still viable, 12 percent of employed blacks worked in agriculture and 25 percent in professional services.

The increase of blacks in the textile and apparel industries, in part, was in response to the defection of a segment of the traditional white labor

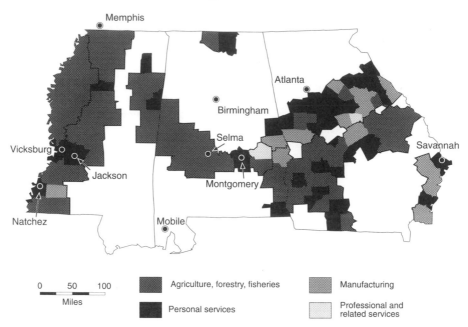

MAP 12.5. Primary types of employment of blacks, 1960 (source of data: U.S. Bureau of the Census 1963a)

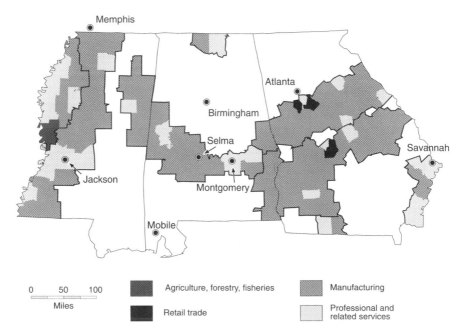

MAP 12.6. Primary types of employment of blacks, 1990 (source of data: U.S. Bureau of the Census 1993)

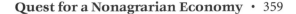

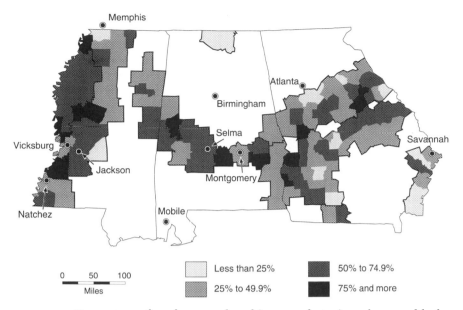

MAP 12.7. Percentage of workers employed in manufacturing who were black, 1990 (source of data: U.S. Bureau of the Census 1993)

force, primarily the adult children of operatives, who opted for high-wage jobs in more sophisticated types of manufacturing. Blacks filtered into textile mills and apparel plants to replace whites who moved into higher-wage jobs (map 12.8). Factories, especially apparel, also filtered into plantation counties where the pools of inexpensive black workers existed. New South industrial strategist Daniel Tompkins would have viewed this process as perfectly logical. Nearly a century earlier he predicted that "possibly, after a long time, when the white operatives have left the coarse work behind, negroes may become successful in this work." By the 1970s, newer types of manufacturing, such as furniture and electrical machinery, also employed large black labor forces in parts of the plantation regions.[46]

Some of the local blacks who rose to political prominence in the South's plantation regions found limitations in what they could accomplish. Although blacks might achieve election to political office, this did not give them the power to solve historical economic and social problems. Such was especially true in some municipalities and counties where blacks assumed political control. Not only have most black leaders not been able to stimulate significant economic development, but black political control sometimes has negative economic effects. If political rhetoric goes to the extent of creating an antiwhite image, this essentially insures that white manage-

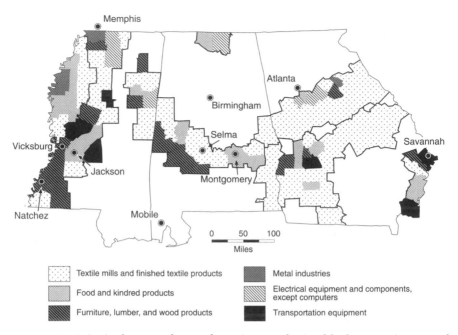

MAP 12.8. Principal types of manufacturing employing blacks, 1990 (source of data: U.S. Bureau of the Census 1993)

ment is not likely to locate a factory or any other type of private job-producing facility in such a place. Much of what black elected officials in the plantation regions have been able to deliver is federally funded social programs. Although housing may have been improved and health problems dealt with to a degree, large numbers of private sector jobs have eluded them. In the plantation regions, the most lucrative jobs for blacks are public sector ones. Increasingly, blacks have sought to wrestle more and more public jobs from whites, including control of the majority-black school districts still directed in some counties and municipalities by white superintendents and school boards.

In the quest for economic development and the creation of new jobs, some white and black county and community leaders in the plantation regions have desperately looked to undesirable facilities, or NIMBYs (Not In My Back Yard) and LULUs (Locally Unpopular Land Uses), including hazardous waste landfills and incinerators, noxious factories, state and federal prisons, and race tracks and other types of gambling enterprises including casinos. Poor, underdeveloped counties in the South with large black populations are among the places in the nation where what is termed "environmental racism" exists.[47] One of the largest hazardous waste landfills in the eastern United States is at Emelle in the Alabama Black Belt.

Other plantation areas, including Jefferson County, Mississippi; Hancock and Taylor Counties, Georgia; and St. John the Baptist Parish, Louisiana, became involved in controversies concerning attempts to locate hazardous waste facilities or noxious factories in them.[48] Federal and state prisons were accepted by economically depressed Hancock County, Georgia, and Marshall County, Mississippi, during the early 1990s, and new federal prisons were proposed by the Clinton administration as one part of an economic solution to the lack of jobs in the alluvial Mississippi Valley.[49]

Greene and Macon Counties, Alabama, among the first places in the plantation regions where blacks assumed political control, were among the first to turn to race tracks with pari-mutuel betting as a source of jobs and tax revenue. In 1976, the black-controlled Greene County Commission granted a company owned largely by local white businessmen and land-holders and headed by the son of legendary University of Alabama football coach Paul "Bear" Bryant the franchise for a greyhound dog race track. A greyhound race track opened in Macon County in 1984. Gambling as economic development in the plantation South moved to a higher, more sophisticated stage in the 1990s with the opening of expensive, flashy casinos built by major national and international corporations in Mississippi counties along the Mississippi River. Tunica, Coahoma, Washington, and Adams were among the counties that legalized betting, with residents wagering that with gambling would come the elusive prosperity.[50] The outcome of such enterprises may be like that in Greene County, which remains an impoverished place. Although generating as much as a two-hundred-thousand-dollar gross some nights in the 1970s and 1980s, by 1994 the race track was operating on borrowed money. None of the profits were saved, and beyond a new courthouse and other public buildings few of the funds were spent to improve the county's infrastructure for economic development.[51]

By the late-1990s, there was increasing evidence that leaders in some counties in the plantation regions no longer were willing to take just any type of facility which would create jobs. That the opposition to undesirable facilities was not split along racial lines but included biracial groups offered hope that some whites and blacks were willing to put the past behind and to cooperate in improvement of the economies and amenities of their communities.

CHAPTER THIRTEEN

Epilogue

❦ ❦ ❦

How, at the conclusion of the twentieth century, should the Southern plantation regions be interpreted? One of the most tempting approaches is to portray them as postcolonial areas. Colony and plantation have been intimately associated in world geography and history since the sixteenth century when European powers embarked on imperial paths, carved out large domains in the tropics and subtropics, and established great commercial farms. The South is the part of the United States which is the most similar to the rest of the world, and the plantation regions are the areas of the South which are most comparable to the new nations that inherited plantation economies.

The social research on the American South which began to emerge in the 1920s quickly tied a colonial concept to the Southern realm. The South was a colony of the North, becoming more so after the Civil War. Rupert Vance thought that the South retained "a colonial economy," and explanation of the realm lay "in the colonial system under which it was founded, . . . the plantation system to which it passed, and the cotton system with its tenancy which prevailed after abolition."[1] C. Vann Woodward and other scholars continued to employ the colony concept, especially for the New South era.[2] Key, however, comprehended that spatial variation existed within the Southern colony. The plantation regions were the areas of the South which were most like the imperial world and, therefore, the colonial nuclei. A few years prior to the 1954 *Brown* decision, Key wrote:

> The hard core of the political South—and the backbone of southern political unity—is made up of those counties and sections of the southern states in which Negroes constitute a substantial proportion of the population. In these areas a real problem of politics . . . is the maintenance of control by a white minority. The situation resembles fundamentally that of the Dutch in the East Indies or the former position of the British in India. Here, in the southern black belts, the

problem of governance is similarly one of control by a small, white minority of a huge, retarded, colored population. And, as in the case of the colonials, that white minority can maintain its position only with the support, and by the tolerance, of those outside—in the home country or in the rest of the United States.[3]

The new underdeveloped countries that emerged from the colonial empires after the Second World War are unable to escape the plantation system and its effects; the same seems true of the plantation regions of the American South. In certain respects, the lower Piedmont, the Black Belt, the Loess Plains, and the alluvial Mississippi Valley have more in common with the former colonies of the Caribbean and Central and South America than with the metropolitan United States. Whether the agricultural system is extinct or viable, the legacy of the plantation lingers in its effects on the economy, the social structure, and the political system. Some scholars maintain that a " 'colonial economy' no longer exists because 'outsiders' have so thoroughly penetrated the South that both the people and the economy have lost their distinct identities."[4] But the impact of outsiders has focused on metropolitan areas, especially on major ones such as Atlanta, Houston, and Dallas–Fort Worth. Large parts of the plantation regions have not shared significantly in the alien invasion, even the one related to new factories. For many, most of the "Egypt land" of the South remains terra incognita, a remote nonmetropolitan domain still shrouded in myth and misapprehension.

The civil rights movement can in certain respects be interpreted as a black nationalist struggle. It was part of the anticolonial endeavors that swept the world following the Second World War. The similarities between the civil rights movement in the American South and independence struggles in world colonies, especially those by blacks in the Caribbean and Africa, did not escape black leaders. "Sure we identified with the blacks in Africa," remembered John Lewis. "They were getting their freedom, and we still didn't have ours in what we believed was a free country."[5] Stokely Carmichael and Charles Hamilton's *Black Power* was the most overt contemporary effort to tie the civil rights movement to anti-imperialism that rocked the colonial world. For Carmichael and Hamilton, the "colony" of the United States was not a place but black people, and colonialism was "institutional racism." "Objective relationship" was what mattered, "not rhetoric (such as constitutions *articulating* equal rights) or geography." Colonial status had distinct political, economic, and social aspects, which operated in different geographical settings, ranging from Southern cotton plantations to the "dark ghettos" of the nation's metropolises. Cheap labor was the principal commodity of the United States' black colony.[6] Like independence leaders in world colonies, civil rights workers in the Southern plantation regions thought that freeing a dependent population from the oppressive agricultural system was fundamental to improvement of lives. "The problem is the depressive, impoverishing, circularly-indebting cycle of the plantation system," wrote two SNCC members who committed themselves to extended work among blacks in Holmes County, Mississippi. "The goal is to release the impoverished from that system and provide them the opportunity of gaining and maintaining their economic freedom."[7]

In certain respects, recent changes in the South's plantation regions are similar to those of new nations that emerged from world empires. A syndrome of economic characteristics, which Latin American scholars of the dependency school term "dependent underdevelopment," distinguishes the new Caribbean nations that emerged from the colonial plantation empires. Political independence has not brought economic independence; limited success has been achieved in economic development and diversification efforts; consumption absorbs most of the gross domestic product; and much recurrent expense is financed by foreign grants and development assistance.[8] Beckford observed that "each type of plantation . . . has its own inherent characteristics that create social diseconomies in plantation society. But a general pattern seems to emerge. In all cases the social diseconomies find ultimate expression in a rather unique combination of resource underutilization alongside underconsumption and poverty among the majority of people in plantation society."[9]

As Third World countries have been unable to escape the plantation system and its effects, so the plantation regions of the American South have not eluded the consequences of the agricultural system. Dependency and underdevelopment are what Phillips and Raper foretold. At the close of the twentieth century, the South's plantation regions still share certain characteristics of the plantation America culture sphere. Despite amelioration of segregation and discrimination, these postcolonial areas continue to have multiracial societies and structured caste lines. Planters, timber companies, factories, and government agencies participate in the control and manipulation of the socioeconomic and political structures. Even newly elected black officials in the South, like ones in the Caribbean, often adopt the image of the whites whom they replaced.[10] Dependency, the trait that traditionally characterized plantation workers, now portrays large areas of the plantation regions. The dependent economy is distinguished by low wages, insufficient economic diversification, shortage of local development

FIG. 13.1. The aftermath of the civil rights movement. A boarded-up pre–Civil War house in the white residential core of Tuskegee, Alabama, in 1987. Charles S. Aiken

FIG. 13.2. The aftermath of the civil rights movement. Abandoned United Methodist Church building in Cruger, Holmes County, Mississippi, falling into ruin, in 1985. After disagreements with the United Methodist Church over desegregation, former members formed an independent Methodist Church and constructed a small building nearby. Most Methodist congregations in Mississippi remained with the United Methodist Church. Charles S. Aiken

capital, creation of few new enterprises, and inadequate economic linkages to the growth sectors of the national and international economies. Large areas of the regions are capable of attracting job-creation facilities only from the bottom sector of the economy or facilities that have social, environmental, or economic stigma. Such conditions exist whether the local plantation economy is still viable or whether it is extinct. The Yazoo Delta has a strong modern agricultural economy and a relatively large underemployed labor force that cannot be absorbed by the agricultural system. In the Alabama Black Belt, the agricultural economy is largely gone, and a new type of local economy that can fully assimilate the underemployed population has not developed.

The postcolonial interpretation of the Southern plantation regions is supported by landscape features. As in former colonial areas, the new settlement pattern of blacks is one in which former plantation workers and their descendants are concentrated in hamlets and towns just beyond the borders of the old estates. The new settlement pattern is similar to that found in the Caribbean and other parts of the Plantation America culture sphere. Crop agriculture may have disintegrated, but many of the large landholdings remain virtually intact, planted to pine trees or lying idle.

Certain landscape elements are in response to desegregation and political empowerment of a historically oppressed people. As blacks gained access to public facilities of whites, the role of private, segregated facilities grew in importance. Not only did the number of private schools increase, but also the number of private country clubs, swimming pools, and other private places of recreation and entertainment for whites grew. As whites began to abandon historical towns that passed into the political control of blacks, certain icons of the past were protected. Among them are buildings that once housed the institutions of a segregated plantation society, including pre–Civil War houses. Often such dwellings are unoccupied and sometimes are boarded up, but they usually are not sold to blacks (fig. 13.1). Examples of places with such dwellings include Tuskegee, Alabama; Natchez and Holly Springs, Mississippi; and Sparta, Georgia. The white population in some towns and in parts of the countryside has declined to the point that even churches, such as Horeb Church in Hancock County, one of the oldest Baptist congregations in Georgia, and Woodstock Presbyterian Church at Philomath in Oglethorpe County, which was organized in 1794, have been disbanded and their edifices forsaken. Among the abandoned church buildings are those of congregations rent asunder over the issue of desegregation, such as the United Methodist Churches at Webb and Cruger, Mississippi, whose former buildings are falling into ruin (fig. 13.2).

The plantation landscapes are also littered with abandoned and modified school buildings, which are among the most overt testimonies to the end of legalized segregation (fig. 13.3). Deserted campuses of Jim Crow–era colleges, such as that of Mississippi Industrial College in Holly Springs, are striking landscape emblems. Massive desegregation caught the badly underfunded Mississippi Industrial in the midst of campus renovations and additions. The all-black college could not compete once the inexpensive Mississippi community colleges were opened to all. The partially completed addition to the gymnasium sits as a stark landscape symbol of the end of racial segregation as partially unfinished planter mansions, such as Longwood in Natchez, stand as monuments to the fall of slavery.

New public buildings that house government services stand in stark contrast to older edifices. In the postcolonial world, the number and quality of government services vary from country to country and, in part, are dependent upon external agencies, such as the United Nations and the World Bank. In the plantation regions of the American South, new public health clinics, welfare and subsidized-housing offices, fire and police stations, and government administrative centers are frequently new structures built with federal funds. Such edifices, from the striking, new modernistic city hall in Tuskegee, Alabama, to the large new health center in the midst of cotton and soybean fields north of Mound Bayou, Mississippi, do not bespeak the poverty that surrounds them (fig. 13.4, 13.5). Also scattered

FIG. 13.3. The landscape of desegregation. Abandoned school building, Hale County, Alabama, in 1987. This building is typical of ones constructed during the 1960s to create "equal" as well as "separate" facilities, but mass desegregation caused most whites of the Black Belt to flee the public schools. Charles S. Aiken

FIG. 13.4. The landscape of the War on Poverty. The new federally funded city hall, Tuskegee, Alabama, in 1987. Charles S. Aiken

across the landscape of the plantation regions are objects associated with new postcolonial economic development efforts. These range from the sign on the side of an abandoned building in Crawfordville, Georgia, advertising industrial sites, to the new catfish processing plant just east of Sunflower, Mississippi, to the gleaming gambling casinos along the Mississippi River, to the huge hazardous waste landfill near Emelle, Alabama, to the dog race track just off Interstate Highway 59 in Greene County, Alabama, to the empty industrial park at Tuskegee, Alabama, to the new regional prison hidden in the hills just south of Sparta, Georgia, to the vacant buildings and empty lots along the wide, well-paved streets of Floyd McKissick's failed dream of Soul City near Durham, North Carolina (figs. 13.6–13.8).

Although the postcolonial concept provides a tempting interpretation for the South's cotton plantation regions, beyond a certain point it becomes a narrow, forced notion that is neither comprehensive nor pragmatic to their interpretation. The plantation regions may have certain analogies to countries that emerged from colonial empires, but, fundamentally, they are not Third World places but places within a great democratic country with one of the world's strongest economies. The British colonies that became the American South began to differ from other world colonies before creation of the United States. George Washington, a political dissenter, the commander of the Continental army, the first president, and the most significant person in forging the new nation, was a Southern

FIG. 13.5. The landscape of the War on Poverty. A new federally funded health center just north of Mound Bayou, Mississippi, in 1987. Charles S. Aiken

FIG. 13.6. Quest for a new economy. A sign advertising Crawfordville, Georgia, in 1992. The relative situation of Crawfordville, a town with many vacant store buildings, changed at the end of the New South era. Charles S. Aiken

FIG. 13.7. Quest for a new economy. New catfish packing plant, Sunflower County, Mississippi, in 1987. The plant was funded with a five-million-dollar HUD block grant to the white-controlled, black ghetto town of Sunflower, which receives a small portion of the profits, and a loan from an insurance company. Charles S. Aiken

slave-owning planter, as was Thomas Jefferson, who authored the Declaration of Independence. Moreover, actual connections between the freedom struggles in world colonies and the civil rights movement in the United States were tenuous. The primary freedom struggle in the United States was fought a century before the civil rights movement over the issue of slavery. In the 1950s and 1960s, American blacks sought independence within the context of a strong democracy and the civil rights guaranteed to citizens by its constitution. Martin Luther King Jr. knew that, "consciously or unconsciously," the American Negro was "caught up by the spirit of the times . . . with his black brothers of Africa and his brown and yellow brothers in Asia, South America and the Caribbean." But he also comprehended that the freedom struggle in the United States was "a special American phenomenon which must be understood in the light of American history."[11]

Another concept for interpretation of the South's plantation regions is that they are among the places whose relative geographical situations within the nation have changed over time. One approach is to view the regions within the context of the transformation of the United States from a rural to an urban nation. At the beginning of the twentieth century, most of the nation's largest companies were related to agriculture and to the extraction of natural resources. American Cotton Oil, American Sugar, and Standard Oil were representative the types of companies which made up the nation's corporate structure. Not until 1920, well into the century, did the urban population surpass the rural. Seventy years later, at the close of the century, only 24 percent of the nation's population was rural and 76 percent urban, and the nation's economy was dominated by manufacturing and service corporations.

The propensity of Americans to believe the agrarian myth and rural blocs in Congress and state legislatures holding disproportionate power long after the majority of the electorate moved to cities, which were not effectively constrained until after the Supreme Court's 1962 *Baker v. Carr* decision, delayed conceptualization of the United States as an urban nation. Even comprehensive studies of American urban geography and urban history were post–Second World War phenomena. In the 1960s, a new spatial concept of what long had been the actual relationship between rural and urban in America began to be accepted. The connections between American cities and their rural hinterlands are no longer conceived in the shadow of agrarian myth and rhetoric. As parts of various types of intricate regional, national, and international systems, many functions of American cities are independent of their immediate hinterlands. Cities, especially the nation's great metropolises, including Chicago, Los Angeles, Minneapolis–St. Paul, Atlanta, and Dallas–Fort Worth, not agricultural regions, are the economic, social, and political heartlands. Where once farmers, planters, and businessmen of small cities and towns and the rural countryside held the economic and political power, the authority and vigor has shifted to medium-size cities and metropolises. During the last half of the twentieth century the plantation regions, like other of the nation's agrarian regions, lost much of their remaining power and influence.[12] Not only are the nation's rural areas, even those with viable agriculture, no longer superior to or even equals of cities, but some, including large parts of the plantation regions, have been left so far behind economically that they are given epithets such as "lagging regions" and "forgotten places."[13]

The filtering of marginal manufacturing from metropolises, the heartland, into lagging rural regions, the hinterland, is assumed to be the primary stimulus of economic development.[14] Whether this concept is completely accurate or not, strategies to assist lagging regions are geared to the industrial filtering idea. Most of the programs to help the persons residing in such regions emphasize education, job training, and improvements in the infrastructure, primarily the construction of roads and highways.[15] The multiplier effect on funds pumped into an area by factories, together with those from transfer payments and monies of government service agencies,

FIG. 13.8. Quest for a new economy. "Don't dump on us." Anti–hazardous-waste-dump sign on a pre–Civil War house in Sparta, Hancock County, Georgia, in 1992. The proposed facility allied whites and blacks of this racially polarized county into two biracial groups that opposed and supported the facility. Charles S. Aiken

is assumed to ripple through the economy and improve the lot of all inhabitants through a trickle-down effect.

One of the most important overlooked aspects of the South's plantation regions conceived as lagging areas is that significant spatial diversity exists among them and within them. Counties vary in such fundamental traits as age structure of their populations and size of their labor forces. Counties also are not equal in their integration with urban America and greater American society.[16] Small and medium-size cities within the plantation regions, including Athens, Georgia; Montgomery, Alabama; and Jackson, Mississippi, are fundamental to urban integration and ameliorated socioeconomic conditions. Most plantation counties that are on the fringes of such cities and large metropolises including Atlanta, Memphis, and Birmingham also have high levels of integration and socioeconomic well-being. Because of its proximity to Athens, rural Oglethorpe County, Georgia, has a high degree of integration. An alien sojourner driving State Highway 22 through the desolate pine forest between Philomath and Lexington might not immediately comprehend the accessibility to metropolitan America. But if her automobile breaks down near the site of the Barrow family's Sylls Fork Plantation, the tiny cordless telephone in her purse and the small computer on the automobile's back seat put her in instant contact with other places in the United States and the world. However, parts of the old plantation regions, even some counties close to metropolises, remain relatively remote from urban America, though they may be spatially close. The areas include the southeastern part of the lower Georgia Piedmont, the western Alabama Black Belt, the southern Yazoo Delta, and a portion of the Natchez district. Not only are such counties isolated, but they are dysfunctional places.

The fundamental problems of the places left behind, including large parts of the cotton plantation regions, are not ones caused primarily by exploitation from without, as is assumed by the postcolonial model, but ones that are a result of leadership failure from within. This failure may be Faulknerian, resting in the inertia and obstructionism of narrow-minded scions of old planter families, such as the Compsons and the McCaslins, and in the Varners and Snopes, greedy, short-sighted descendants of poor whites who arose within the leadership vacuum. In part, the failure may also rest within the black community. Political empowerment did not guarantee that aggressive and effective black leaders would emerge in every place in the plantation South. Although legacy of discrimination, inferior segregated schools, and denied political participation still cast their shadows over the plantation counties, increasingly blacks have had more chances to become masters of their own fate.

No simple solution exists for the social and economic problems that persist in the plantation regions. In the short term, most areas of these regions will continue to need significant financial assistance from federal and state governments. In the long term, the underlying solution to their problems, in large part, lies in that which the former slaves and white Southerners who dissented with the New South race creed looked to, education. Public education in most counties and municipalities of the plantation South still has a long way to go to achieve even average national standards. Not only does the historical problem of chronic underfunding persist, but improvements must be made within a national context that has witnessed many questionable, ineffectual practices sold since 1970 under the guise of education and teaching reform. Education improvements in the plantation regions depend, in part, on changes in prevailing philosophies of public education at the national level.

For many persons, blacks and whites, the plantation South remains a depressing place, among the least desirable in the nation in which to live and hardly a place to visit. Many young persons still view leaving the place of their birth, not as going away from home, but as escape. Yet, others, even some well-educated blacks and whites, tell you that the plantation South is the best place in the United States in which to live. They speak of family and friends, of the beauty of the countryside, of recreational opportunities, and of the friendliness of the people. But what most do not usually perceive, even those who are loners, is that which is really the most important to them. In a plantation society, even one that is rapidly receding, you cannot be anonymous, for everyone is known. And you do not have to concern yourself with creating your place in that society, for from birth it is largely defined for you.

At the end of the twentieth century, race relations in the plantation regions are improved significantly from what they were at midcentury. The codes of segregation and conduct dictated by a plantation society are disappearing, replaced not by true integration but by a mutual tolerance of one race for the other and by new forms of de facto segregation. The veil separating the races is not destroyed but merely adjusted and rehung. The failure of members of the two races to communicate effectively with one another, a failure during the New South era which Faulkner attempted to capture in *Go Down, Moses*, persists in the Modern South in a deceptively altered form. Martin Luther King Jr.'s dream that one day the sons of former slaves and the sons of former slave owners would be able to sit down together at the table of brotherhood appears to have come to pass in that whites and blacks now sit around tables to discuss political, economic, and, sometimes, social matters.[17] But at the conclusion of the twentieth century, the brotherhood of King's connotation hardly exists, and, unfortunately, lack of real communication and understanding between the races persists, not just in the old Egypt land of the South but throughout America.

Notes

CHAPTER ONE
OVERVIEW OF THE
SOUTHERN PLANTATION

1. Phillips 1925, 738–39.
2. Phillips 1903, 233–34. For critiques of Ulrich Bonnell Phillips and his work, see Singal 1982, 35–57; Roper 1984; Dillon 1985; Smith and Inscoe 1990. On large industrial farms as plantations, see Gregor 1962, 1965.
3. Galloway 1977; Curtin 1990, 3–11.
4. Land 1969, 1–10, 29–31; Thompson 1975, 7–18.
5. Wagley 1960.
6. Aiken 1971a.
7. Merle Prunty Jr. argued that the Southern plantation is distinguished by six elements: a landholding large enough to be distinguishable from that of a family farm; a distinct division of labor and management; agricultural specialization; a large input of cultivation power per unit of area; a distinctive settlement form and spatial organization that reflect centralized control, especially control of cultivating power; and location in an area of the South with a plantation tradition (Prunty 1955, 460). Although specific details of each of the six elements changed through time, the six elements themselves have been consistent. Discussions of the definition of *plantation* are also in Taylor 1954a; Gregor 1962; Jones 1968; Graham and Floering 1984.
8. Curtin 1990, 4.
9. Beard and Beard 1927, 2:269; Morison and Commager 1930, 627; Shugg 1939, 234–36; Taylor 1954a.
10. Jacques 1869, 133.
11. Bruce 1900, 59.
12. Phillips 1903, 235–36.
13. Ransom and Sutch 1977, 56.
14. Ibid., 334–35.
15. Phillips 1910, 37.
16. Russell 1913, 17–18; Gray 1933; Taylor 1954b.
17. Hart 1982, 9–10.
18. U.S. Bureau of the Census 1913a, 6:878.
19. Quoted in Stone 1908, 90–91.
20. Prunty 1955, 1962; Aiken 1978.
21. Stampp 1956, 27–33.
22. U.S. Census Office, 1860, Population Schedules of the Eighth Census. Microfilm rolls 30, 150, 630, 1231. The 1860 slave census schedules list not only the number of slaves for each slaveholder but also the number of slave houses. The percentage of slaveholders in the sample areas with five or more houses, the minimum number of single-household dwellings assumed in later definitions to qualify a landholding as a plantation, ranged from 22.8 percent in Oglethorpe County, Georgia, to 60.1 percent in St. Helena Parish, South Carolina, and the proportion of slave houses owned from 55.8 percent in Oglethorpe to 91.9 percent in St. Helena. In addition, masters with five

or more houses owned 56.7 percent of the slaves in Oglethorpe County, 74.2 percent in the southern division of Macon County, Alabama, 89.5 percent in Tunica County, Mississippi, and 89.9 percent in St. Helena Parish.

23. Aiken 1973.

24. Olmsted 1856, 88, 111, 416, 420; 1860, 35, passim.

25. Works Progress Administration, Georgia Narratives, vol. 4, pt. 2, 253–54. Hunter was probably referring to the Billups plantation in nearby Morgan County rather than in Oglethorpe.

26. Olmsted 1860, 140.

27. Because large slaveholders employed one or more overseers, slave quarters with a house for the overseer could be built away from the owner's dwelling. For efficient management and help with discipline, the size of a slave village usually was limited to between 50 and 100 slaves. Sometimes, when the population grew to more than 100, the slaves were divided between the original landholding and a newly created plantation (Olmsted 1860, 28). In Tunica County, Mississippi, the 160 slaves of J. T. Watkins lived on two plantations. One had 90 slaves and seventeen houses, and the other 70 slaves and ten houses. William Henry Heyward owned 386 slaves and ninety-eight slave houses in Prince William Parish, Beaufort District, South Carolina. The dwellings were in groups of fifteen, twenty-five, twenty-eight, and thirty on four plantations (Census Office, 1860, Population Schedules of the Eighth Census. Microfilm rolls 603 and 1231). In his will, Isaac Franklin left detailed instruction for the creation of "three additional negro quarters or sets or clusters of houses, in the same style and plan as those on [his] other plantations" on his 8,700-acre tract in West Feliciana Parish, Louisiana, as the slave

population of his estate grew (in Stephenson 1938, 152).

28. That in the 1850s many slaves still lived in such houses is revealed in the interviews with former slaves conducted by the Writers' Project of the Works Progress Administration during the 1930s. Leah Garret, who was a slave on a Georgia plantation, recalled, "Us lived in a long house dat had a flat top and little rooms like mule stalls, just big enough for you to git in and sleep. Dey warn't no floors in dese rooms and neither no beds" (Works Project Administration, 1937, vol. 12, Ga., pt. 1, 15). The fire hazard posed by a poorly constructed chimney is divulged in an interview with Peter Clifton, who was a slave on a plantation on the lower Piedmont of South Carolina. "Us live in a log house wid a plank floor and a wooden chinney, dat was always ketching afire" (Works Project Administration, 1937, vol. 1, 205).

29. A planter in 1850 argued against multiroom houses for more than one family because he believed that the crowding was "unhealthy," contributed to "contention," and promoted "immorality between the sexes" (Breeden 1980, 121).

30. Ibid., 114–39. Sixteen- and 18-foot square rooms of slave houses were common. The room dimensions diffused from the British Isles to colonial America, where they were employed in the houses of various classes.

31. Olmsted 1860, 421–22; Stephenson 1938, 97; Herman 1981, 11–13.

32. Rehder 1978, 217–34.

33. Works Progress Administration, Alabama Narratives, vol. 6, pt. 1, 238; Mississippi Narratives, pt. 2, vol. 7, Supp. Series 1, 525.

34. Johnson 1934, 13–14; Thompson 1975, 252–53.

35. Quoted in Ford 1983, 34.

36. Reid 1965, 47, 48.

37. Shlomowitz 1982.
38. Barrow (Sr.) Papers 1866–81; Barrow Jr. 1881, 831.
39. Brooks 1914, 37–54.
40. Ransom and Sutch 1977, 65–78; Wayne 1983, 110–49; Wiener 1978, 3–73; Flynn 1983, 57–83; Davis 1982, 89–120.
41. Flynn 1983, 3; Reid 1973.
42. A South Carolinian 1877, 678; Somers 1871, 120.
43. Barrow Jr. 1881.
44. Ibid., 83–85. Barrow (Sr.) Papers, 1866–81.
45. Wayne 1983, 135.
46. Works Progress Administration. Alabama Narratives, vol. 1, 172; Mississippi Narratives, vol. 8, pt. 3, 989.
47. Herman 1981, 203.
48. Barrow Jr. 1881, 832.
49. E. King 1879, 303; Thompson 1940, 252; Prunty 1955, 470–71.
50. Bond 1970, 21.
51. Washington 1928, 29–30.
52. Barrow Jr. 1881, 835.
53. Davis 1982, 74.
54. Banks 1905, 82.
55. Flynn 1983, 85–86.
56. Unidentified planter, 1912, Brooks Papers, Inquiries I.
57. Range 1954, 85.
58. Flynn 1983, 87–88.
59. Ibid., 90.
60. Welch 1943, addendum.
61. Kousser 1974, 14–15, 60–61.
62. Ibid., 16–17.
63. Kirwan 1951, 58–64.
64. Kousser 1974, 166–67, 246–47.
65. Constitution of the United States, 1787, with Articles and Amendments as of 1890, in Aiken Papers.
66. Key 1949, 533–54; Kousser 1974, 239; Constitution of the State of Mississippi 1890, Adopted Nov. 1, A.D. 1890, in Aiken Papers.
67. Woodward 1951, 337–38; Kousser 1974, 249–50.
68. Key 1949, 517.
69. Odum 1936, 111.
70. Bond 1970, 3–126; Anderson 1988, 4–32.
71. Bond 1970, 84–126; Anderson 1988, 148.
72. Bond 1970, 153.
73. Percy 1941, 298.
74. Quoted in Kousser 1974, 250.
75. Bledsoe 1942, 2; Percy 1941, 282.
76. Phillips 1904, 257–58.

CHAPTER TWO
FROM OLD SOUTH TO
NEW SOUTH PLANTATION

1. Welch 1943, addendum. Before the end of the New South era, Bureau of the Census officials realized that sharecroppers were not legal tenants, but the agency continued to enumerate them through the 1959 Census of Agriculture (U.S. Bureau of the Census 1947, 131).
2. Brooks 1914, 52–53.
3. Phillips 1910, 37.
4. Brooks 1914, 52, 63.
5. John Dozier Pou, 1912, February 15 letter to Robert P. Brooks, in Brooks Papers, Inquiries I.
6. Brooks 1914, 51–52.
7. Phillips 1903, 67–68.
8. Johnson and Turner 1930, 11.
9. Raper 1936, 207.
10. Wright 1986, 48–49.
11. Stone 1908, 90. According to Paul S. Taylor, it was Stone's criticism of the way in which data was collected on Southern agriculture, counting tenant farms but not plantations, which caused the Bureau of the Census to conduct a special investigation of plantation farming as part of the 1910 census (Taylor 1954a, 141).
12. U.S. Bureau of the Census 1916, 13.
13. U.S. Bureau of the Census 1914, 878–79; 1916, 13–16, 29–33. The eleven former Confederate states had 14,584 "wage-labor plantations." Because most sugarcane plantations were operated with

hired workers rather than tenants, Louisiana had 2,230 wage-labor plantations that paid an average $5,843 for labor in 1909. (See also Taylor 1954a; Gregor 1965).

14. Although state data are available on the number of plantations in 1910, no county data appear to have survived the congressional purge of the 1910 census of agriculture.

15. U.S. Bureau of the Census 1916, 20–26.

16. Georgia Plantation Schedules, 1911, in Brooks Papers.

17. E.g., Boeger and Goldenweiser 1916; Brannen 1924; Johnson and Turner 1930; Long 1931; Hartman and Wooten 1935; Woofter et al. 1936; Holley, Winston, and Woofter, 1940.

18. Woofter et al. 1936, xvii–xix.

19. Ibid., xix–xxii, 3–5, 17–20.

20. U.S. Bureau of the Census [1948], v–vi.

21. Faulkner 1957, 316.

22. E.g., Mandle 1978, 52–70.

23. E.g., Clark 1944, 1946; Woodward 1951, 179–85; Ransom and Sutch 1977.

24. Aiken 1973.

25. Layman 1867, 208; Grady 1881, 728.

26. Smith 1952, 26.

27. Roper 1903, 239–40; U.S. Bureau of the Census 1906, 12.

28. Nixon 1915, 22–23.

29. Eliot 1927, 20–23; Hicks 1931, 186–204, 415–16.

30. Newton and Workman 1919, 427–32.

31. Johnson 1978.

32. *International Cotton Seed Products Directory* 1923, 294–308.

33. U.S. Census Office 1865, 739; U.S. Bureau of the Census 1913b, 449; Nixon 1930.

34. *International Cotton Seed Products Directory* 1923, 43, 66.

35. Lösch 1967, 10–11, 365–67; Moulton 1931.

36. Hoover 1948, 49.

37. Moorhouse and Cooper 1920, 28.

38. Aiken 1973.

39. Anthony 1976b.

40. Barrow 1881, 832.

41. Hill 1915, 53–55. Georgia Plantation Schedules, 1911, in Brooks Papers. Comprehension of a transition from a first to a second post–Civil War generation of housing of plantation blacks illuminates numerous contemporary references to the alteration. Writing of a "tall brown man, a hard worker" who was a tenant on a southwest Georgia plantation about 1900, W. E. B. Du Bois observed, "This distressingly new board house is his, and he has just moved out of yonder moss-grown cabin with its one square room" (1903, 116).

42. Brooks 1914, 94–95.

43. Aiken 1971a.

44. Ibid.

45. U.S. Census Office 1864, 64–69; Harper 1922, 226–27.

46. Phillips 1908, 212–51.

47. Russell 1988; *Atlanta Resurgens* 1971.

48. Brandfon 1967.

49. Stone 1908, 86; Langsford and Thibodeaux 1939, 13.

50. Hudson 1982.

51. Faulkner 1942, 340–41.

52. Vance 1935, 266.

53. Stone 1908, 87.

54. U.S. Bureau of the Census [1948].

55. Range 1954, 83; Earl 1992.

56. Cohn 1948, 12.

57. Palmer 1981; Cobb 1992, 277–305.

58. Johnson, Embree, and Alexander 1935; Johnson 1941, 17–32.

CHAPTER THREE
THE DEMISE OF THE PLANTATION

1. Mandle 1978, 82–83.

2. Baker 1926, 470–71.

3. Sauer 1941, 18–22.

4. The area described is based on Raper's studies of Greene County, Georgia,

which adjoins Oglethorpe (Raper 1936, 1943).

5. Raper and Reid 1941, 3–17.

6. Lange and Taylor 1939, 18–20, 39–42.

7. Johnson and Turner 1930, 6–7, 15.

8. Ibid., 11.

9. Gibson 1941, 16.

10. Raper 1936, 3–4.

11. Billings 1979, 216; Wiener 1978, 71–73, 222–27.

12. Barnett 1931.

13. Russell 1982, 42; Colley 1926a. The Ivan Allen family was prominent in the leadership of Atlant from the 1890s over the next century. Ivan Allen Jr. was the city's progressive mayor during the critical part of the civil rights era and helped lead the city into the Modern South. One member of the family was so detached from his heritage that when he discovered a journalist was writing the Allens, he reportedly exclaimed: "My God, you're going to find out we owned slaves, aren't you?" (Pomerantz 1996a, 1996b).

14. Meinig 1986, 184–85.

15. Quoted in Doyle 1982, 16. Statements about leadership and the decline of planter families, especially ones made by their members, must be accepted with discretion. In 1932, William Alexander Percy charmed and baited Hortense Powdermaker into believing that he "was one of the few remaining aristocrats in Mississippi": "Most all the others were either dead, in the poor house, or in New York" (Powdermaker 1966, 141).

16. The 1936 study directed by T. J. Woofter found "considerable turnover in plantation ownership." Only 21 percent of the 450,000 acres that made up the 534 cotton plantations studied across the South had been owned twenty-five or more years, and 41 percent had been owned less than ten (Woofter et al. 1936, 22–23, 199).

17. Rozier 1988, 295–99. Frank White and Moses Harris, 1912, Brooks Papers, Inquiries I.

18. Robert P. Brooks, 1911, Reports on Georgia Plantation Districts, Brooks Papers, District III, 1–3; District X, 3.

19. Raper 1936, 101–3.

20. Ibid., 104.

21. Hartman and Wooten 1935, 13, 62–70.

22. Ely and Galpin 1919; Spillman 1919.

23. For most censuses, standing-rent tenants are enumerated as cash tenants. In this study, standing-rent tenants are combined with cash tenants for censuses in which standing-rent tenants are enumerated separately.

24. Hoffsommer 1934, 29:6–7; Daniel 1985, 81.

25. Raper 1936, 148.

26. Ibid., 148.

27. Buchanan 1959, 12.

28. U.S. Department of Agriculture 1922, 350. Only 10 percent of the United States cotton crop was lost to the insect in 1915, but the destruction jumped to 31 percent in 1921. By 1928 the loss had declined to 14 percent (U.S. Department of Agriculture 1930, 198).

29. Soule 1920; Lundy 1922.

30. Acree 1912, 31–34.

31. Raper 1936, 67.

32. U.S. Bureau of the Census, *Cotton Production in the United States*, 1915, 10–11; 1922, 11; 1924, 11–12.

33. Raper 1936, 216–17.

34. Hartman and Wooten 1935, 51–52.

35. Wheeler 1927.

36. Immigration Needed 1922; Raper 1943, 161.

37. Prunty and Aiken 1972. The seven-county area of Greene, Hancock, Morgan, Oglethorpe, Putnam, Taliaferro, and Wilkes produced 170,900 five-hundred-pound bales of cotton in 1918 but only 59,200 in 1929. Production never rose above the 71,800 bales harvested in 1930. At the end of the Second

World War, the area produced only 36,400 bales (U.S. Bureau of the Census, *Cotton Production in the United States*, 1918–46).

38. Johnson and Turner 1930, 13; Hartman and Wooten 1935, 78.
39. Johnson and Turner 1930, 11, 21–22.
40. Raper 1936, 185–91; Hartman and Wooten 1935, 80.
41. The Negro Exodus 1922.
42. Quoted in The Negro Exodus 1922.
43. Gantt 1922.
44. Hoffsommer 1934, 29:7; Raper 1936, 214; Johnson and Turner 1930, 12.
45. Johnson and Turner 1930, 12, 25.
46. Quoted in Bunche 1973, 183.
47. Colley 1926b.
48. Wheeler 1927.
49. Raper 1936, 221. Sixteen plantations were reported in Greene County, Georgia, in the 1940 plantation census. Only 11.6 percent of the county's 911 tenant farms were parts of plantations (U.S. Bureau of the Census [1948], 8).
50. Johnson and Turner 1930, 29–30.
51. Margaret Mitchell, 1936, July 10 letter to Henry Steele Commager, in Mitchell-Marsh Papers, box 16, file 22.
52. Pyron 1991, 252–53.
53. Silk 1989.
54. Mitchell 1936, 859.
55. Caldwell 1932, 15–20, 231–32.
56. U.S. Bureau of the Census [1948], 10.
57. Superior Court Clerk's Office, Oglethorpe County, Georgia, 1932. The ownership history is traced for the larger tract after 1942.
58. Ibid., 1944; 1946; 1947; 1954. Deed books FFF, HHH, 3-J, 3-K, 3-R. In 1991, land in Oglethorpe County, Georgia, had an average value of $375 per acre, exclusive of timber (interview with Oglethorpe County, Georgia, tax assessor by author, June 23, 1992).
59. Smith and Phillips 1925, 238.
60. Bartram 1791, 56–57.
61. The Yazoo Delta, a developing planta-

tion region with significant numbers of cash tenants, was an exception. The percentages of cash tenants appear to have been low in mature plantation regions and high in both new and declining regions.
62. Webb 1968, 24.
63. U.S. Bureau of the Census, *Cotton Production in the United States*, 1908–18; Welch 1943, 13.
64. U.S. Bureau of the Census 1901, 545; 1902, 432; 1932a, 1310; 1932b, 1082–83.
65. Davis, Gardner, and Gardner 1941.
66. Ibid., 276–80; U.S. Bureau of the Census [1948], 12.
67. Davis, Gardner, and Gardner 1941, 285–89, 329–34.
68. Hinds 1916, 27.
69. U.S. Bureau of the Census 1912, 47–59; 1932, 99–103. Black Belt population figures do not include those for Montgomery County.
70. Hinds 1916, 61.
71. Cleland 1920, 378; Boles 1930.
72. Stoney 1941, 16.
73. Johnson 1934, 10–12; U.S. Bureau of the Census 1913b, 34; *Cotton Production in the United States*, 1915, 8; 1935, 8.
74. Johnson 1934, 12.
75. U.S. Bureau of the Census 1912, 58; 1952, 2–86; *Cotton Production in the United States*, 1915, 8; 1948, 11.
76. Rubin 1951, 11, 185–86.

CHAPTER FOUR
MECHANIZATION OF
THE PLANTATION

1. Prunty and Aiken 1972.
2. Gray 1958, 54; Hilliard 1972.
3. Gray 1958, 9–32.
4. Mississippi Crop and Livestock Reporting Service 1955, 112–13.
5. Conversion of a Cotton Plantation 1945, 5.
6. Brodell and Ewing 1948, 27.
7. Gaines and Crowe 1950, 3.

8. Tallahatchie Leasing Cooperative Association, 1940, two February 27 bid invitations and contracts. FSA Records, Cooperative associations, Miss. 1935–54, box 374.
9. Prunty and Aiken 1972, 298–99.
10. Street 1957, 107–29.
11. Ibid., 121–22.
12. Program for Pickers 1936.
13. Aiken 1973.
14. Ibid., 213; U.S. Bureau of the Census 1946a, 7–9.
15. U.S. Dept. of Agriculture, Consumer and Marketing Service, Cotton Division, summary sheets of the 1956–57 Census of Cotton Gin Equipment.
16. Agricultural Extension Agents, 1967, Letters to Charles S. Aiken, Aiken papers.
17. Goosing the Cotton 1963; Growers "Goosing" Their Cotton Should Know Safe Pesticides 1964.
18. Agricultural Extension Agents, 1965, Letters to Charles S. Aiken, Aiken papers. Porter 1962.
19. Mandle 1978, 96–97.
20. Prunty 1955.
21. Barton and McNeely 1939, 8–12; Welch 1943, 43–51; Conversion of a Cotton Plantation 1945; Pedersen and Raper 1954.
22. Pedersen and Raper 1954; Prunty 1962.
23. LeRay, Wilber, and Crowe 1960, 11–13; Prunty 1962, 164–65.
24. McClinton 1938, 48–50; Baker 1983, 155–56.
25. U.S. Bureau of the Census 1948, 14. The definition of *plantation* used by Prunty (1955) was employed to identify plantations in Tate County. The 168 plantations in Tate County in 1940 averaged 594 acres and 8.8 tenants. The large reduction in the number of plantations between 1940 and 1961 occurred because of the decline of cotton acreage owing to federal programs. In addition, with the advent of mechanization, the number of resident laborers on many of the county's larger farms dropped below the minimum of five needed to meet the definition of plantation (U.S. Bureau of the Census [1948], 2; Aiken 1962).
26. Prunty and Aiken 1972.
27. Johnson and Turner 1930, 5.
28. White 1951, 21–23.
29. Prunty and Aiken 1972.
30. Fisher 1970; Aiken 1971b.
31. Only ten articles on American farm tenancy are listed in *Reader's Guide* from 1920 through 1934. The number increased dramatically in 1935, and 174 are listed from 1935 through 1944. Interest in farm tenancy subsided after 1944. Only eleven articles on the topic are listed from 1945 through 1960.
32. Conrad 1965.
33. Caldwell 1951, 150–55.
34. James Agee, 1938, Journal, in Agee-McDowell Papers, box 1, folder 2, notebook 2, 20. *Tobacco Road* and *The Grapes of Wrath* are the best known of the sharecropper novels that were published from the 1920s into the 1940s. The genre also includes Dorothy Scarborough's *In the Land of Cotton* (1923) and George Sessions Perry's *Hold Autumn in Your Hand* (1941). (See Sonnichsen 1969). The principal characters in the sharecropper novels are members of white tenant-farm families.
35. Caldwell and Bourke-White 1974, vii–viii.
36. Agee and Evans 1966, inside front cover.
37. Maharidge and Williamson 1989, 70.
38. James Agee, 1938, Journal, Agee-McDowell Papers, box 1, folder 2, notebook 2, 20.
39. Conrad 1965, 42–44, 54–55.
40. Ibid., 54–58.
41. Ibid., 49–62.
42. That mechanization did not play a major role in the elimination of tenant

farmers during the first half of the Great Depression decade is revealed by *The Collapse of Cotton Tenancy*. The principal purpose of the small book was to awaken Americans to the plight of the South's landless farmers. Although an entire chapter was devoted to the impact of the AAA on farm tenancy, only four pages discussed mechanization. The impact of tractors and mechanical harvesters were imminent problems, not current ones (Johnson, Embree, and Alexander 1935, 43–44). In his 1938 M.A. thesis on the Percy family's Trail Lake plantation, McClinton described the replacement of tenants by machinery in the Yazoo Delta as "a relatively new trend" (6).

43. Quoted in Meltzer 1978, 163–80.
44. Taylor 1938a, 1938b.
45. MacLeish 1938, 59.
46. Lange and Taylor 1939, 40–42, 73.
47. Meltzer 1978, 202–3. Kehl demonstrated that significant parallels exist between passages in *The Grapes of Wrath* and Lange's pictures in MacLeish's *Land of the Free* (Kehl 1974).
48. Steinbeck 1939, 44–48, 157.
49. Cantor 1969, 3–17; Petcotte 1976.
50. Mitchell 1979, 38–40; Cantor 1969, 23.
51. U.S. Executive Office of the President 1937, President's Committee on Farm Tenancy; Baldwin 1968, 157–192; Grubbs 1971, 136–161; Mitchell 1979, 232–233.
52. Mitchell 1979, 171.
53. Missouri Sharecroppers, 1939, January report to the president, Wallace Papers, box 3117. Cantor 1969, 64, 127–29; Lange and Taylor 1939, 38.
54. Quoted in Cantor 1969, 157.
55. Henry A. Wallace, 1939, March 6 letter to J. E. Rankin; June 7 letter to Mrs. Roosevelt, Wallace Papers, box 3117. Eleanor Roosevelt caustically responded to Wallace's patronizing letter: "Should we be developing more industries and

services? Should we practice birth control or drown the surplus population?" Eleanor Roosevelt, 1939, June 15 note to Henry Wallace, Wallace Papers, box 3117.

56. Street 1957, 175–201.
57. Johnson 1947, 37; Dickins 1949, 5; Pedersen and Raper 1954, 8.
58. Mandle 1978, 84–97.
59. Grant 1972; Henri 1975; Grossman 1989.
60. Estimated net intercensal migration was computed by components of change method. U.S. Bureau of the Census 1975, 95.
61. Pedersen and Raper 1954, 26.
62. Maier 1969, 6–7, 120–12; James 1968; Bryles 1968.
63. The wage for harvesting is based on the amount paid for picking 100 pounds of seed cotton.
64. Taylor 1938b, 861–62.
65. Cohn 1950.

CHAPTER FIVE
THE WORLD OF PLANTATION BLACKS

1. U.S. Bureau of the Census 1975, 12, 22.
2. Johnson, Embree, and Alexander 1935, 14.
3. Holley, Winston, and Woofter 1940, 58.
4. Quoted in Freidel 1965, 8.
5. Quoted in Breeden 1980, 124.
6. Woofter et al. 1936, 224.
7. Ibid., 224.
8. Ibid., 100, 228.
9. Johnson 1934, 93.
10. McClinton 1938, 59–62.
11. Farm Security Administration, 1941, appraisal report Mahoney property, FSA Records, Cooperative Associations, Miss. 1935–54, Tallahatchie Leasing Association, box 377. Faulkner 1942, 341.
12. Jordan 1974, 3–25.
13. Cell 1982, 169–70.

14. Myrdal 1944, 54–55, 75–78.
15. Du Bois 1903, 181–82.
16. Myrdal 1944, 60–61.
17. Sellers 1973, 10.
18. Johnson et al. 1941, 277–80.
19. Quoted in Maharidge and Williamson 1989, 239.
20. Powdermaker 1966, 191.
21. Percy 1941, 298.
22. Phillips 1918, 313–14.
23. Raper 1936, 371.
24. Ibid., 23.
25. Du Bois 1918, 722–23.
26. Percy 1941, 299 300. Percy does not identify Dollard by name but refers to him as "a learned gentleman from Yale, who was psychoanalyzing the whole Negro population of a neighboring town." According to Daniel Patrick Moynihan, Dollard was among the first to associate aggression with frustration (Moynihan 1988, vii). See also Dollard et al. 1939, 22, 152–95. In a 1964 interview with Robert Penn Warren, Martin Luther King Jr. stated this same idea to explain why militant blacks had begun to attack him. "Unwittingly, these angry blacks were transferring to him their 'bitterness toward the white man'" (Peak 1987, 130).

    Percy's reaction to Dollard's answer is interesting in light of his just having written that although blacks and whites "believe they have an innate and miraculous understanding of one another . . . the sober fact is we understand one another not at all" (1941, 299–300).
27. Daniels 1938, 173; Wyatt-Brown 1994, 268.
28. Johnson 1941, 283–89.
29. Greenland to Open Monday 1941; Raper 1943, 338–39.
30. Raper 1943, 338.
31. New Greenland Has a Gala Opening 1941.
32. Wynes 1967; Dykeman and Stokely 1962.
33. Arthur F. Raper, 1934, Raper Papers, October 1–4 field notes, Putnam Co., Ga., 1–2; 7–8.
34. Mississippi Department of Education 1931, 98–105, 163–75; Raper 1936, 307.
35. Mississippi Department of Education 1931, 22. SCLC, 1966, Wilcox County: New structures versus old problems. King Papers, ser. 1, subser. 2, box 28, 3.
36. For example, in 1910 the Mississippi legislature passed a law that permitted the organization of consolidated rural school districts with the power to tax and to issue bonds (Mississippi Department of Education 1931, 9).
37. Ibid., 51–54.
38. Raper 1936, 313.
39. Anderson 1988, 186.
40. William B. Eskridge, interview in The Youth of the Rural Organizing and Cultural Center 1991, 181.
41. Moody 1968, 10.
42. Mississippi State Department of Education 1931, 74–75, 88–91; Raper 1936, 330–31.
43. Frank D. Alexander, 1944, Cultural Reconnaissance Survey of Coahoma County, Mississippi. Original uncensored manuscript for Bureau of Agricultural Economics, USDA, which contains material objected to by Mississippi congressman Jamie Whitten, in Raper Papers, 21–22.
44. Alexander 1944, 21–22.
45. Johnson 1941, 119.
46. Woofter et al. 1936, 10–11. The Woofter study found that plantations operated exclusively by white tenants were more common on the upper Piedmont and in the middle Tennessee Valley than in the other Southern regions (196).
47. Raper and Reid 1941, 20.
48. Rice 1968.
49. Harold Rose's discussions of territoriality and residential segregation in large southern cities give insights into the characteristics of residential segrega-

tion in smaller southern municipalities (Rose 1970; 1971, 1–13).

50. Aiken 1977, 5.

51. The research for Powdermaker's study of Indianola was conducted before Dollard's. Dollard's study, which was published first, has overshadowed Powdermaker's. *After Freedom* has a better description of Indianola's black subculture, while *Caste and Class in a Southern Town* treats race relations in greater depth (Rudwick 1968, ix). Dollard considered the "small industrial section" with cotton gins and a cotton seed oil mill at the eastern end of the business district a fourth spatial unit that was distinct from the business district (Dollard 1937, 4).

52. Powdermaker 1939, 9–14.

53. Dollard 1937, 2–3.

54. Raper 1936, 111–15; Herman 1981, 109–41.

55. Raper 1936, 111.

56. The same pattern of land ownership by blacks developed in the Caribbean colonies following emancipation. Whites continued to own plantations, and the only land available to blacks was in the hills and poor soil areas (Lowenthal 1972, 58–59).

57. Kirby 1969.

58. Faulkner 1959, 398–99.

59. John Lewis, interview in Rains 1983, 72.

60. Woodson 1930, 112, 123–31.

61. Ibid., 122–23; Herman 1981, 219–45; Hamilton 1991, 43–98.

62. Russell 1977.

63. T. C. Miller, 1938, memorandum to J. H. Wood, Cooperative Activities on Gee's Bend Project, FSA Records, Cooperative Associations, Alabama, box 2. Industrial and Rural Workers 1942.

64. A. D. Stewart, 1943, October 4 memorandum to R. W. Hudges. FSA Records, Cooperative Associations 1935–54, Mileston Co-op, Mississippi, box 343. C. C. Coats and J. M. Turner, 1940, Oc-

tober 18 appraisal report for the Holzman farm, FSA Records, Cooperative Associations 1935–54, Tallahatchie Cooperative Leasing Association, Mississippi, box 374. Thomas R. Vaughn, 1942, January 1 memorandum to E. B. Whitaker, Purchase of land by Tallahatchie Cooperative Leasing Association, FSA Records, Cooperative Associations 1935–54, Tallahatchie Leasing Association, box 377.

65. Kousser 1974, 224–26.

66. Key 1949, 5, 224–26, 503–17.

67. Ibid., 5–6, 666.

68. Ibid., 5–12; Black and Black 1987, 3–15.

CHAPTER SIX
MOBILIZATION

1. Morris 1984, 12–16, 77–99, 120–38, 195–224; Peake 1987, 37–64; Meier and Rudwick 1973, 72–98.

2. Rice 1968; Lawson 1976, 23–54.

3. Morris 1984, 77–99; Peake 1987, 42–49.

4. Baker 1960, 4.

5. Carson 1981, 9–20.

6. Brady 1955, 48.

7. Kluger 1976, 3–17.

8. Ibid., 346–66.

9. Ibid., 586–94; Newman 1994, 426–44.

10. King 1958, 201.

11. Tumin et al. 1958, 149–52; Bartley 1969, 12–20.

12. Key 1949, 666; Carter 1992.

13. King 1958, 201.

14. Other organizations formed to preserve segregation included the Defenders of State Sovereignty and Individual Liberties in Virginia, the States' Rights Council of Georgia, the Patriots of North Carolina, and the North Carolina Association for Preservation of the White Race (Wilhoit 1973, 48–51).

15. U.S. House 1967, 3–4; Chalmers 1976, 1–21; Whitehead 1970, 11–18.

16. U.S. House 1967, 5–11; Chalmers 1976, 22–342; Whitehead 1970, 18–21.
17. Klu [*sic*] Klux Klan, 1963, "The Freedom Fighter" (December), mimeograph, Lowenstein Papers, ser. 6, subser. 1, folder 144. U.S. House 1967, 11, 17–62.
18. U.S. House 1967, 60–62.
19. Ibid., 133–34.
20. McMillen 1971, 18–43, 119.
21. Brady 1955.
22. Martin 1957, 34, 37, 110, 155.
23. Carter 1959, 212.
24. Martin 1957, 155.
25. Powdermaker 1966, 142.
26. Frank Prial, 1965, October 11 memorandum to Edgar May and Robert Clampitt, Shelby, Bibb, Chilton, Perry, and Hale Counties, OEO Records, Inspection Reports 1945–47, box 1, Ala.
27. McMillen 1971, 362.
28. Research to measure the relationship between white oppression of blacks and black political participation employing Ku Klux Klan membership has shown no significant correlation. The lack of such a relationship has been misinterpreted as a result of local civil rights organizations that curtail Klan activity (e.g., Matthews and Prothro 1963, 164–66; Stewart and Sheffield 1987).
29. Percy 1941, 231–32.
30. U.S. v. Original Knights of the Ku Klux Klan, 250 F. Supp. 330, 335 (E. D. La. 1965).
31. Al Krumlauf, 1965, OEO Records, Inspection Reports 1964–67, box 21, Ga., 2; U.S. House 1967, 151.
32. Let's Be Careful, 1965, "Benton County Freedom Train," vol. 3, September 5, mimeograph, OEO Records, Inspection Reports 1964–67, box 40, Miss. The United Klans of America's small White People's March for Freedom klavern at Ashland was the only one in Benton County (U.S. House 1967, 153).
33. Cagin and Dray 1988; U.S. House 1967, 122.
34. Bartley 1969, 103–5.
35. McMillen 1971, 314–18.
36. Wilhoit 1973; Carter 1992, 21–23.
37. McMillen 1971, 319–42.
38. Brady 1955, 84.
39. McMillen 1971, 360.
40. Frank Prial, 1965, October 11 memorandum to Edgar May and Robert Clampitt, , 13.
41. Quoted in Huie 1955; October 12 letter to Roy Wilkins, Huie Papers, Emmett Till file. Whitfield 1988, 51–69.
42. Silver 1964, 29–30; Berry 1973, 52; Smith 1990.
43. Smith 1990; Nossiter 1991; Massengil 1994; Vollers 1995.
44. Silver 1964, 22.
45. A. A. Berle, 1965, September 10 memorandum to Inspector General OEO, SERA Corporation, OEO Records, Inspection Reports 1964–65, box 40, Miss.
46. Burk 1984, 253–66; Blumberg 1984, xvii-xxiii.
47. Burk 1984, 253–77.
48. Quoted in Schlesinger 1978, 234.
49. Civil Rights Act of 1957, Public Law 85–110, Stat. 634, U.S.C. at 703 (1957).
50. U.S. Department of Justice, Report of the Attorney General 1962, 161.
51. Claude Screws, sheriff of Baker County, Georgia, had beaten to death Robert Hall, a black man, on the courthouse lawn in full view of witnesses. No state murder charges were brought, but in a trial initiated by federal prosecutors, an all-white jury convicted Screws under a Reconstruction-era statute of depriving Hall of his civil rights. In overturning the conviction, the United States Supreme Court ruled that the prosecution had shown that Screws had intended to kill Hall but had failed to prove that Screws had attacked Hall with the intention of depriving him of

his civil rights. Screws v. the United States, 1495 U.S. (1944); Branch 1988, 408–9.

Between September 1, 1958, and July 1, 1964, the Justice Department filed forty-five court suits under Section 1971(a) but only sixteen under Section 1971(b). U.S. Department of Justice, Report of the Attorney General 1964, 172.

52. Doar and Landsberg 1975, 893.

53. Ibid., 898, 971–72; U.S. Department of Justice, Report of the Attorney General 1959, 180–86.

54. Lawson 1976, 205.

55. Doar and Landsberg 1975, 955.

56. U.S. Department of Justice, Report of the Attorney General 1958, 182; 1959, 192–93; Lawson 1976, 206.

57. U.S. Department of Justice, Report of the Attorney General 1959, 193–94.

58. Martin 1957, 105–14; U.S. Commission on Civil Rights 1959, 77; Norrell 1986, 92–110.

59. Doar and Landsberg 1975, 898; U.S. Commission on Civil Rights 1959, 69–97.

60. The Justice Department discovered that at least half of the registration cards of enrolled whites contained the same types of technical defects as the cards of the purged blacks.

61. Lawson 1976, 250–61; Brauer 1977, 68–88; Sorensen 1965, 478–79.

62. Lawson 1976, 261; Brauer 1977, 112–19.

63. U.S. Department of Justice, Report of the Attorney General 1961, 165–66; 1962, 160–61; 1963, 180–81.

64. Branch 1988, 408–10; Doar and Landsberg 1975, 909–12; Hampton and Fayer 1990, 142–43.

65. Doar and Landsberg 1975, 906.

66. Alabama Council on Human Relations, 1960, January newsletter, mimeograph, King Papers, ser. 1, subser. 2, box 28. U.S. Department of Justice, Report of the Attorney General 1964, 173.

67. U.S. Department of Justice, Report of the Attorney General 1963, 183; Wirt 1970, 95–96; Garrow 1978, 31.

68. United States v. Ike Shankle and the State of Mississippi, final judgement and decree, F.D.C., N. Dist. Miss. Civil No. D-C-45-61 (1964), in SNCC Papers, subgroup A, ser. 4, box 18.

69. John M. Straud, 1964, August 22, Sardis, Mississippi, Report, MFDP Papers, box 25.

70. Wirt 1970, 96–108.

71. Bruce Gordon, 1963, November 9 Selma, Alabama, field report, SNCC Papers, subgroup A, ser. 3, box 94. Burke Marshall, 1963, October 11 letter to Julian Bond, SNCC Papers, subgroup A, ser. 4, box 16.

72. Bruce Gordon, 1963, November 9 Selma, Alabama, field report. U.S. Department of Justice, Report of the Attorney General 1964, 174.

73. Bruce Gordon, 1963, November 9 Selma, Alabama field report, 5–6.

74. Quoted in Wirt 1970, 13–14; quoted in Garrow 1978, 34.

75. Wirt 1970, 322–23.

76. Chestnut 1990, 176, 185.

77. Halberstam 1993, 436–55.

78. Marshall 1964, 23; U.S. Department of Justice, Report of the Attorney General 1964, 172.

79. David, Kendall, 1964, July 23 report on Tate County, Mississippi, SNCC Papers, subgroup A, ser. 3, box 22. Davis 1965.

80. SNCC, 1963, Minutes of the Executive Committee meeting, December 27–31, SNCC Papers, subgroup A, ser. 3, box 6.

81. Frank Prial, 1965, October 11 memorandum to Edgar May and Robert Clampitt, 2. A. P. Boynton, 1963, October 8 letter to Martin Luther King Jr. King Papers, subgroup A, ser. 1, subser. 1, box 21. Garrow 1981, 224.

CHAPTER SEVEN
CONFRONTATION

1. South Enters New Era 1960.
2. Quoted in Analavege 1966b.
3. Morris 1984, 134.
4. Branch 1988, 233; Peake 1987, 149.
5. Sellers 1973, 46.
6. Zinn 1965, 124.
7. Doar and Landsberg 1975, 892–97, 958–62; Branch 1988, 487–88.
8. Hampton and Fayer 1990, 98.
9. Zinn 1965, 62–64; Hampton and Fayer 1990, 140–41.
10. Branch 1988, 228–33.
11. Hayden 1962, 7–8; Forman 1972, 224–25.
12. Robert Moses, 1961, Aug. 3 letter to Wyatt T. Walker. MFDP Papers, Box 23.
13. Hayden 1962, 9–11; Zinn 1965, 68.
14. Hayden 1962; Zinn 1965, 62–78; Carson 1981, 45–50.
15. Quoted in Zinn 1965, 79.
16. Zinn 1965, 79–86; Carson 1981, 77–78.
17. Myles Horton and the Highlander Folk School were vital to the civil rights movement. Many of the persons who became principal leaders of the struggle were trained at Highlander. SNCC regularly sent its workers to the school for seminars and additional training.
18. Quoted in Hampton and Fayer 1990, 179–80.
19. SNCC, 1963, Minutes of the Executive Committee meeting. December 27–31, 28–30.
20. Quoted in Carson 1981, 98.
21. Peake 1987, 112.
22. Brauer 1977, 230–67.
23. Wicker 1963.
24. Civil Rights Act of 1964, Public Law 88–352, Stat. 241, U.S.C. at 287 (1964); Lawson 1976, 298–99; Whalen and Whalen 1985, 239–42.
25. SNCC, 1963, Mississippi Staff Report, circa September, Operation Mississippi, SCLC Papers, subgroup E, ser. 1,

subser. 3, box 141. SNCC, 1964, Minutes of the March 29 Executive Committee meeting, SNCC Papers, subgroup A, ser. 3, box 6.
26. Sutherland 1965, 65–67; Harris 1982, 60–89.
27. SNCC, 1964, Minutes of the March 29 Executive Committee meeting, 2; Sutherland 1965, 64–67; McAdam, 1988, 66–86.
28. Barrett 1965; Williams 1988, 213–18. Governor Ross Barnett had secretly agreed with President John Kennedy and Attorney General Robert Kennedy to a charade in which he would stage capitulation to federal authority. On the night before Meredith was to be registered, Barnett announced on television to Mississippians: "I urge all Mississippians and instruct every state officer under my command to do everything in their power to preserve peace and to avoid violence in any form. Surrounded on all sides by the armed forces and oppressive power of the United States of America, my courage and my convictions do not waver. My heart still says, 'Never,' but my calm judgment abhors the bloodshed that would follow. . . . I know that we are completely surrounded by armed forces and that we are physically overpowered. I know that our principles remain true, but we must at all odds preserve the peace and avoid bloodshed" (quoted in Barrett 1965, 146; McMillen 1971, 345–46).

As Barnett spoke, a riot at the university was well under way with many of the participants apparently responding to his televised speech on September 13 in which he warned that "a solemn hour . . . the moment of our greatest crisis since the War Between the States" was at hand. "There is no case in history where the Caucasian race has survived social integration. . . .

We must either submit to the unlawful dictate of the federal government or stand up like men and tell them 'Never'" (quoted in Williams 1988, 215).

29. Quoted in Orfield 1969, 118.

30. E.g., Whitehead 1970, 312–13.

31. U.S. Commission on Civil Rights 1968b, 222–56.

32. Garrow 1978, 179–81.

33. Forman 1972, 265–66.

34. SCLC, 1965, Proposed budget and program, Summer Community Organization and Political Education Program (SCOPE), King Papers, ser. 1, subser. 2, box 34, 1.

35. James L. Bevel, 1964, April 13 memorandum to Martin Luther King Jr., Non-Violent vs. Brinkmanship, King Papers, ser. 1, subser. 2, box 28. Eric Kindberg and Ann Kindberg, 1965, September 1, Proposed Alabama Project and Budget, King Papers, ser. 1, subser. 2, box 28.

36. James L. Bevel, 1964, Program for action in Alabama, King Papers, ser. 1, subser. 2, box 28, 5–6.

37. Branch 1988, 81–87; Niebuhr 1932, 250–54; King 1958, 97–99. Curiously, the relationship between the civil rights movement in the 1950s and 1960s and the long, but successful, woman suffrage movement, which ended with passage of the Nineteenth Amendment to the United States Constitution in 1919, has never been seriously explored. By its very nature, the suffrage movement, which involved thousands of mothers, daughters, and wives, was "nonviolent." In its final stages, female leaders resorted to such scandalous "direct action" tactics as public marches. The most infamous was what began as a large orderly, but unsanctioned, parade in Washington the day before Woodrow Wilson's inauguration in 1913. On Pennsylvania Avenue, drunken ruffians engaged in a planned attack on the marchers in front of police, who provided no protection. Deliberate confrontational protests by women outside the White House day after day led to the jailing of participants and eventually to police using sadistic brute force against those who were arrested.

38. King 1958, 102.

39. Garrow 1978, 220–23; King quoted in Powledge 1964.

40. SCLC, 1965, Proposed budget and program, Summer Community Organization and Political Education Program (SCOPE), 1, 16. Jack Nelson, interview, in Martha McKay and Al Krumlauf, 1965, September 20 memorandum to Edgar May and Robert Clampitt, Atlanta, the bloodless war, OEO Records, Inspection Reports 1964–67, box 19, Fla.-Ga.

41. James L. Bevel, 1964, Program for action in Alabama, 3–4.

42. Eric Kindberg and Ann Kindberg, 1965, September 1, Proposed Alabama Project and Budget, 5.

43. Martin Luther King, Jr., 1964, Assignments, King Papers, subgroup A, ser. 1, subser. 1, box 21.

44. Chestnut 1990, 188.

45. Laurie Pritchett, interview in Hampton and Fayer 1990, 105–6; Williams 1988, 163–83.

46. James L. Bevel, 1964, Program for action in Alabama, 3–4.

47. SNCC, 1963, Minutes of the Executive Committee meeting, December 27–31, 28–30, 7.

48. Chestnut 1990, 188.

49. John R. Lewis and Silas Norman Jr., 1965, March 7 letter to Martin Luther King Jr., SNCC Papers, subgroup A, ser. 1, box 2.

50. Stevens 1964, 1.

51. Herbers 1965a. Joseph Smitherman, interview in Hampton and Fayer 1990,

216–17. The argument can be made that after the Selma campaign, when Martin Luther King Jr. turned the thrust of the civil rights movement northward, he selected Chicago for the same reasons that he picked Birmingham and Selma. In Mayor Richard Daley, Chicago had a public official who eventually would permit police to go too far. Ironically, King's assassination in Memphis in 1968, which led to burning and looting in Chicago and other cities, is what instigated Daley's infamous shoot-to-kill order for arsonists. Daley's effort to use force to maintain order at any cost culminated in the clash between police and antiwar demonstrators outside the building where the Democratic National Convention was being held in August 1968. The event was televised live to the nation.

52. SCLC, 1964, December 31 media release, King Papers, subgroup A, ser. 1, box 21, A9; Herbers 1965b.

53. Herbers 1965b.

54. Herbers 1965c.

55. Martin Luther King Jr., 1965, February 19 telegram to Nicholas Katzenbach, King Papers, subgroup A, ser. 1, subser. 1, box 21.

56. SCLC, 1965, Alabama Chronology, King Papers, subgroup A, ser. 1, subser. 1, box 21. Herbers 1965d.

57. James Orange and Charlier Love, 1965, June 10 summary report on . . . SCLC . . . activity . . . within the past six months, King Papers, subgroup A, ser. 1, subser. 2, box 28, 38; Jones 1966, 357–58.

58. John R. Lewis and Silas Norman Jr., 1965. March 7 letter to Martin Luther King Jr.

59. Abernathy 1989, 327–29. According to Abernathy, "Martin had been receiving an excessive number of death threats as the result of the Selma campaign. . . .

Finally, he agreed to go back to Atlanta on Friday, rest on Saturday, take up his pastoral responsibilities on Sunday, and then come back to Selma Monday morning."

60. James Orange and Charlie Love, 1965, June 10 summary report on . . . SCLC . . . activity . . . within the past six months, 38; Reed 1965a; Jones 1966, 357–58, 38. Garrow 1978, 67–77.

61. Garrow 1978, 63–67, 133–35.

62. Quoted in Reed 1965b; Reed 1965c.

63. Voting Rights Act of 1965. Public Law 89–110, Stat. 437, U.S.C. at 486 (1965); U.S. Department of Justice, Report of the Attorney General 1966, 182–88.

64. U.S. Department of Justice, Report of the Attorney General 1966, 187–88.

65. Ibid., 111–12.

66. Sellers 1973, 151–59; Lawson 1985, 105.

67. Carson 1981, 161–74, 175–228.

68. Street 1966; Lawson 1985, 49–50.

69. Himel 1966; Cook 1966a; Lawson 1985.

70. The Meredith march was not the last one in the rural South, but no subsequent march involved as many people or attracted a similar amount of media attention. In April 1969 several hundred blacks began a two-week, 250-mile trek from Asheville, North Carolina, to the state capitol at Raleigh to protest the death sentence a seventeen-year-old black received for robbing and killing a white merchant. A short article about the march was buried on a back page of the *New York Times*.

71. Quoted in Dunbar 1969, 40.

72. Fruhman 1966; Pollard 1966a, 1966b; Roberts 1966b; Sellers 1973, 160–69. The march also demonstrated that for Southern rural blacks Martin Luther King Jr. was the undisputed leader of the civil rights movement and that almost two years before his assassination King was well along the way on his passage into legend. Although Mississippi was SNCC, not SCLC, territory and the

cosmopolitan King appeared out of place in his new walking shoes and new wide-brimmed hat, he was the leader with the commanding presence. Cleveland Sellers was especially captivated by King's status among poverty-stricken blacks who waited along the line of march:

> From the very beginning of the march, poor blacks along the route were awestruck by Dr. King's presence. They had heard about him, seen him on television, but had never expected to see him in person. . . .
>
> The same incredible scene would occur several times each day. The blacks along the way would line the side of the road, waiting in the broiling sun to see him. As we moved closer, they would edge out onto the pavement, peering under the brims of their starched bonnets and tattered straw hats. As we drew abreast someone would say, "There he is! Martin Luther King!" This would precipitate a rush of two, sometimes as many as three thousand people. (Sellers 1973, 164)

73. Tom Levin, 1965, July 13 CDGM discussion paper, The pacifist poor in the War on Poverty, OEO Records, ser. 4, box 27, 5, CDGM file.

74. U.S. Bureau of the Census 1946b, 123–31; 1967, 260–67. Claude Ramsay, 1965, August 16 AFL-CIO report on the Delta farm strike, SNCC Papers, subgroup A, ser. 3, box 6, 1–2.

75. Is Meredith Right? 1966.

76. Doar and Landsberg 1975, 918, 980.

77. U.S. Department of Justice, Report of the Attorney General 1960, 191–92.

78. Ibid.; Doar and Landsberg 1975, 912–15; Forman 1972, 116–27.

79. Baker 1961; U.S. Department of Justice, Report of the Attorney General 1961,

174–75; Historic Summer in Tennessee 1964.

80. In Tennessee the Issue Is Economic, 1963.

81. Margaret Lauren, 1965, May 7 letter to friends of SNCC, SNCC Papers, subgroup P, ser. 8, box 50.

82. Leonard Mitchell, 1965, November 24 memorandum to Martin Luther King Jr., King Papers, ser. 1, subser. 2, box 28.

83. Forman 1972, 292–91; Mills 1993, 6–42, 315–20.

84. "Secret" Crisis in the Delta 1966.

85. U.S. House 1941, 279.

86. A study of Yazoo Delta plantations conducted in 1957 found a significant shortage in adults between ages twenty-four and thirty-five (LeRay, Wilber, and Crowe 1960, 15–16).

87. Shaw, Mississippi: New Sounds in the Delta, 1965, SNCC Papers, subgroup A, ser. 3, box 6. Hilton 1969, 72–76; Howell, 1969, 36–38.

88. Claude Ramsay, 1965, August 16 AFL-CIO report on the Delta farm strike, 1–2.

89. Quoted in Hilton 1969, 72–75.

90. Minimum Wage Law Nears 1967.

CHAPTER EIGHT
THE WAR ON POVERTY

1. When Robert Moses went to McComb, Mississippi, in 1961, he quickly discovered that his suit, white shirt, and tie hindered his efforts to establish rapport with poor blacks. Moses changed his dress to denim bib overalls, an inexpensive western shirt, and work boots. Others followed his lead, and soon the casual outfits, usually with jeans substituted for overalls, became the official uniforms of civil rights workers in the rural South. The civil rights activists took the uniforms back to Northern and Western college and university campuses as status symbols

of social involvement and dissension. They soon revolutionized dress on campuses across the nation, diffusing even to those in the South (Cagin and Dray 1988, 170; McAdam 1988, 142).

2. Don Hamer, 1964, The plantation of Mr. John Aker Hayes, in McAdam 1988, 870.

3. Aiken 1990.

4. Piven and Cloward 1977, 264–67.

5. MacDonald 1977, 1–8.

6. Hoffsommer 1934, 2:2.

7. Myrdal 1944, 356–59. The name of the Aid to Dependent Children program was changed in 1962 to Aid to Families with Dependent Children (Gordon 1994, 1).

8. In 1940, most employed blacks and whites in the North were covered under the Social Security Act. In the South, where large numbers of both races were still engaged in farming, approximately 60 percent of black and almost 50 percent of white workers were not covered (Myrdal 1944, 1280).

9. Piven and Cloward 1977, 134–37; Shaw, Mississippi, 1965, SNCC Papers, subgroup A, ser. 3, box 6.

10. Dunbar 1969, 44–47.

11. Carson 1981, 79–80; Raines 1983, 241–43; Cagin and Dray 1988, 227–30.

12. Gordon 1994; Furstenburg 1994.

13. Galbraith 1958; Levitan 1969, 12–13.

14. Harrington 1962, 1–2; Levitan 1969, 11–15.

15. Levitan 1969, 14–18.

16. Brauer 1977, 46–51, 58; Forman 1972, 130.

17. Levitan 1969, 17–18.

18. Economic Opportunity Act of 1964. Public Law 88–452, Stat. 508, U.S.C. at 585 (1964).

19. Donovan 1967, 36; Levitan 1969, 45–46.

20. Yarmolinsky 1969, 45–49. Although the poverty programs of the Economic Opportunity Act were innovative, they also were limited. The programs were designed to have the greatest impact on blacks and on children. The programs also were biased in favor of large cities. Forty percent of the poor were rural, but the Economic Opportunity Act lacked programs that aggressively addressed rural poverty (Donovan 1967, 97).

21. Batteau 1990, 7–8.

22. Levitan 1969, 59–60.

23. Yarmolinsky 1969, 45–49.

24. Pavin and Cloward 1977, 273.

25. Samuel F. Yette, 1966, July 15 memorandum to Sargent Shriver: Yesterday's White House meeting with civil rights coordinators, OEO Records, ser. 2, box 11, Civil Rights file. Yette was concerned about the apparent negative perception the Commission on Civil Rights held concerning OEO's civil rights efforts. "In the past" the commission "has considered our comparatively small civil rights staff as little evidence of sincere efforts," but "there is not necessarily a direct relationship between the two."

26. John Dean, 1965, December 8 memorandum to Earl Redwine, Gainesville-Hall County Economic Opportunity Organization, 2, OEO Records, Inspection Reports 1964–67, box 20, Ga.

27. James L. Keeney, 1967, May 15 letter to William H. Stuckey Jr., OEO Records, Inspection Reports 1964–67, box 20, Ga.

28. Markham Ball, 1965, September 27 memorandum to deputy and associate directors, division chiefs and deputies of VISTA, VISTA policies on civil rights activities of volunteers, OEO Records, ser. 3, box 19, VISTA file.

29. Thomas Levin, 1965, The Pacifist Poor in the War on Poverty, OEO Records, ser. 4, box 27, CDGM file.

30. Martha McKay, 1965, November 9 memorandum to Edgar May and

Robert Clampitt, Chattahoochee-Flint, OEO Records, Inspection Reports 1965–67, box 21, Ga.

31. U.S. Senate 1967, vol. 8, 2337–41.

32. Martha McKay, 1965, November 9 memorandum to Edgar May and Robert Clampitt.

33. Martha McKay, 1965, December 3 memorandum to Edgar May and Robert Clampitt, Northeast Georgia, OEO Records, Inspection Reports 1964–67, box 20, Ga. Such statements by rural whites to civil rights and OEO workers were frequently made in jest and were especially humorous to a wily entrapper if the interviewer failed to comprehend the baiting.

34. Martha McKay, 1965, December 16 memorandum to Edgar May and Robert Clampitt, Oconee Planning and Development Commission, OEO Records, Inspection Reports 1964–67, box 20, Ga. An OEO evaluation of Atlanta stated: "Climate, . . . plus development as a transportation and distribution center, have made it what it is. Some residents add luck and determination on the part of the power structure (one is tempted to call it collusion) to foster and preserve Atlanta's 'image'—as a progressive, modern city; businesslike, efficient, handling all its own problems with its own creative talent and resources. It is not added but is evident that whatever problems are admitted to will be solved on the terms of the power structure and within their set limits. The integration which has been allowed comes under this dictum." Martha McKay and Al Krumlauf, 1965, September 20 memorandum to Edgar May and Robert Clampitt.

35. Martha McKay, 1965, December 14 memorandum to Edgar May and Robert Clampitt, West Central Georgia, OEO Records, Inspection Reports 1965–67, box 20, Ga.; Carter 1992, 16–23.

36. U.S. Senate, 1967, 1151–84. Martha McKay, 1965, October 25 memorandum to Edgar May and Robert Clampitt, Chattahoochee-Flint, OEO Records, Inspection Reports, 1965–67, box 21, Ga.. Five of the seven counties were in the lower Piedmont plantation region and two in the white yeoman farmer upper Piedmont.

37. Robert Martin, 1966, July 20 memorandum to Edgar May, Lowndes County, Alabama, OEO Records, Inspection Reports 1964–67, box 2, Ala.-Arz.

38. Samuel F. Yette, 1965, January 14 memorandum to Sargent Shriver, Negroes in the Alabama CAP program, OEO Records, ser. 4, box 23, Ala. CAP.

39. Orzell Billingsly, 1965, August 4 letter to Sargent Shriver signed by Willis L. Gulley et al., OEO Records, ser. 4, box 24, Ala. CAP.

40. Orzell Billingsly, 1965, August 8 letter to Sargent Shriver signed by Willis L. Gulley et al. Robert Martin, 1966, July 20 memorandum to Edgar May.

41. Tersh Boasberg, 1965, August 16 memorandum to Sargent Shriver, Operation Dixie, OEO Records, ser. 2, box 12, Ala. CAP, 1–5.

42. Ibid.

43. Tersh Boasberg, 1965, circa September note to Samuel Yette, OEO Records, ser. 2, box 12, Ala. CAP.

44. Marie Carl, 1965, September 22 investigation of the Johnson-Mayberry Conflict in the Macon County Community Action Program, OEO Records, ser. 4, box 23, Ala. CAP.

45. Robert W. Saunders, 1966, August 23 memorandum for file, Dallas County, Alabama, Grant, OEO Records, ser. 4, box 23, Ala. CAP.

46. Myrdal 1944, 727–29.

47. John Cook et al., 1965, October 29 let-

ter to Samuel Yette; Tersh Boasberg, 1965, August 19 memorandum to Samuel Yette, OEO Records, ser. 4, box 23, Ala. CAP.

48. Frank Prial, 1965, November 1 memorandum to Edgar May et al., Area 2, Huntsville-Madison County, OEO Records, Inspection Reports 1964–67, box 1, Ala.

49. Tate Williams, 1966, May 26 letter to the Commission on Civil Rights, OEO Records, ser. 4, box 35, Ala. CAP.

50. Joseph Bradford, 1966, July 13 memorandum to Theodore Berry, Civil Rights Inquiry re OEO-USDA programs in Alabama Blackbelt, OEO Records, ser. 4, box 23, Ala. CAP. Another method by which OEO officials could " 'sneak' a grant by the Governor of Alabama" was to fund local Community Action Programs under Title III-B of the Economic Opportunity Act, which provided for a "national emphasis" directly from the Washington office. A national emphasis program for seasonal farm workers was used to fund CAPs in Lowndes and Wilcox Counties. Theodore Berry, 1966, July 27 memorandum to Sargent Shriver, Position on Grant to Lowndes Co., OEO Records, Inspection Reports 1964–67, box 2, Ala.-Arz.

51. U.S. Commission on Civil Rights 1968a, 40.

52. Office of Economic Opportunity, 1966, Mississippi Programs as of March 4, OEO Record, ser. 4, box 24, Miss. CAP; Jack Gonzales, 1965, circa September memorandum, Child Development Group of Mississippi, OEO Records, ser. 4, box 27, CDGM file. The first OEO program to reach most of Mississippi's counties was rural loans, a small assistance program that benefited only a few of the state's legions of poor.

53. William R. Seward, 1966, April 20

memorandum to Edgar May, STAR, Inc., OEO Records, Inspection Reports 1964–67, box 40, Miss.

54. Greenberg 1969, 32–33.

55. Ibid., 18–19, 668–98; U.S. Senate 1967, vol. 4.

56. William R. Seward, 1966, March 17 memorandum to Edgar May and Robert Clampitt, Bolivar County, Mississippi; March 3, 1966, memorandum to Edgar May and Robert Clampitt, Preliminary Report, Bolivar County, Mississippi, OEO Records, Inspection Reports 1964–67, box 40, Miss.

57. William R. Seward, 1966, March 17 memorandum to Edgar May and Robert Clampitt, Bolivar County, Mississippi.

58. Martha McKay, 1965, November 9 memorandum to Edgar May and Robert Clampitt.

59. Mosley and Williams 1967, 3–8, 26.

60. Office of Economic Opportunity, 1966, Mississippi Programs as of March 4; Mosley and Williams 1967, 20–22.

61. Office of Economic Opportunity, 1966, June 24 Inspector's Field Report, Coahoma Opportunities, OEO Records, Inspection Reports 1964–67, box 40, Miss.; 1966, Mississippi Programs as of March 4; Mosley and Williams 1967, 41–71.

62. Samuel F. Yette, 1966, July 15 memorandum to Sargent Shriver, Yesterday's White House meeting with civil rights coordinators; Martha McKay, 1965, November 9 memorandum to Edgar May and Robert Clampitt, 7.

63. Phillips 1928, 31.

64. Young quoted in Mosley and Williams 1967, 17, 30–32.

65. Tersh Boasberg, 1966, March 8 memorandum to Samuel Yette et al., Mississippi Delta, OEO Records, ser. 4, 5, 6, box 4.

66. Halberstam 1993, 452–54. Breland

quoted in Huie, 1955, Oct. 8 letter to Roy Wilkins. Huie Papers, Emmett Till file.

67. Roberts 1966a.
68. Mosley and Williams 1967, 31; William R. Seward, 1966, March 3 memorandum to Edgar May and Robert Martin.
69. Frank Prial, 1965, October 11 memorandum to Edgar May and Robert Clampitt, 7.
70. Quoted in Couto 1991, 106–7.
71. Deputy Administrator Consumer Food Programs, 1966, March 7 memorandum to deputy assistant inspector, Leflore County Needy Family Program, OEO Records, ser. 4, box 27, Miss. CAP.
72. National Association for the Advancement of Colored People, 1965, Statement on Economic Advancement Adopted at the 1965 National Convention, OEO Records, ser. 2, box 12; Carmichael and Hamilton 1967, 183.
73. Rural Organizing and Cultural Center 1991, 77; Peter Berle, 1965, August 18 memorandum to William Haddad and Robert Clampitt, Clarksdale, Mississippi, OEO Records, Inspection Reports 1964–67, box 40, Miss., 3.
74. Cauto 1991, 106.
75. Charles Collins, 1967, September 25 memorandum to Edgar May, Allegations Re Crawfordville Enterprises, OEO Records, Inspection Reports 1964–67, box 20, Ga., 8–9.
76. William R. Seward, 1966, January 28 memorandum to Robert Martin, Bacon Co. Homemaker Program, OEO Records, Inspection Reports 1964–67, box 20, Ga.
77. James Heller, 1965, October 21 memorandum to Samuel Yette, Natchez, Mississippi, OEO Records, ser. 4, box 27, Miss. CAP; Mississippi Freedom Democratic Party, 1966, The Case against Sunflower Co. Progress, OEO Records, ser. 4, box 28, Miss. CAP. All charges of

prejudice and illegal activity by Community Action agencies in the plantation regions were not factual. The OEO's investigation of CENLA Community Action Committee in Alexandria, Louisiana, found that although local whites took advantage of factions among blacks, there was "no outright racial discrimination on the board or the staff" (Michael Coleman, 1966, April 26 memorandum to Edgar May, CENLA Community Action Committee, OEO Records, ser. 4, 5, 6, box 34).
78. Donovan 1967, 83–87; Jack Gonzales, [1965], Child Development Group of Mississippi, OEO Records, ser. 4, box 27, CDGM file.
79. Dunbar 1990, 48.
80. Kittredge 1986.
81. Mosley and Williams 1967, 34–36, 95.
82. Quoted in Mosley and Williams 1967, 34.
83. Robert Martin, 1966, October 17 memorandum to Edgar May, Project Help, OEO Records, Inspection Reports 1964–67, box 39, Minn.-Miss.; Martha McKay, 1965, December 14 memorandum to Edgar May and Robert Clampitt.
84. U.S. Senate 1987, 4:1159–60.

CHAPTER NINE
SCHOOL DESEGREGATION

1. Brown v. Board of Education of Topeka, Kansas, 347 U.S. (1954).
2. Wisdom 1975, 141.
3. Mississippi Department of Education 1931, 83.
4. Martin 1957, 56.
5. Brown v. Board of Education of Topeka, Kansas, 349 U.S. 294, 299–301 (1955).
6. Martin 1957, 4.
7. McMillen 1971, 15–16.
8. Moore 1959.
9. Southern Regional Council 1963. Most of the large public high schools for

blacks across the plantation South were "training schools," which were to teach "practical" manual skills such as farming, carpentry, and masonry. This controversial concept of education was promoted after the Civil War by Booker T. Washington. Many of the whites who continued to champion this dated concept of education after the Second World War believed that blacks either did not have the intellectual ability to master academic subjects beyond an elementary level or should not be permitted to master them in order to keep blacks submissive. By the 1950s, many blacks viewed the concept of "training school" with disdain (see Nossiter 1988).

10. Bartley 1969, 115–27.

11. Southern Education Reporting Service 1961, 5–6, 15–16, 25.

12. Read 1975, 31.

13. Briggs v. Elliott, 132 F. Supp. 776, 777 (E.D.S.C. 1955).

14. The Bureau of the Census defines the South as the eleven former Confederate states together with Kentucky, Delaware, Maryland, West Virginia, and Oklahoma.

15. U.S. Commission on Civil Rights 1964b, 1–2, 287–92.

16. Mosley and Williams 1967, 2.

17. U.S. Commission on Civil Rights 1964b, 132–36.

18. Martin 1957, 170.

19. Civil Rights Act of 1964. Public Law 88-352, Stat. 241, U.S.C. §IV, sec. 407, at 295; §VI, sec. 602, at 301 (1964).

20. Elementary and Secondary Education Act of 1965. Public Law 89-10, Stat. 27 U.S.C. at 29; Orfield 1969, 2–4, 36–46.

21. Quoted in Orfield 1969, 63.

22. U.S. Commission on Civil Rights 1967, 10–11.

23. Ibid., 11–12; Orfield 1969, 85–92.

24. U.S. Commission on Civil Rights 1967, 11–12.

25. Samuel Yette, 1965, August 31 memorandum to Holmes Brown, Integration in Head Start, OEO Records, ser. 4, box 23, Ala., Head Start file; Edstrom 1966.

26. OEO Records, 1965, December 8 media release, Freedom of choice disallowed for Head Start. 1966, August 25 media release, Twenty-five summer Head Start sponsors rejected. Ruby G. Martin, 1966, September 21 memorandum to Samuel Yette, Discrepancies between OE and OEO on compliance by schools, OEO Records, ser. 2, box 12, Office of Education file.

27. U.S. Commission on Civil Rights 1966, 51–52.

28. U.S. Commission on Civil Rights 1964b, 290–92; 1967, 104–6.

29. Southern Education Reporting Service, 1967; U.S. Commission on Civil Rights 1967, 25.

30. U.S. Commission on Civil Rights 1966, 34–35. Because integration often resulted in demotion and the firing of black principals and teachers, by the late 1960s it was common for civil rights attorneys to request federal courts to assure protection of black school employees' jobs in desegregation orders (Galphin 1970, A4).

31. U.S. Commission on Civil Rights 1967, 88.

32. Ibid., 47–51, 66, 88.

33. Ibid., 87–89.

34. Foster 1965, 61. Harold Howe, 1967, January memorandum to superintendents and boards of education, OEO Records, ser. 2, box 12, Office of Education file.

35. Singleton v. Jackson Municipal Separate School District, 348 F. 2d 729 (5th Cir. 1965); Singleton v. Jackson Municipal Separate School District, 355 F. 2d 855 (5th Cir. 1966); Read 1975, 20–23.

36. United States v. Jefferson County Board of Education, 372 F. 2d 836, 867–69, 896 (5th Cir. 1966); Read 1975, 16–

28. Because of the sweeping nature of the decision, an *en banc* rehearing attended by all of the twelve judges of the Fifth Circuit was held, and the opinion of *Jefferson* was adopted by the court. United States v. Jefferson County Board of Education, 380 F. 2d 385 (5th Cir. 1967).

37. Green v. County School Board of New Kent County, 391 U.S. 430, 431 (1968). The racially diverse rural settlement pattern does not mean that concentrations of dwellings of blacks or whites did not exist or that the traditional segregated settlement pattern did not prevail in municipalities.

38. Green v. County School Board of New Kent County, 391 U.S. 430, 435–42 (1968).

39. Read 1975, 29–32.

40. Herbers 1969.

41. Edelman 1973, 39–40.

42. Munford 1973, 12–14; Wilkinson 1979, 18–20.

43. Alexander v. Holmes County Board of Education, 396 U.S. 19, 20–21 (1969).

44. Read 1975, 30–31.

45. Carter v. West Feliciana Parish School Board, 396 U.S. 290, 290 (1970).

46. Nordheimer 1969; U.S. Commission on Civil Rights 1967, 69.

47. Read 1975, 32; Wilkinson 1979, 122.

48. Quoted in Nordheimer, 1969.

49. Southern Education Reporting Service 1961, 5–6, 15–16, 25; 1967, 21. Southern Regional Council, 1963, August 19 report L-45, Public Education in Mississippi, SCLC Papers, acc. 1, ser. 1, subser. 3, box 139.

50. McMillen 1971, 298–302; U.S. Commission on Civil Rights 1967, 70; Southern Education Reporting Service 1967, 6, 21; Williams 1967.

51. Wooten 1970.

52. Quoted in Auchmutey 1984b.

53. Munford 1973, 23–24. The Supreme Court's *Alexander* decision refers to thirty-three school districts, but according to Munford there were only thirty (13).

54. U.S. Commission on Civil Rights 1967, 76; Mississippi Department of Education, Bureau of Management Systems, 1989, Public School Enrollment 1989–90: End of First Month, 48–49, Jackson, Mississippi.

55. Galphin 1970; Wooten 1970; Nevin and Bills 1976, 47–59.

56. Bell 1975, 351; Cimons 1982, 6F.

57. Wooten 1970, 34L.

58. McNeal v. Tate County School District, 460 F. 2d, 568 (5th Cir., 1972).

59. U.S. Bureau of the Census 1992b, 60; Porter and Nehrt 1977, 89.

60. Prattville, Alabama, Chamber of Commerce, 1990, Origins of Households Moving to Autauga County, Alabama, in 1990. Computer file.

61. The 1990 census figures on number of pupils by race enrolled in public school systems, which are based on a sample, do not agree with the actual enrollment figures of the Alabama Department of Education. U.S. Bureau of the Census 1993, 304–19. Alabama Department of Education, 1990, School Enrollment by Ethnic Group for First Month 1990–91, Montgomery, Ala.

62. Agee and Evans 1941. The Macon County, Alabama, school system was almost 90 percent black in 1990–91, but Notasulga High, located in the northern part of the county near the boundary between the Upland South and the Lowland South, was 57 percent black. Black and white Notasulga parents united to fight the majority-black Macon County school board's proposal to abolish the "truly integrated" 258-student facility by consolidating it into a new central high school in Tuskegee (Nossiter 1991).

63. Alabama Department of Education 1990.

64. Ibid. In the late 1980s, residents of Meriwether County, Georgia, became embroiled in a dispute over consolidation of three small public high schools into a new comprehensive one. Two of the three schools were majority black (76% and 86%), and one was majority white (67%). The new consolidated school would be majority black (White 1989b).

CHAPTER TEN
THE RIGHT TO VOTE—
AN ILLUSIVE BLACK POWER

1. Quoted in Lawson 1985, 21.
2. U.S. Department of Justice, Report of the Attorney General 1968, 112; King 1967, 112.
3. Katzenbach 1966, 59.
4. Lawson 1985, 37–39.
5. U.S. Commission on Civil Rights 1968, 222–23.
6. Ibid., 224–25, 246–47.
7. Quoted in Sellers 1973, 166–69.
8. Ibid., 157.
9. Karim 1971, 131–32.
10. Garrow 1981; Tompkins 1993.
11. G. C. Moore, 1968, February 29 memorandum to W. C. Sullivan, Counterintelligence program, black nationalists–hate groups, racial intelligence. G. C. Moore, 1970, April 15 memorandum to W. C. Sullivan, Counterintelligence program, black nationalist–hate groups, racial intelligence, Stokely Carmichael, in U.S. Senate 1976, 386–92, 767–69.
12. Tompkins 1993.
13. G. C. Moore, 1968, March 26 memorandum to W. C. Sullivan, Counterintelligence program, black nationalist-hate groups, racial intelligence, Washington spring project, in U.S. Senate 1976, 788–90.
14. Tompkins 1993.
15. Van Deburg 1992, 9.
16. U.S. Commission on Civil Rights 1968b, 222–23; Watters and Cleghorn 1967, 300.
17. Joint Center for Political and Economic Studies 1994, xxiii.
18. U.S. Commission on Civil Rights 1968b, 19–20.
19. Rice 1971.
20. Parker 1990, 30–33; House Deplores Sniper Shooting 1966.
21. Quoted in U.S. Commission on Civil Rights 1968b, 27.
22. Parker 1990, 72.
23. Smith 1964, 280–300.
24. Parker 1990, 53.
25. Hester 1982, 841–15.
26. Parker 1990, 153.
27. Voting Rights Act Amendments of 1982. Public Law 97-295, Stat. 96, U.S.C. at 131 (1982).
28. South Carolina v. Katzenbach, 383 U.S. 301 (1966); Allen v. State Board of Education, 393 U.S. 544 (1969); Perkins v. Matthews, 400 U.S. 379 (1971); Bell, Krane, and Lauth 1982, 51–58; Grofman, Handley, and Niemi 1992, 24–25.
29. Georgia v. United States, 411 U.S. 526 (1973); White v. Regester, 412 U.S. 755 (1973); Zimmer v. McKeithen, 485 F. 2d 1297 (5th Cir., 1973); Scher and Button 1984, 37–38; Grofman, Handley, and Niemi 1992, 30–34.
30. City of Mobile v. Bolden, 446 U.S. 55 (1980); Voting Rights Act Amendments of 1982. Public Law 97-205, Stat. 96, U.S.C. at 131 (1982); Grofman, Handley, and Niemi 1992, 38–41.
31. Thornburg v. Gingles, 478 U.S. 30, 2766–67 (1986).
32. Shaw v. Reno, 509 U.S., 113 S.Ct. 2816, 125 L.Ed. 2d 511 (1993).
33. Watters and Cleghorn 1967, 267.
34. Ibid., 300.
35. Quoted in Campbell 1972, 39.
36. Lawson 1985, 101–2.
37. Frank Prial, 1965, October 11 memorandum to Edgar May and Robert

Clampitt; Salamon 1973; Lemann 1991, 327–30.

38. Joint Center for Political Studies 1973, xvii; Joint Center for Political and Economic Studies 1994, xxiii.

39. Hanks 1987, 68; Negroes Take Over 1968.

40. Chapman 1969; Jenkins 1969; Newberry 1989.

41. Nossiter 1990; Turque and Manly 1990; Haddad 1990.

42. Sikora 1969; quoted in Pool Area Ransacked 1969; quoted in Lanker 1989, 223.

43. African American Members of Congress 1993.

44. Holmes 1994.

45. African American Members of Congress 1993; Joint Center for Political and Economic Studies 1990; 1994; Mapping the Challenges. The Second Mississippi Congressional District, the Yazoo Delta, was represented by an appointed white in 1993. Mike Epsy resigned to accept the office of secretary of agriculture in the Clinton administration. The Alabama congressional district that includes Montgomery is among the seven.

46. Karlan 1995.

47. Carter 1993.

48. Bush v. Vera, 1996 WL 315857 (U.S.); Shaw v. Hunt, 1996 WL 315870 (U.S.); Abrams v. Johnson 1997 WL 331802 (U.S.).

CHAPTER ELEVEN
NEW SETTLEMENT PATTERNS

1. Aiken 1985, 393–95.

2. U.S. Department of Agriculture, FmHA 1986, 11–13. In nonmetropolitan areas, the FmHA has authority in municipalities with population up to twenty thousand, and to an agency-defined "rural boundary" in the urban fringes of those with populations of twenty thousand and more.

3. White Oak is large enough to be shown as an unnamed place on the Helena 1:250,000 topographic map (U.S. Geological Survey 1977).

4. Mrs. H. M. Calloway, interview with author, September 4, 1985, Philomath, Ga.

5. Barrow 1881, 833.

6. Aiken 1987.

7. Cumming 1974, 24.

8. E.g., Souther v. City of Knoxville, Tenn., No. 91773-2 (Ch. Knox Co., Tenn., Filed March 25, 1987).

9. E.g., City of Pleasant Grove, Ala. v. U.S., 55 U.S.L.W. 4133 (U.S. 1987); U.S. v. Town of Indian Head, Md., No. R86-964 (D. Md., March 25, 1986); U.S. v. City of Bessemer, Ala., No. CV-83-P-3050-S (D. Ala., August 20, 1985); Dotson v. City of Indianola, Miss., No. GC80-220-WK-0 (D. Miss., June 4, 1985).

10. For the 1980 census, Mississippi was one of five states that contracted with the Bureau of the Census for the preparation of block statistics, including racial composition of the population, for all rural and urban territory (U.S. Bureau of the Census 1982, 6–7). A spatially detailed population data set exists for all incorporated places in the Yazoo Delta and for the countryside surrounding them for 1980 as well as for 1990. In small municipalities the census blocks are defined in the same manner as in large urban areas, primarily by using streets as boundaries. Roads, streams, utility lines, and other features are employed as boundaries for census blocks in the rural countryside. Rural blocks vary in size and shape and range from less than an acre to several square miles.

11. Sutherland 1965, 39.

12. U.S. Bureau of the Census 1963b, 26–38.
13. U.S. Executive Office of the President 1968–1976; 1977–1981; U.S. Bureau of the Census 1983a, 26–111 to 26–127.
14. Auchmutey 1984a.
15. Aiken 1987, 567–69.
16. Residuals from regression analysis in which black fringe population was the dependent variable and black municipal population the independent were used to identify municipalities with populations of five hundred or more which significantly underbounded their black populations in 1980. The twelve with positive standardized residuals of 0.50 and greater, which included Indianola, Tunica, and Belzoni, were analyzed in detail (Aiken 1987, 567–74).
17. Hathorn 1988; Christion 1991; Cawthon 1991.
18. U.S. Dept of Agriculture, FmHA 1986; U.S. Dept. of Housing and Urban Development 1986.
19. Crigler and Heaton v. Kincade, No. 7704 (Ch. Coahoma Co., Miss., October 20, 1985).
20. Aiken 1987, 574–76; Mississippi State Tax Commission 1986, 1990.
21. Aiken 1987, 574–76.
22. Dotson v. City of Indianola, Miss., No. GC80-220-WK-0 (D. Miss., June 4, 1985); City of Pleasant Grove, Ala. v. United States, 55 U.S.L.W. 4133 (U.S. 1987).
23. Aiken 1990.
24. Rose 1971, 6. Rose defined an "all-Negro" town as one having a "nonwhite" population of more than 95 percent (Rose 1965, 362).
25. E.g., Morrill 1965; Rose 1970, 1971, 1972.
26. Drake and Cayton 1962.
27. The index of dissimilarity, one of the most widely employed measures of segregation, revealed growth in segregation among the municipalities of the Yazoo Delta between 1950 and 1980. On a scale of 0 for complete integration to 100 for complete segregation, the index rose from 12.8 to 18.4. The indices are low because most of the region's municipalities have relatively large percentages of blacks and whites. The increase in segregation was greater among the smaller municipalities, the 52 with populations less than five thousand, than among the entire group. The index rose from 22.3 in 1950 to 31.1 in 1980 (Aiken 1990).
28. Aiken 1990.
29. U.S. Bureau of the Census 1981, 18; 1991, 45.
30. Morrill 1965; Rose 1970, 1971.
31. Nine of the twelve sample municipalities in the Yazoo Delta had federally funded housing (Aiken 1990).
32. Johansen and Fuguitt 1984, 85–135.
33. For the fifty-two Yazoo Delta municipalities with populations of less than five thousand in 1980, the correlation between per capita retail sales and percentage of the population which was black in 1980 was $-.58$ ($\alpha = .05$). Average per capita sales for the fifty-eight municipalities was $3,880 in 1980. Average per capita retail sales was less than $3,880 in all but two of the twelve municipalities with a black population of 75 percent or more in 1980. Between 1980 and 1988, eleven of the twelve municipalities experienced declines in actual retail sales, and all had declines measured in constant dollars (Mississippi State Tax Commission 1980, 1988).
34. Mississippi State Tax Commission 1970, 1980, 1990.
35. Pred 1963; Rose 1971, 72–81; Aiken 1990, 233–38.
36. Pennington and Butchko 1978, 1–7; Bartram 1791, 57–59.
37. Quoted in Ahlgren 1967.

CHAPTER TWELVE
QUEST FOR A NONAGRARIAN
ECONOMY

1. Raper 1936, 3–4, 171.
2. Ibid., 406.
3. Phillips 1906, 798.
4. Phillips 1910.
5. Harrington 1984, 209.
6. MDC 1986; Southern Governors' Association 1986; Southern Growth Policies Board 1986; Wooten 1988; Walker 1989; Aiken 1995.
7. U.S. Bureau of the Census 1947, 144–46. "Nonwhite" is employed by the Bureau of the Census for historical consistency. Summary racial data are not available before 1900, and Negroes were not enumerated separately in all censuses after 1900. "Nonwhite" includes Negroes, Native Americans, Chinese, Japanese, and other "nonwhite" races (ibid., 144–45).
8. E.g., Sinclair 1986; Black Farmers Must Battle 1989.
9. Carnoy 1994, 13–33.
10. California Governors Commission on the Los Angeles Riots 1965, 3–4.
11. Brown 1970, 101.
12. Freeman 1970, 7.
13. E.g., Alter 1993.
14. Tilly 1968; Fogelson and Hill 1968, 235, 243; Feagin and Hahn 1973, 6–12. The popular explanation for the migration of blacks from the South, the mechanization of agriculture, has never been seriously challenged. That agricultural mechanization was the single important cause of black migration from the rural South persists in both popular and professional literature. Donald Grubbs attempted to make his *Cry from the Cotton*, an examination of the Southern Tenant Farms Union of the 1930s, relevant to the urban problems of the 1960s by asserting, "Through the Agricultural Adjustment Administration, Franklin Roosevelt gave Southern planters the means and the incentive to substitute machines and underemployed casual labor for their tenants. . . . Yesterday, through ignorance or greed, the propertied drove the propertyless off the land; tomorrow, Harlem and Watts and the South Side will be burning" (Grubbs 1971, xi).

Using data for the Yazoo Delta, Richard Day statistically demonstrated that four stages in the adoption of new technology for cotton production created a two-stage push of sharecroppers from the land. Day thought that the results were "reasonably representative of southern agriculture . . . as a whole" (Day 1967, 443).

Robert Fogelson and Robert Hill (1968) concluded that "the 1960s riots were a manifestation of race and racism in the United States, a reflection of the social problems of modern black ghettos, a protest against the essential conditions of life there, and an indicator of the necessity for fundamental changes in American Society" (243).
15. Fay Bennett, 1965, September 7 National Sharecroppers Fund memorandum on city riots and rural poverty, OEO Records, ser. 2, box 12.
16. E.g., Smith 1974, 142–44.
17. For discussion of agrarianism and the agrarian myth, see Johnston 1940 and Aiken 1975.
18. Comptroller General of the United States 1969, 39.
19. T. C. Miller, 1938, March 15 memorandum to J. H. Wood, Cooperative Activities on Gee's Bend Project. C. B. Bladwin, 1938, June 9 letter to Mile Perkins, Proposed Loan to Gee's Bend Cooperative Association. Boradman N. Ivey, 1952, April 16 letter to Julian Brown, Gee's Bend Cooperative Asso-

ciation, FSA Records, Cooperative Associations, Ala., box 2. Charles S. Aiken, 1994, The Agrarian Myth and the Migration of Blacks to Metropolitan America, paper read at the national meeting, Association of American Geographers, Chicago, in Aiken Papers.

20. Callahan 1987.

21. Farmers Home Administration, 1969, January 16 status report on Southwest Alabama Farmers Cooperative Association, OEO Records, Department of Agriculture to Civil Service Commission, box 1.

22. Brown 1988, 191; Mitchell 1921, 14–15.

23. Rowan 1970, 55; Knox 1983.

24. Gregg 1941, 48.

25. Quoted in Cash 1941, 181.

26. Cell 1982, 127–30.

27. Shaw 1988; Morgan 1988; Rowan 1970, 54, 62.

28. Cash 1941, 205–6.

29. Charles S. Aiken, 1973, Migration from the Cotton Fields: Tractors v. Sharecroppers during the Great Depression, paper read at the annual meeting, Southeastern Division, Association of American Geographers, Boone, N.C., in Aiken Papers. Many of the white sharecroppers in 1930 were on the northern margins or beyond the fringes of the plantation regions.

30. Bosworth 1941, 6.

31. Davis, Gardner, and Gardner 1941, 261; Cobb 1993, 116.

32. Dunbar 1969, 11.

33. Dollard 1937, 4; Davis, Gardner, and Gardner 1941, 255–61; Rubin 1951, 8.

34. Twelve Southerners 1930.

35. Quoted in Vesey 1993.

36. Cell 1982, 126–27.

37. That Atlanta and Birmingham, though only 200 miles apart and segregated, were fundamentally different types of cities is illustrated by the SCLC's strategy. The organization had its headquarters in Atlanta, where its leaders felt reasonably safe. The 1963 "Project C" campaign was launched in Birmingham, where King believed that a violent confrontation would occur.

38. Cobb 1993, 5–34.

39. Rubin 1951, 8, 73–76.

40. Thatcher 1977.

41. Cobb 1993, 256; U.S. Bureau of the Census 1992b, 290.

42. Allan, Kaufman, and Underwood 1972, 207–10.

43. Lonsdale 1969; Stuart 1983.

44. Hathorn 1989; Mississippi State Tax Commission 1990.

45. Bell 1973, 742–83.

46. Tompkins 1899, 110; Johnson 1990.

47. Siler 1991; Bronstein 1993.

48. Moore 1992; Charlier 1992; Mantius 1990; McCarthy 1993; Cheakalos 1992.

49. Whitt 1989. Calvin L. Beale, 1992, Penitential Development: Prisons, Population, and Jobs in Nonmetro America, Paper presented at the Southern Demographic Association meeting, Charleston, S.C.

50. Skinner 1976; Christensen 1984; Campbell 1993a, 1993b; Cooper 1994.

51. Smith 1996.

CHAPTER THIRTEEN
EPILOGUE

1. Vance 1935, 467–68.

2. E.g., Woodward 1951, 291–320; Wright 1986, 156–97, 269–74; Boles 1995, 337–449.

3. Key 1949, 5.

4. Wright 1986, 270.

5. Quoted in Williams 1988, 139.

6. Carmichael and Hamilton 1967, 2–32.

7. Henry J. Lorenzi and Susan Sadoff Lorenzi, 1966, January 4 report, Housing—Holmes County [Miss.], Kenney Papers, app. B, reel 7, 2.

8. Peters 1992, 139–40.
9. Beckford 1972, 177.
10. Rozier 1982, 187–96; Peters 1992, 126–65.
11. King 1967, 169–70.
12. Black and Black 1987, 21–49, 175–94.
13. Layson and Falk 1993.
14. Berry, B. 1973, 1:6–17.
15. Brown and Warner 1989.
16. Cleland and Fontanez 1994.
17. King 1963b, 219.

Bibliography

MANUSCRIPT COLLECTIONS

Agee-McDowell Papers. Papers of James Agee and David McDowell. Special Collections, Hoskins Library, University of Tennessee, Knoxville.

Aiken Papers. Materials in possession of Charles S. Aiken.

Barrow Papers. Papers of David C. Barrow. Hargrett Rare Book and Manuscript Library, University of Georgia Libraries, Athens.

Barrow Jr. Papers. Papers of David C. Barrow Jr. Hargrett Rare Book and Manuscript Library, University of Georgia Libraries, Athens.

Brooks Papers. Papers of Robert Preston Brooks. Hargrett Rare Book and Manuscript Library, University of Georgia Libraries, Athens.

FSA Records. Records of the United States Farm Security Administration and the United States Farmers Home Administration. Record Group 96, National Archives, Suitland, Maryland.

Hargrett Papers. Papers of the Hargrett Family. Hargrett Rare Book and Manuscript Library, University of Georgia Libraries, Athens.

Huie Papers. Papers of William Bradford Huie. Libraries, Ohio State University, Columbus.

Kenney Papers. Papers of Michael Kenney and the Student Nonviolent Coordinating Committee. Library, University of North Carolina, Chapel Hill (microform).

King Papers. Papers of Martin Luther King Jr. Acc. No. 215. King Center, Atlanta, Georgia.

Lowenstein Papers. Papers of Allard Kenneth Lowenstein. Southern Historical Collection, Wilson Library, University of North Carolina, Chapel Hill.

MFDP Papers. Papers of the Mississippi Freedom Democratic Party. King Center, Atlanta, Georgia.

Mitchell-Marsh Papers. Papers of Margaret Mitchell. Hargrett Rare Book and Manuscript Library, University of Georgia Libraries, Athens.

National Cotton Council Files. Vertical Files, Library, National Cotton Council of America, Memphis, Tennessee.

OEO Records. Records of the United States Office of Economic Opportunity and the United States Community Services Administration. Record Group 381, National Archives, Washington, D.C.

Raper Papers. Papers of Arthur F. Raper. Southern Historical Collection, Wilson Library, University of North Carolina, Chapel Hill.

SCLC Papers. Papers of the Southern Christian Leadership Conference. Acc. No. 1. King Center, Atlanta, Georgia.

SNCC Papers. Papers of the Student Nonviolent Coordinating Committee. Acc. No. 52. King Center, Atlanta, Georgia.

U.S. Census Office. 1860. Population Schedules of the Eighth Census of the United States. National Archives, Washington, D.C. (microform).

Wallace Papers. Papers of Henry A. Wallace. Correspondence of the United States Secretary of Agriculture, 1906–70. Record Group 16, National Archives, Washington, D.C.

UNPUBLISHED RECORDS

Alabama State Department of Education.
Chamber of Commerce, Prattville, Alabama.
Chancery Clerk's Office, Bolivar County, Mississippi.
Chancery Clerk's Office, Coahoma County, Mississippi.
Chancery Clerk's Office, Holmes County, Mississippi.
Chancery Clerk's Office, Tallahatchie County, Mississippi.
Chancery Clerk's Office, Tate County, Mississippi.
Chancery Clerk's Office, Tunica County, Mississippi.
Mississippi State Department of Education.
Superior Court Clerk's Office, Clarke County, Georgia.
Superior Court Clerk's Office, Greene County, Georgia.
Superior Court Clerk's Office, Hancock County, Georgia.
Superior Court Clerk's Office, Morgan County, Georgia.
Superior Court Clerk's Office, Oglethorpe County, Georgia.
Tax Assessor's Office, Oglethorpe County, Georgia.
U.S. Department of Agriculture. Agricultural Stabilization and Conservation Service. Tate County, Mississippi Office.
U.S. Department of Agriculture. Consumer and Marketing Service, Cotton Division, Memphis, Tennessee.
U.S. Department of Agriculture. Farmers Home Administration. Bolivar, Coahoma, Holmes, Humphreys, Sunflower, Tunica, and Washington Counties, Mississippi Offices.
U.S. Department of Agriculture. Farmers Home Administration. Mississippi District Offices 3, 4, and 5.
U.S. Department of Housing and Urban Development. Jackson, Mississippi, Area Office.
U.S. Department of Housing and Urban Development. Office of Community Planning and Development. Washington, D.C.

U.S. Department of Housing and Urban Development. Office of the Assistant Secretary for Housing–Federal Housing Commissioner. Washington, D.C.

U.S. Department of Housing and Urban Development. Region IV Office. Atlanta, Georgia.

U.S. Department of Justice. Civil Rights Division. Washington, D.C.

U.S. Federal District Court of the Northern District of Mississippi. Clarksdale and Greenville Clerks' Offices.

U.S. Federal Emergency Relief Administration

SECONDARY SOURCES

Abernathy, Ralph David. 1989. *And the Walls Came Tumbling Down*. New York: Harper & Row.

Acree, W. G. 1912. History of the Mexican Boll Weevil and Its March. *Georgia Agricultural Quarterly* 5 (January): 31–34.

African American Members of Congress throughout History. 1993. *Voting Rights Review* (Spring): 17–18.

Agee, James, and Walker Evans. 1941. *Let Us Now Praise Famous Men*. Boston: Houghton Mifflin.

——. 1966. *Let Us Now Praise Famous Men*. New York: Ballantine Books.

Ahlgren, Frank R. 1967. The South's Poor. *Commercial Appeal*, October 9, 6.

Aiken, Charles S. 1962. Transitional Plantation Occupance in Tate County, Mississippi. M.A. thesis, Department of Geography, University of Georgia.

——. 1971a. Examination of the Role of the Eli Whitney Cotton Gin in the Origin of the United States Cotton Regions. *Proceedings of the Association of American Geographers* 3:5–9.

——. 1971b. The Fragmented Neoplantation: A New Type of Farm Operation in the Southeast. *Southeastern Geographer* 11 (April): 43–51.

——. 1973. The Evolution of Cotton Ginning in the Southeastern United States. *Geographical Review* 63 (April): 196–224.

——. 1975. Expressions of Agrarianism in American Geography: The Cases of Isaiah Bowman, J. Russell Smith, and O. E. Baker. *Professional Geographer* 27 (February): 19–29.

——. 1977. Faulkner's Yoknapatawpha County: Geographical Fact into Fiction. *Geographical Review* 67 (January): 1–21.

——. 1978. The Decline of Sharecropping in the Lower Mississippi River Valley. In *Man and Environment in the Lower Mississippi River Valley*, vol. 19 of *Geoscience and Man*, edited by Sam B. Hilliard, 151–65. Baton Rouge: School of Geoscience, Louisiana State University.

——. 1982. The Image of the Plantation in Southern Fiction: The Case of William Faulkner. In *Proceedings, Tall Timbers Ecology and Management Conference, February 22–24, 1979*, 189–202. Thomasville, Ga.: Tall Timbers Research Station.

———. 1985. New Settlement Pattern of Rural Blacks in the American South. *Geographical Review* 75 (October): 383–404.

———. 1987. Race as a Factor in Municipal Underbounding. *Annals of the Association of American Geographers* 77 (December): 564–79.

———. 1990. A New Type of Black Ghetto in the Plantation South. *Annals of the Association of American Geographers* 80 (June): 223–46.

———. 1995. The Rural South : A Historic View. In *The Changing American Countryside: Rural People and Places*, edited by Emery N. Castle, 318–38. Lawrence: University Press of Kansas.

Alabama Department of Education. 1990. Enrollment by Sex by Ethnic Group for First Month. November 16. Montgomery. Computer print copy.

Allan, Leslie, Eileen K. Kaufman, and Joanna Underwood. 1972. *Pollution in the Pulp and Paper Industry*. Cambridge: MIT Press.

Alter, Jonathan. 1993. Elitism and the Immigration Backlash. *Newsweek*, July 26, 35.

Analavege, Robert. 1966a. Delta People Organize: 30,000 Will Be Jobless. *Southern Patriot* 24 (June): 1, 4.

———. 1966b. Which Way in Grenada? *Southern Patriot* 24 (August): 1, 3.

Anderson, James D. 1988. *The Education of Blacks in the South, 1860–1935*. Chapel Hill: University of North Carolina Press.

Anderson, James R. 1973. *A Geography of Agriculture in the United States' Southeast*. Geography of World Agriculture No. 2. Budapest: Akadémia: Kiadó.

———. 1982. Plantation Agriculture in the Middle Suwannee Basin of Florida, 1825–1850. In *Proceedings, Tall Timbers Ecology and Management Conference, February 22–24, 1979*, 21–45. Thomasville, Ga.: Tall Timbers Research Station.

Anthony, Carl. 1976a. The Big House and the Slave Quarters, Part I, Prelude to New World Architecture. *Landscape* 20 (Spring): 8–18.

———. 1976b. The Big House and the Slave Quarters, Part II, African Contributions to the New World. *Landscape* 20 (Autumn): 9–15.

Atlanta Resurgens. 1971. Atlanta: First National Bank of Atlanta.

Auchmutey, Jim. 1984a. In Every Way, the South Is Not as Poor as in 1964. *Atlanta Constitution*, June 26, A10.

———. 1984b. Tunica: Poorest County in America. *Atlanta Constitution*, June 26, A1, A11.

Baker, Ella J. 1960. Bigger than a Hamburger. *Southern Patriot* 18 (May): 4.

———. 1961. Tent City: Freedom's Front Line. *Southern Patriot* 19 (February): 1, 4.

Baker, Lewis. 1983. *The Percys of Mississippi: Politics and Literature in the New South*. Baton Rouge: Louisiana State University Press.

Baker, O. E. 1926. Agricultural Regions of North America: The Basis of Classification. *Economic Geography* 2 (October): 459–93.

Baldwin, Sidney. 1968. *Poverty and Politics: The Rise and Decline of the Farm Security Administration*. Chapel Hill: University of North Carolina Press.

Ball, Howard, Dale Krane, and Thomas P. Lauth. 1982. *Compromised Compliance: Implementation of the Voting Rights Act*. Westport, Conn.: Greenwood Press.

Banks, Enoch M. 1905. *The Economics of Land Tenure in Georgia*. New York: Columbia University Press.

Barnett, Frank Willis. 1931. Thriving Little City [Eutaw, Alabama] Has Many Industries and Professional Men. *Birmingham News*, October 18. Ala.-Miss. page.

Barrett, Russell H. 1965. *Integration at Ole Miss*. Chicago: Quadrangle Books.

Barrow, David C., Jr. 1881. A Georgia Plantation. *Scribner's Monthly* 21 (April): 830–36.

Bartley, Numan V. 1969. *The Rise of Massive Resistance: Race and Politics in the South during the 1950s*. Baton Rouge: Louisiana State University Press.

Barton, G. T., and J. G. McNeely. 1939. *Recent Changes in Farm Organization in Three Arkansas Plantation Counties*. Preliminary Report. Fayetteville, Ark.: University of Arkansas, Agricultural Experiment Station.

Bartram, William. 1791. *Travels through North & South Carolina, Georgia, East & West Florida*. Philadelphia: James & Johnson.

Batteau, Allen W. 1990. *The Invention of Appalachia*. Tucson: University of Arizona Press.

Beale, Calvin L. 1992. Penitential Development: Prisons, Population, and Jobs in Nonmetro America. Paper presented at annual meeting, Southern Demographic Association.

Beard, Charles A., and Mary R. Beard. 1927. *The Rise of American Civilization*. 2 vols. New York: Macmillan.

Beckford, George L. 1972. *Persistent Poverty: Underdevelopment in Plantation Economies of the Third World*. New York: Oxford University Press.

Bell, Derrick A., Jr. 1973. *Race Racism and American Law*. Boston: Little, Brown.

———. Waiting on the Promise of *Brown*. *Law and Contemporary Problems* 39 (Spring): 341–73.

Berry, Brian J. 1973. *Growth Centers in the American Urban System, 1960-1970*. 2 vols. Cambridge, Mass.: Ballinger.

Berry, Jason. 1973. *Amazing Grace: With Charles Evers in Mississippi*. New York: Saturday Review Press.

Billings, Dwight B. 1979. *Planters and the Making of a "New South": Class, Politics, and Development in North Carolina, 1865–1900*. Chapel Hill: University of North Carolina Press.

Birt, Geoffrey. 1953. Wilcox School Survey Shows Over $3 Million Outlay Needed. *Mobile Press Register*, March 17, 1.

Black, Earl, and Merle Black. 1987. *Politics and Society in the South*. Cambridge: Harvard University Press.

Black Farmers Must Battle Difficult Odds to Survive. 1989. *Tuscaloosa News*, May 7, 11A.

Bledsoe, O. F. 1942. Sharecropping with the Government and Private Individuals. National Cotton Council of America, Memphis, Tenn. Library, Tenancy file.

Blumberg, Rhoda Lois. 1984. *Civil Rights: The 1960s Freedom Struggle*. Boston: Twayne Publishers.

Boeger, E. A., and E. A. Goldenweiser. 1916. *A Study of the Tenant Systems of Farming in the Yazoo-Mississippi Delta*. Bulletin 639. Washington, D.C.: U.S. Department of Agriculture.

Boles, John B. 1995. *The South through Time: A History of an American Region*. Englewood Cliffs, N.J.: Prentice-Hall.

Boles, W. J. 1930. Greene County Ranks High in Agriculture: Old Ante-Bellum Plantation Homes Still Stand in the Rich Farming Belt. *Birmingham News*, January 25, 1.

Bond, Horace Mann. 1970. *The Education of the Negro in the American Social Order*. New York: Octagon Books. Original edition, New York: Prentice-Hall, 1934.

Bosworth, Karl A. 1941. *Black Belt Country: Rural Government in the Cotton Country of Alabama*. University: Bureau of Public Administration, University of Alabama.

Brady, Tom P. 1955. *Black Monday*. Jackson, Miss.: Citizens Councils of America.

Branch, Taylor. 1988. *Parting the Waters: America in the King Years, 1954-1963*. New York: Simon & Schuster.

Brandfon, Robert L. 1967. *Cotton Kingdom of the New South: A History of the Yazoo Mississippi Delta from Reconstruction to the Twentieth Century*. Cambridge: Harvard University Press.

Brannen, C. O. 1924. *Relation of Land Tenure to Plantation Organization*. Bulletin 1269. Washington, D.C.: U.S. Department of Agriculture.

Brauer, Carl M. 1977. *John Kennedy and the Second Reconstruction*. New York: Columbia University Press.

Breeden, James A., ed. 1980. *Advice among Masters: The Ideal in Slave Management in the Old South*. Westport, Conn.: Greenwood Press.

Brodell, A. P., and J. A. Ewing. 1948. *Use of Tractor Power, Animal Power, and Hand Methods in Crop Production*. FM-69. Washington, D.C.: U.S. Department of Agriculture, Bureau of Agricultural Economics.

Bronstein, Scott. 1993. Are Poor Blacks Targets of Waste Facility Site Decisions? *Atlanta Constitution*, February 22, A1.

Brooks, Robert Preston. 1914. *The Agrarian Revolution in Georgia, 1865–1912*. Bulletin of the University of Wisconsin. No. 639. History Series, Vol 3. Madison, Wisconsin.

Brown, Calvin S. 1962. Faulkner's Geography and Topography. *PMLA* 78:652–59.

Brown, Chris. 1988. The Old Mills at Cedar Shoals. In *History of Newton County, Georgia*. Covington: Newton County Historical Society, 189–201.

Brown, David L., and Mildred Warner. 1989. Persistent Low Income Areas in the United States: Some Concepts and Challenges. In *National Rural Studies Committee: A Proceedings*. Second Annual Meeting. Corvallis: Western Rural Development Center, Oregon State University.

Brown, Lester R. 1970. *Seeds of Change*. New York: Praeger.

Bruce, Philip A. 1900. Social and Economic Revolution in the Southern States. *Contemporary Review* (July): 59.

Bryles, Mark. 1968. Farm Labor Incentive Plans. *Delta Farm Press*, December 12, 3.

Buchanan, R. Ogilvie. 1959. Some Reflections on Agricultural Geography. *Geography* 44:1–13.

Bunche, Ralph J. 1973. *The Political Status of the Negro in the Age of FDR*. Edited by Dewey W. Grantham. Chicago: University of Chicago Press.

Burk, Robert Fredrick. 1984. *The Eisenhower Administration and Black Civil Rights*. Knoxville: University of Tennessee Press.

Cagin, Seth, and Philip Dray. 1988. *We Are Not Afraid: The Story of Goodman, Schwerner, and Chaney and the Civil Rights Campaign for Mississippi*. New York: Bantam Books.

Caldwell, Erskine. 1932. *Tobacco Road*. New York: Grosset and Dunlap.

———. 1951. *Call It Experience: The Years of Learning How to Write*. New York: Duell, Sloan, and Pearce.

Caldwell, Erskine, and Margaret Bourke-White. 1937. *You Have Seen Their Faces*. New York: Modern Age Books.

———. 1974. *Tobacco Road*. Savannah, Ga.: Beehive Press. Expanded edition.

California Governor's Commission on the Los Angeles Riots. 1965. *Violence in the City—An End or a Beginning?*. Los Angeles: n.p.

Callahan, Nancy. 1987. *The Freedom Quilting Bee*. Tuscaloosa: University of Alabama Press.

Campbell, David. 1972. The Lowndes County Freedom Organization: An Appraisal. *New South* 37 (Winter): 37–42.

Campbell, Sarah C. 1993a. Harrah's Cleared for Two Casino Licenses. *Commercial Appeal*, October 22, A1, A12.

———. 1993b. Regulation or Market to Limit Casinos? *Commercial Appeal*, May 26, A1, A9.

Cantor, Louis. 1969. *A Prologue to the Protest Movement: The Missouri Sharecropper Roadside Demonstration of 1939*. Durham, N.C.: Duke University Press.

Carmichael, Stokely, and Charles V. Hamilton. 1967. *Black Power: The Politics of Liberation in America*. New York: Vintage Books.

Carnoy, Martin. 1994. *Faded Dreams: The Politics and Economics of Race in America*. New York: Cambridge University Press.

Carson, Clayborne. 1981. *In Struggle: SNCC and the Black Awakening of the 1960s*. Cambridge: Harvard University Press.

Carter, Dan T. 1995. *The Politics of Rage: George Wallace, the Origins of the New Conservatism, and the Transformation of American Politics*. New York: Simon & Schuster.

Carter, Hodding, III. 1959. *The South Strikes Back*. New York: Doubleday.

Carter, Jimmy. 1992. *Turning Point: A Candidate, State, and a Nation Come of Age*. New York: Time Books.

Carter, Selwyn. 1993. No Second Black Congressional District for Alabama. *Voting Rights Review* (Fall): 5–6.

Cash, W. J. 1941. *The Mind of the South*. New York: Alfred A. Knopf.

Cawthon, Raad. 1991. Better Times on Sugar Ditch: Help Comes to a Poor County. *Atlanta Journal/Atlanta Constitution*, March 23, A3.

Cell, John W. 1982. *The Highest Stage of White Supremacy: The Origins of Segregation in South Africa and the American South*. New York: Cambridge University Press.

Chalmers, David M. 1976. *Hooded Americanism: The History of the Ku Klux Klan*. New York: New Viewpoints.

Chapman, Joseph. 1969. Greene County Elects Six Negro Candidates. *Birmingham Post-Herald*, July 30, 1.

Charlier, Tom. 1992. Two Counties: Give Us Your Tired Refuse, We're Poor. *Commercial Appeal*, October 26, A1.

Cheakalos, Christina. 1992. A Thin Green Line: Louisiana Town Fights a Chemical Giant. *Atlanta Journal/Atlanta Constitution*, February 23, B4, B5.

Chestnut, J. L., Jr. 1990. *Black in Selma: The Uncommon Life of J. L. Chestnut, Jr.* New York: Farrar, Straus and Giroux.

Christensen, Mike. 1984. Alabama Towns Try to Join Dog Track Gravy Train. *Atlanta Journal/Atlanta Constitution*, February 5, 1A, 16A.

Christion, Cornell. 1991. Tunica's Sugar Ditch Fades into the Past. *Commercial Appeal*, March 8, B1.

Cimons, Marlene. 1982. Dual School Systems in South Thrive. *Los Angeles Times*, March 1, F1, F6.

Clark, Thomas D. 1944. *Pills, Petticoats, and Plows: The Southern Country Store*. New York: Bobbs-Merrill.

———. 1946. The Furnishing and Supply System in Southern Agriculture since 1865. *Journal of Southern History* 12 (February): 36–37.

Classics in the Making. 1992. Cape Cod, Mass.: C. M. L. Co.

Cleland, Charles L., and Will Fontanez. 1994. Rurality Scores for U.S. Counties, 1994. Bulletin 689. Knoxville: Agricultural Experiment Station, University of Tennessee.

Cleland, Herdman F. 1920. The Black Belt of Alabama. *Geographical Review* 10 (December): 375–87.

Cobb, James C. 1992. *The Most Southern Place on Earth: The Mississippi Delta and the Roots of Regional Identity*. New York: Oxford University Press.

———. 1993. *The Selling of the South: The Southern Crusade for Industrial Development, 1936–1990*. 2d ed. Urbana: University of Illinois Press.

Cohn, David. L. 1948. *Where I Was Born and Raised*. Boston: Houghton Mifflin.

———. 1950. Lament for the South That Is No More. *New York Times Magazine*, January 22, 14, 41.

Colley, Marion. 1926a. Historic Danburg, the Center of a Community of Skilled Farmers. *Washington (Ga.) News-Reporter*, December 15, 1.

——. 1926b. How Wilkes County Farmers Are Solving Their Problems. *Washington (Ga.) News-Reporter*, December 3, 1.

Conrad, David Eugene. 1965. *The Forgotten Farmers: The Story of Sharecroppers in the New Deal.* Urbana: University of Illinois Press.

Conversion of a Cotton Plantation. 1945. *Cotton and Cotton Oil Press* 46 (October 20): 5–9.

Cook, K. W. 1966a. King Leads Marchers on Meredith's Route. *Commercial Appeal*, June 8, 1.

——. 1966b. Negroes Urged to Leave Vietnam. *Commercial Appeal*, June 10, 8.

Cooper, Helene. 1994. Mixed Blessing: Southern Town's Past Haunts Golden Future for Gambling Riches. *Wall Street Journal*, June 22, A1, A6.

Couto, Richard A. 1991. *Ain't Gonna Let Nobody Turn Me Round: The Pursuit of Racial Justice in the Rural South.* Philadelphia: Temple University Press.

Cumming, Joe. 1974. Greene County, Alabama: Hope for the Future. *Southern Voices* 1 (March–April): 22–30.

Curtin, Philip D. 1990. *The Rise and Fall of the Plantation Complex: Essays in Atlantic History.* New York: Cambridge University Press.

Dallas County [Alabama] Citizens Council. 1963. Advertisement. *Selma Times-Journal*, June 9, 3.

Daniel, Pete. 1985. *Breaking the Land: The Transformation of Cotton, Tobacco, and Rice Cultures since 1880.* Urbana: University of Illinois Press.

Daniels, Jonathan. 1938. *A Southerner Discovers the South.* New York: Macmillan.

Davis, Adell. 1965. Testimony. U.S. Commission on Civil Rights. *Hearings Held in Jackson, Miss., February 16–20, 1965.* Vol. 1, *Voting.* Washington, D.C.: GPO, 111–20.

Davis, Allison, Burleigh B. Gardner, and Mary R. Gardner. 1941. *Deep South: A Social Anthropological Study of Caste and Class.* Chicago: University of Chicago Press.

Davis, Ronald L. F. 1982. *Good and Faithful Labor: From Slavery to Sharecropping in the Natchez District, 1860–1890.* Westport, Conn.: Greenwood Press.

Day, Richard H. 1967. The Economics of Technological Change and the Demise of the Sharecropper. *American Economic Review* 62:427–49.

Demerath, N. J., III, Gerald Marawell, and Michael T. Aiken. 1971. *Dynamics of Idealism.* San Francisco: Jossey-Bass.

Dickens, Dorothy. 1949. *The Labor Supply and Mechanized Cotton Production.* Bulletin 463. State College, Miss.: Mississippi Agricultural Experiment Station, Mississippi State College.

Dillon, Merton L. 1985. *Ulrich Bonnell Phillips: Historian of the Old South.* Baton Rouge: Louisiana State University Press.

Doar, John, and Dorothy Landsberg. 1975. The Performance of the FBI in Investigating Violation of Federal Laws Protecting the Right to Vote, 1960–1967. In U.S. Senate, Select Committee to Study Governmental Operations with Respect to Intelligence Activities, *Hearings.* Vol. 6, *Federal Bureau of Investigation.* 94th Cong., 1st sess., 888–991. Washington, D.C.: GPO, 1976.

Dollard, John. 1937. *Cast and Class in a Southern Town*. New Haven: Yale University Press.

Dollard, John, et al. 1939. *Frustration and Aggression*. New Haven: Yale University Press.

Donovan, John C. 1967. *The Politics of Poverty*. New York: Pegasus.

Doyle, Don Harrison. 1982. Urbanization and Southern Culture: Economic Elites in Four New South Cities (Atlanta, Nashville, Charleston, Mobile) c.1865–1910. In *Toward a New South?: Studies in Post–Civil War Southern Communities*, edited by Orville Vernon Burton and Robert C. McMath Jr., 11–36. Westport, Conn.: Greenwood Press.

Drake, St. Clair, and Horace R. Cayton. 1962. *Black Metropolis*. 2 vols. New York: Harper & Row.

Du Bois, W. E. Burghardt. 1903. *The Souls of Black Folk: Essays and Sketches*. Chicago: A. C. McClurg.

———. 1918. Review of Ulrich Bonnell Phillips' *American Negro Slavery*. *American Political Science Review* 12 (November 1918): 722–26.

Dunbar, Anthony. 1969. *The Will to Survive: A Study of a Mississippi Plantation Community*. Atlanta: Southern Regional Council.

———. 1990. *Delta Time: A Journey through Mississippi*. New York: Pantheon Books.

Dykeman, Wilma, and James Stokely. 1962. *Seeds of Southern Change: The Life of Will Alexander*. Chicago: University of Chicago Press.

Earl, Joe. 1992. It's So Hard to Say Good Buy. *Atlanta Constitution*, October 29, E1, E4.

Edelman, Marian Wright. 1973. Southern School Desegregation, 1954–1953: A Judicial-Political Overview. *Annals of the American Academy of Political and Social Science* 407 (May): 32–42.

Edstrom, Eve. 1966. Rights Commission Cites HEW's Failure to Enforce Compliance. *Washington Post*, February 24, A2.

Eliot, Clara. 1927. *The Farmer's Campaign for Credit*. New York: D. Appleton.

Elwell, L. D. 1979. Farmers Home Administration Rural Housing Programs. *Family Economics Review*, no. 2:17–20.

Ely, Richard T., and Charles J. Gilpin. 1919. Tenancy in an Ideal System of Landownership. *American Economic Review Supplement* 9 (March): 180–213.

Faulkner, William. 1942. *Go Down, Moses*. New York: Random House.

———. 1957. *The Town*. New York: Random House.

———. 1959. *The Mansion*. New York: Random House.

Feagin, Joe R., and Harlan Hahn. 1973. *Ghetto Revolts: The Politics of Violence in American Cities*. New York: Macmillan.

Federico, Ronald C. 1984. *The Social Welfare Institution: An Introduction*. Lexington, Mass.: D. C. Heath.

Fisher, James S. 1970. Federal Crop Allotment Program and Responses by Individual Farm Operators. *Southeastern Geographer* 10 (November): 47–58.

Flynn, Charles L., Jr. 1983. *White Land, Black Labor: Caste and Class in Late Nineteenth-Century Georgia.* Baton Rouge: Louisiana State University Press.

Fogelson, Robert M., and Robert B. Hill. 1968. Who Riots? A Study of Participation in the 1967 Riots. In National Advisory Commission on Civil Disorders, *Supplemental Studies,* 235–43. Washington, D.C.: GPO.

Ford, Arthur M. 1973. *Political Economies of Rural Poverty in the South.* Cambridge, Mass.: Ballinger.

Ford, Lacy. 1983. Labor and Ideology in the South Carolina Up-Country: The Transition to Free-Labor Agriculture. In *The Southern Enigma: Essays on Race, Class, and Folk Culture,* edited by Walter J. Fraser Jr. and Winfred B. Moore Jr., 25–41. Westport, Conn.: Greenwood Press.

Forman, James. 1972. *The Making of Black Revolutionaries: A Personal Account.* New York: Macmillan.

Foster, G. W., Jr. 1965. Title VI: Southern Education Faces the Facts. *Saturday Review,* March 20, 60–61, 76.

Freedman, Leonard. 1969. *Public Housing, the Politics of Poverty.* New York: Holt, Rinehart and Winston.

Freeman, Orval L. 1970. Review of Lester R. Brown's *Seeds of Change. War on Hunger, a Report from the Agency of International Development* 4, no. 5: 6–8.

Freidel, Frank. 1965. *F.D.R. and the South.* Baton Rouge: Louisiana State University Press.

Friedman, Lawrence M. 1968. *Government and Slum Housing.* Chicago: Rand McNally.

Fruhman, Barton. 1966. Rights Leaders Flay Emotions. *Commercial Appeal,* June 8, 21.

Furstenberg, Frank F., Jr. 1994. Women and Children Last. Review of Linda Gordon's *Pitied but Not Entitled: Single Mothers and the History of Welfare, 1890–1935. New York Times Book Review,* October 16, 46.

Gaines, James P., and Grady B. Crowe. 1950. *Workstock vs. Tractors in the Yazoo-Mississippi Delta.* Bulletin 470. State College, Miss: Agricultural Experiment Station, Mississippi State College.

Galbraith, John K. 1958. *The Affluent Society.* Boston: Houghton Mifflin.

Galloway, J. H. 1977. The Mediterranean Sugar Industry. *Geographical Review* 67 (April): 177–92.

Galphin, Bruce. 1970. School Battle for South Not Yet Won. *Washington Post,* January 2, A1, A4.

Gantt, Larry. 1922. Letter to the *Madison (Ga.) Madisonian,* December 22, 6.

Garrow, David J. 1978. *Protest at Selma: Martin Luther King, Jr., and the Voting Rights Act of 1965.* New Haven: Yale University Press.

———. 1981. *The FBI and Martin Luther King, Jr., from "Solo" to Memphis.* New York: W. W. Norton.

Genovese, Eugene D. 1974. *Roll, Jordan, Roll: The World the Slaves Made.* New York: Pantheon Books.

Gibson, J. Sullivan. 1940. The Western Black Belt—Its Geographic Status. *Annals of the Association of American Geographers* 30 (March): 57–58.

———. 1941. The Alabama Black Belt and Its Geographic Status. *Economic Geography* 17 (January): 1–23.

Good, Paul. 1968. *Cycle to Nowhere.* U.S. Commission on Civil Rights. Clearinghouse Publication No. 14. Washington, D.C.: GPO.

Goosing the Cotton. 1963. *Time,* April 26, 60.

Gordon, Linda. 1994. *Pitied but Not Entitled: Single Mothers and the History of Welfare, 1890–1935.* New York: Free Press.

Gottlieb, Peter. 1987. *Making Their Own Way: Southern Blacks' Migration to Pittsburgh.* Urbana: University of Illinois Press.

Grady, Henry W. 1881. Cotton and Its Kingdom. *Harper's New Monthly Magazine* 63 (October): 719–34.

Graham, Edgar, and Ingrid Floering. 1984. *The Modern Plantation in the Third World.* New York: St. Martin's.

Grant, Robert B. 1972. *The Black Man Comes to the City: A Documentary Account from the Great Migration to the Great Depression, 1915–1930.* Chicago: Nelson-Hall Co.

Gray, Lewis Cecil. 1933. *History of Agriculture in the Southern States to 1860.* 2 vols. Washington, D.C.: Carnegie Institution.

Gray, R. B. 1958. *Development of the Agricultural Tractor in the United States.* St. Joseph, Mich.: American Society of Agricultural Engineers.

Greenberg, Polly. 1969. *The Devil Has Slippery Shoes: A Biased Biography of the Child Development Group of Mississippi.* Toronto: Collier-Macmillan Canada.

Greenland to Open Monday. 1941. *Greensboro (Ga.) Herald-Journal,* January 17, 1.

Gregg, William. 1941. *Essays on Domestic Industry.* Graniteville, S.C.: Graniteville Co.

Gregor, Howard F. 1962. The Plantation in California. *Professional Geographer* 14 (March): 1–4.

———. 1965. The Changing Plantation. *Annals of the Association of American Geographers* 55 (June): 221–38.

Gregor Sebba, ed. 1952. *Georgia Studies: Writings of Robert Preston Brooks.* Athens: University of Georgia Press.

Grier, George W. 1967. The Negro Ghettos and Federal Housing Policy. *Law and Contemporary Problems* 32:550–60.

Grofman, Bernard, Lisa Handley, and Richard G. Niemi. 1992. *Minority Representation and the Quest for Voting Equality.* New York: Cambridge University Press.

Groh, George. 1972. *The Black Migration: The Journey to Urban America.* New York: Weybright and Talley.

Grossman, James R. 1989. *Land of Hope: Chicago, Black Southerners, and the Great Migration.* Chicago: University of Chicago Press.

Growers "Goosing" Their Cotton Should Know Safe Pesticides. 1964. *Cotton Trade Journal,* June 5, 60.

Grubbs, Donald H. 1971. *Cry from the Cotton: The Southern Tenant Farmers' Union and the New Deal*. Chapel Hill: University of North Carolina Press.

Haddad, Charles. 1990. For Albany, a Difficult Rebirth. *Atlanta Journal/Atlanta Constitution*, May 27, H1, H3.

Halberstam, David. 1993. *The Fifties*. New York: Villard Books.

Hamilton, Kenneth M. 1991. *Black Towns and Profit: Promotion and Development in the Trans-Appalachian West, 1877–1915*. Urbana: University of Illinois Press.

Hampton, Henry, and Steve Fayer. 1990. *Voices of Freedom: An Oral History of the Civil Rights Movement from the 1950s through the 1980s*. New York: Bantam Books.

Hanks, Lawrence J. 1987. *The Struggle for Black Political Empowerment in Three Georgia Counties*. Knoxville: University of Tennessee Press.

Harper, Roland M. 1922. Development of Agriculture in Upper Georgia from 1890 to 1920. *Georgia Historical Quarterly* 6 (September): 226–27.

Harrington, Michael. 1962. *The Other America: Poverty in the United States*. New York: Macmillan.

———. 1984. *The New American Poverty*. New York: Holt, Rinehart and Winston.

Harris, David. 1982. *Dreams Die Hard*. New York: St. Martin's.

Harris, J. William. 1982. Plantations and Power: Emancipation on the David Barrow Plantations. In *Toward a New South?: Studies in Post–Civil War Southern Communities*, edited by Orville Vernon Burton and Robert C. McMath Jr., 246–64. Westport, Conn.: Greenwood Press.

Hart, John Fraser. 1981. Migration to the Blacktop: Population Redistribution in the South. *Landscape* 25, no. 3:15–19.

———. 1982. The Role of the Plantation in Southern Agriculture. In *Proceedings, Tall Timbers Ecology and Management Conference, February 22–24, 1979*, 1–19. Thomasville, Ga.: Tall Timbers Research Station.

Hartman, W. A., and W. H. Wooten. 1935. *Georgia Land Use Problems*. Bulletin 191. Experiment, Ga.: University System of Georgia, Georgia Experiment Station.

Hathorn, Clay. 1988. Sugar Ditch Deal Disputed: Twenty-one Families Left in a Bind. *Commercial Appeal*, October 5, A1, A4.

———. 1989. Hill Counties Enriched by Furniture Industry. *Commercial Appeal*, March 13, A1.

Hayden, Tom. 1962. *Revolution in Mississippi*. New York: Students for a Democratic Society.

Henri, Florette. 1975. *Black Migration: Movement North, 1900–1920*. New York: Anchor Press/Doubleday.

Herbers, John. 1965a. Dr. King Opens Alabama Drive. *New York Times*, January 3, 1, 20.

———. 1965b. Voting Is Crux of Civil Rights Hopes. *New York Times*, February 14, E5.

———. 1965c. Negroes Beaten in Alabama Riot. *New York Times*, February 19, 1, 29.

———. 1965d. Dr. King, Back in Alabama, Calls for a March on Capitol to Push Voting Drive. *New York Times*, February 23, 16.

——. 1969. Ambiguous Moves on Desegregation. *New York Times*, April 6, E7.

Herman, Janet Sharp. 1981. *The Pursuit of a Dream*. New York: Oxford University Press.

Hester, Kathryn Healy. 1982. Mississippi and the Voting Rights Act, 1965–1982. *Mississippi Law Journal* 52:803–76.

Hicks, John D. *The Populist Revolt*. 1931. Minneapolis: University of Minnesota Press.

Hill, W. B. 1915. *Rural Survey of Clarke County, Georgia with Special Reference to the Negroes*. Bulletin of the University of Georgia No. 15. Phelps-Stokes Fellowship Studies No. 2. Athens: University of Georgia.

Hilliard, Sam B. 1972. The Dynamics of Power, Recent Trends in Mechanization on the American Farm. *Technology and Culture* 13:1–24.

——. 1984. *Atlas of Antebellum Southern Agriculture*. Baton Rouge: Louisiana State University Press.

Hilton, Bruce. 1969. *The Delta Ministry*. London: Collier-Macmillan.

Himel, Ramon. 1966. Shouted Warning, Blasts Crack Stillness on Road. *Commercial Appeal*, June 7, 1.

Hinds, W. E. 1916. *Boll Weevil in Alabama*. Bulletin 188. Auburn: Alabama Polytechnic Institute, Agricultural Experiment Station.

Historic Summer in Tennessee. 1964. *Southern Patriot* 22 (June): 4.

Hoffsommer, Harold. 1934. Rural Problem Areas Survey, Cotton Growing Regions (Areas) of the Old South. Report No. 2, Dallas County, Alabama; Report No. 17, Leflore County, Mississippi; Report No. 29, Meriwether County, Georgia. Division of Research and Statistics, Federal Emergency Relief Administration. Washington, D.C. (mimeograph).

Holley, William C., Ellen Winston, and T. J. Woofter. 1940. *The Plantation South, 1934–1937*. Division of Research, Works Projects Administration. Monograph 22. Washington, D.C.: GPO.

Holmes, Steven A. 1994. Experts Say Redistricting Added to the GOP Surge. *New York Times*, November 14, 16.

Hoover, Edgar. 1948. *The Location of Economic Activity*. New York: McGraw-Hill.

House Deplores Sniper Shooting. 1966. *Commercial Appeal*, June 9, 24.

Howell, Leon. 1969. *Freedom City: The Substance of Things Hoped For*. Richmond, Va.: Knox Press.

Hudson, John. 1982. The Yazoo-Mississippi Delta as Plantation Country. In *Proceedings, Tall Timbers Ecology and Management Conference, February 22–24, 1979*, 66–110. Thomasville, Ga.: Tall Timbers Research Station.

Huie, William Bradford. 1956. The Shocking Story of Approved Killing in Mississippi. *Look*, January 24, 46–49.

Immigration Needed. 1922. *Madison (Ga.) Madisonian*, December 15, 10.

Industrial and Rural Workers on FSA Homesteads. 1942. *Monthly Labor Review* 54 (February): 379–80.

In Tennessee, the Issue Is Economic. 1963. *Southern Patriot* 21 (September): 4.

International Cotton Seed Products Directory, 1922–1923. 1923. Dallas: Cotton and Cotton Oil News.

Is Meredith Right? 1966. *New York Times,* June 5, 78.

Jacques, D. H. 1869. Future of Farming. *Rural Carolinian* 1 (November): 111–23.

James, W. F. 1968. Ways to Get and Keep Labor. *Delta Farm Press,* February 22, 42.

Jenkins, Ray. 1969. Majority Rule in the Black Belt: Greene County, Alabama. *New South* 24 (Fall): 60–67.

Johansen, Harley E., and Glenn V. Fuguitt. 1984. *The Changing Rural Village in America: Demographic and Economic Trends since 1950.* Cambridge, Mass.: Ballinger.

Johnson, Charles S. 1934. *Shadow of the Plantation.* Chicago: University of Chicago Press.

———. 1941. *Growing Up in the Black Belt: Negro Youth in the Rural South.* Washington, D.C.: American Council on Education.

Johnson, Charles S., Edwin R. Embree, and W. W. Alexander. 1935. *The Collapse of Cotton Tenancy: Summary of Field Studies and Statistical Surveys, 1933–1935.* Chapel Hill: University of North Carolina Press.

Johnson, Charles S., et al. 1941. *Statistical Atlas of Southern Counties: Listing and Analysis of Socio-Economic Indices of 1104 Southern Counties.* Chapel Hill: University of North Carolina Press.

Johnson, Merrill. 1990. A Survey-Based Analysis of Race and Manufacturing Plant Location in the Nonmetropolitan South. *Southeastern Geographer* 30 (November): 79–93.

Johnson, O. M., and Howard A. Turner. 1930. The Old Plantation Piedmont Cotton Belt. U.S. Department of Agriculture, Bureau of Agricultural Economics, Division of Land Economics. Washington, D.C. (mimeograph).

Johnson, Oscar. 1947. Will the Machine Ruin the South? *Saturday Evening Post,* May 31, 36–37, 94–95, 98.

Johnson, Paul. 1978. Diffusion of Fertilizer in the Corn Belt, 1939–1969. M.S. thesis, Department of Geography, University of Tennessee, Knoxville.

Johnston, Paul H. 1940. Old Ideals versus New Ideals in Farm Life. In *Farmers in a Changing World, USDA Yearbook of Agriculture, 1940.* Washington, D.C., GPO.

Joint Center for Political Studies. 1973. *National Roster of Black Elected Officials.* Vol. 3. Washington, D.C.: Joint Center for Political Studies.

Joint Center for Political and Economic Studies. 1994. *Black Elected Officials: A National Roster, 1993.* Washington, D.C.: Joint Center for Political and Economic Studies.

Jones, Bill. 1966. *The Wallace Story.* Northport, Ala.: American Southern Publishing Co.

Jones, William O. 1968. Plantations. In *International Encyclopedia of the Social Sciences,* edited by David L. Sills, 12:154–59. New York: Macmillan and Free Press.

Jordan, Winthrop D. 1968. *White over Black: American Attitudes toward the Negro, 1550–1812*. Chapel Hill: University of North Carolina Press.

———. 1974. *The White Man's Burden: Historical Origins of Racism in the United States*. New York: Oxford University Press.

Karim, Benjamin Iman, ed. 1971. *Four Speeches by Malcolm X*. New York: Seaver.

Karlan, Pamela S. 1995. The Supreme Court and Voting Rights: Bizarre Districts or Bizarre Decisions? *Southern Changes* 17 (Fall/Winter): 8–13.

Katzenbach, Nicholas. 1966. A Lesson in Responsible Leadership. *New South* 21 (Spring): 55–60.

Kehl, D. G. 1974. Steinbeck's "String of Pictures" in *The Grapes of Wrath*. *Image* (March): 1–10.

Kepple, David, Oliva Barton, and Brett Guge. 1979. Wilcox, a Study in White Flight. *Birmingham News*, August 21, 2.

Kester, Howard. 1936. *Revolt among the Sharecroppers*. New York: Covici-Friede.

Key, V. O., Jr. 1949. *Southern Politics in State and Nation*. New York: Alfred A. Knopf.

King, Edward. 1879. *The Great South*. Hartford, Conn.: American Publishing Co.

King, Martin Luther, Jr. 1958. *Stride toward Freedom: The Montgomery Story*. New York: Harper & Row.

———. 1963a. *Why We Can't Wait*. New York: Harper & Row.

———. 1963b. I Have a Dream. In *A Testament of Hope: The Essential Writings and Speeches of Martin Luther King, Jr.*, edited by James Melvin Washington, 217–20. New York: HarperCollins.

———. 1967. *Where Do We Go from Here: Chaos or Community*. New York: Harper & Row.

Kirby, Jack T. 1969. Clarence Poe's Vision of a Segregated "Great Rural Civilization." *South Atlantic Quarterly* 68 (Winter): 27–38.

Kirby, John B. 1980. *Black Americans in the Roosevelt Era: Liberalism and Race*. Knoxville: University of Tennessee Press.

Kirwan, Albert D. 1951. *Revolt of the Rednecks: Mississippi Politics, 1876–1925*. Lexington: University of Kentucky Press.

Kittredge, Kevin. 1986. Blacks Fallow in Job Fields: Heirs of Sharecropper System Find Poverty Is Their Portion. *Commercial Appeal*, May 5, A1, A8-A9.

Kluger, Richard. 1976. *Simple Justice: The History of Brown v. Board of Education and Black America's Struggle for Equality*. New York: Alfred A. Knopf.

Knox, Margaret. 1983. Mill That Is Cobb Town's Lifeblood Closing Down. *Atlanta Journal/Atlanta Constitution*, August 21, 1B, 5B.

Kousser, J. Morgan. 1974. *The Shaping of Southern Politics: Suffrage Restriction and the Establishment of the One-Party South, 1880–1910*. New Haven: Yale University Press.

———. 1984. The Undermining of the First Reconstruction: Lessons for the Second. In *Minority Vote Dilution*, edited by Chandler Davidson, 27–45. Washington, D.C.: Howard University Press.

Lamb, Robert B. 1963. *The Mule in Southern Agriculture*. University of California Publications in Geography. Vol. 15. Berkeley: Department of Geography, University of California Press.

Land, Aubrey C., ed. 1969. *Bases of the Plantation Society*. Columbia: University of South Carolina Press.

Lange, Dorothea, to Roy Stryker. 1937. Letter, June 3. In Meltzer, Milton. 1978. *Dorothea Lange: A Photographer's Life*. New York: Farrar, Straus & Giroux.

Lange, Dorothea, and Paul Schuster Taylor. 1939. *An American Exodus: A Record of Human Erosion*. New York: Reynal and Hitchcock.

Langsford, E. L., and B. H. Thibodeaux. 1939. *Plantation Organization and Operation in the Yazoo-Mississippi Delta Area*. U.S. Department of Agriculture. Technical Bulletin 682. Washington, D.C.: GPO.

Lanker, Brian. 1989. I Dreamed a World. *National Geographic* 176 (August): 206–25.

Lawson, Steven F. 1976. *Black Ballots: Voting Rights in the South, 1944–1969*. New York: Columbia University Press.

———. 1985. *In Pursuit of Power: Southern Blacks and Electoral Politics, 1965–1982*. New York: Columbia University Press.

———. 1991. *Running for Freedom: Civil Rights and Black Politics in America since 1941*. Philadelphia: Temple University Press.

Layman, Joseph B. 1867. Cotton Planting. In *Report of the Commissioner of Agriculture for the Year 1866*. Washington, D.C.: GPO, 208.

Lemann, Nicholas. 1991. *The Promised Land: The Great Black Migration and How It Changed America*. New York: Alfred A. Knopf.

LeRay, Nelson L., George L. Wilber, and Grady B. Crowe. 1960. *Plantation Organization and the Resident Labor Force, Delta Area, Mississippi*. Bulletin 606. State College, Miss.: Mississippi Agricultural Experiment Station, Mississippi State College.

Lesher, Stephan. 1994. *George Wallace: American Populist*. Reading, Mass.: Addison-Wesley.

Levitan, Sar A. 1969. *The Great Society's Poor Law: A New Approach to Poverty*. Baltimore: Johns Hopkins University Press.

Lewis, Peirce F. 1976. *New Orleans: The Making of an Urban Landscape* Cambridge, Mass.: Ballinger.

Long, Lewis E. 1931. *Farm Power in the Yazoo-Mississippi Delta*. Bulletin 295. Mississippi Agricultural Experiment Station. State College: Mississippi A&M College.

Lonsdale, Richard E. 1969. Barriers to Rural Industrialization in the South. *Proceedings of the Association of American Geographers* 1:84–88.

Lösch, August. 1967. *The Economics of Location*. Translated from the second edition. New York: John Wiley.

Lowenthal, David. 1972. *West Indian Societies*. New York: Oxford University Press.

Lundy, W. A. 1922. Calcium Arsenate in Dusting for Boll Weevils. *Georgia Agriculturist* 1 (November): 15–18.

McAdam, Doug. 1981. *Political Process and the Development of Black Insurgency, 1930–1970*. Chicago: University of Chicago Press.

———. 1988. *Freedom Summer*. New York: Oxford University Press.

McCarthy, Rebecca. 1993. "I Cannot Be Bought for a Landfill." *Atlanta Constitution*, November 16, C1.

McClinton, Raymond. 1938. A Social-Economic Analysis of a Mississippi Delta Plantation [Trail Lake Planting Company]. M.A. thesis, Department of Sociology, University of North Carolina, Chapel Hill.

MacDonald, Dwight. 1963. Our Invisible Poor. *New Yorker*, January 19, 82–92.

MacDonald, Maurice. 1977. *Food Stamps and Income Maintenance*. New York: Academic Press.

MacLeish, Archibald. 1938. *Land of the Free*. New York: Harcourt Brace.

McMillen, Neil R. 1971. *The Citizens' Council: Organized Resistance to the Second Reconstruction, 1954–1964*. Urbana: University of Illinois Press.

Maharidge, Dale, and Michael Williamson. 1989. *And Their Children after Them*. New York: Pantheon Books.

Maier, Frank H. 1969. An Economic Analysis of Adoption of the Mechanical Cotton Picker. Ph.D. diss., Department of Economics, University of Chicago.

Mandle, Jay R. 1978. *The Roots of Black Poverty: The Southern Plantation Economy after the Civil War*. Durham, N.C.: Duke University Press.

Mantius, Peter. 1990. Miller Reverses Stand on Taylor Hazardous Waste Site. *Atlanta Constitution*, April 20, D1.

Mapping the Challenges. 1997. *Voting Rights Review* (Summer): 10–11.

Marks, Jack. 1955. Senatobia Meeting Boosts Segregation. *Commercial Appeal*, August 14, I7.

Marshall, Burke. 1964. *Federalism and Civil Rights*. New York: Columbia University Press.

Martin, John Bartlow. 1957. *The Deep South Says "Never"*. New York: Ballantine Books.

Massengil, Reed. 1994. *Portrait of a Racist: The Man Who Killed Medgar Evers*. New York: St. Martin's.

Matthews, Donald R., and James W. Prothro. 1963. Social and Economic Factors and Negro Voting Registration in the South. *American Political Science Review* 57:24–44.

MDC. 1986. *Shadows in the Sunbelt*. Chapel Hill, N.C.: MDC.

Meier, August, and Elliott Rudwick. 1973. *CORE: A Study in the Civil Rights Movement, 1942–1968*. New York: Oxford University Press.

Meinig, D. W. 1986. *The Shaping of America: A Geographical Perspective on Five Hundred Years of History*. Vol. 1, *Atlanta America, 1492–1800*. New Haven: Yale University Press.

Meltzer, Milton. 1978. *Dorothea Lange: A Photographer's Life*. New York: Farrar, Straus & Giroux.

Mills, Kay. 1993. *This Little Light of Mine: The Life of Fannie Lou Hamer*. New York: Dutton.

Minimum Wage Law Nears. 1967. *Delta Farm Press*, January 5, 1, 14.

Mississippi Crop and Livestock Reporting Service. 1955. *Base Book of Mississippi Agriculture*. Jackson, Miss.: Mississippi Crop and Livestock Reporting Service.

Mississippi Department of Education. 1931. *Twenty Years of Progress, 1910–1930, and A Biennial Survey Scholastic Years 1929–30 and 1930–31 of Public Education in Mississippi*. Bulletin 67. Jackson, Miss.: Mississippi Department of Education.

Mississippi State Tax Commission. 1970, 1986, 1990. *Service Bulletins*. Jackson, Miss.: Mississippi State Tax Commission.

Mitchell, Broadus. 1921. *The Rise of the Cotton Mills in the South*. Baltimore: Johns Hopkins University Press.

Mitchell, H. L. 1979. *Mean Things Happening in This Land: The Life and Times of H. L. Mitchell, Co-Founder of the Southern Tenant Farmers Union*. Montclair, N.J.: Allanheld, Osmun.

Mitchell, Margaret. 1936. *Gone with the Wind*. New York: Macmillan.

Moody, Anne. 1968. *Coming of Age in Mississippi*. New York: Dial Press.

Moore, Charles. 1959. Hancock Gets State Aid: Small County Builds Big Schools. *Atlanta Journal/Atlanta Constitution*, August 23.

Moore, Gary. 1992. The Nation's Waste Dump? Alabama Bears Influx of Toxins. *Atlanta Constitution*, June 1, A2.

Moorhouse, L. A., and M. R. Cooper. 1920. *The Cost of Producing Cotton*. U.S. Department of Agriculture. Bulletin 896. Washington, D.C.: GPO.

Morgan, Wynette. 1988. The Porter Family. In *History of Newtown County, Georgia*, 880–82. Covington: Newton County Historical Society.

Morison, Samuel E., and Henry Steele Commager. 1930. *The Growth of the American Republic*. New York: Oxford University Press.

Morrill, Richard L. 1965. The Negro Ghetto: Problems and Alternatives. *Geographical Review* 45 (July): 340–61.

Morris, Aldon D. 1984. *The Origin of the Civil Rights Movement: Black Communities Organizing for Change*. New York: Free Press.

Mosley, Donald C., and D. C. Williams Jr. 1967. *An Analysis and Evaluation of a Community Action Anti-Poverty Program in the Mississippi Delta*. State College, Miss.: Mississippi State University, College of Business and Industry.

Moulton, E. S. 1931. *Cotton Production and Distribution in the Gulf Southwest*. U.S. Department of Commerce, Domestic Commerce Series. No. 49. Washington, D.C.: GPO.

Moynihan, Daniel Patrick. 1988. Foreword to reprint of *Caste and Class in a Southern Town*, by John Dollard. Madison: University of Wisconsin Press. Originally published 1937; New Haven: Yale University Press.

Munford, Luther. 1973. White Flight from Desegregation in Mississippi. *Integrated Education: Minority Children in Schools* 11 (May–June): 12–26.

Myrdal, Gunnar. 1944. *An American Dilemma: The Negro Problem in America*. New York: Harper & Row.

National Center for Political and Economic Studies. 1991. *Black Elected Officials: A National Roster, 1990*. Washington, D.C.: National Center for Political and Economic Studies.

National Cotton Council. 1947. *Mechanization of the Cotton Belt*. Memphis, Tenn.: National Cotton Council.

———. [1948] *This Is Cotton*. Memphis Tenn.: National Cotton Council.

Negroes Take Over in Hancock Voting. 1968. *Atlanta Constitution*, November 8, 1.

The Negro Exodus. 1922. *Madison (Ga.) Madisonian*, December 22, 6.

Nevin, David, and Robert E. Bills. 1976. *The Schools That Fear Built: Segregationist Academies in the South*. Washington, D.C.: Acropolis Books.

Newberry, Paul. 1989. Dallas Commission: New Color, Old Problems. *Tuscaloosa News*, June 5, 5.

New Greenland Has Gala Opening. 1941. *Greensboro (Ga.) Herald-Journal*, January 24, 1.

Newman, Roger K. 1994. *Hugo Black: A Biography*. New York: Pantheon Books.

Newton, Roy L., and James M. Workman. 1919. Cotton Warehousing—Benefits of an Adequate System. *Yearbook of the United States Department of Agriculture, 1918*. Washington, D.C.: GPO.

Nixon, H. C. 1930. The Rise of the American Cottonseed Oil Industry. *Journal of Political Economy* 38 (February): 74–75.

———. 1938. *Forty Acres and Steel Mules*. Chapel Hill: University of North Carolina Press.

Nixon, Robert L. 1915. *Cotton Warehouses: Storage Facilities Now Available in the South*. U.S. Department of Agriculture. Bulletin 216. Washington, D.C.: GPO.

Nordheimer, Jon. 1969. South Learning to Live with Desegregation. *New York Times*, November 10, L1, L39.

Norrell, Robert J. 1986. *Reaping the Whirlwind: The Civil Rights Movement in Tuskegee*. New York: Vintage Books.

Nossiter, Adam. 1988. Black Parents in Alabama Trying to Preserve Vestige of Segregation: School That Isolated Them Now Seen as Potent Symbol. *Atlanta Journal/Atlanta Constitution*, December 25, A10.

———. 1990. Selma's Mayor Walking a Fine Line. *Atlanta Journal/Atlanta Constitution*, May 27, H1, H3.

———. 1991. Alabama School Fights for Its Racial Mix. *Atlanta Journal/Atlanta Constitution*, June 30, A3.

Oates, Stephen B. 1982. *Let the Trumpet Sound: The Life of Martin Luther King, Jr.* New York: Harper & Row.

Odum, Howard W. 1936. *Southern Regions of the United States*. Chapel Hill: University of North Carolina Press.

Olmsted, Frederick Law. 1856. *A Journey in the Seaboard Slave States*. New York: Dix & Edwards.

———. 1860. *A Journey in the Back Country*. New York: Burt Franklin.

Orfield, Gary. 1969. *The Reconstruction of Southern Education: The Schools and the 1964 Civil Rights Act*. New York: Wiley-Interscience.

Palmer, Robert. 1981. *Deep Blues*. New York: Viking Press.

Parker, Frank R. 1990. *Black Votes Count: Political Empowerment in Mississippi after 1965*. Chapel Hill: University of North Carolina Press.

Peake, Thomas R. 1987. *Keeping the Dream Alive: A History of the Southern Christian Leadership Conference from King to the Nineteen-Eighties*. New York: Peter Lang.

Pedersen, Harald A., and Arthur F. Raper. 1954. *The Cotton Plantation in Transition*. Bulletin 508. State College: Mississippi State College, Agricultural Experiment Station.

Pennington, Maryland, and Thomas Russell Butchko. 1978. Philomath [Georgia] Historic District. U.S. Department of the Interior, National Park Service. National Register of Historic Places Inventory—Nomination Form. Georgia Department of Natural Resources, Historic Preservation Section, Atlanta.

Percy, William Alexander. 1941. *Lanterns on the Levee: Recollections of a Planter's Son*. New York: Alfred A. Knopf.

Perry, George Sessions. 1941. *Hold Autumn in Your Hand*. New York: Viking. 1950 reprint. New York: Whittlesey House.

Petcotte, Michael. 1976. Buffalo Island: A Cultural-Agricultural Island in the St. Francis Basin of Arkansas. M.S. thesis, Department of Geography, University of Tennessee, Knoxville.

Peters, Donald C. 1992. *The Democratic System in the Eastern Caribbean*. New York: Greenwood Press.

Phillips, Ulrich Bonnell. 1903. The Economics of the Plantation. *South Atlantic Quarterly* 2 (July): 231–36.

———. 1904. The Plantation as a Civilizing Factor. *Sewanee Review* 12:257–67.

———. 1906. The Origin and Growth of the Southern Black Belts. *American Historical Review* 11 (July): 798–816.

———. 1908. *A History of Transportation of the Eastern Cotton Belt to 1860*. New York: Columbia University Press.

———. 1910. The Decadence of the Plantation System. *Annals of the American Academy of Political and Social Science* 35 (January–June): 37–41.

———. 1918. *American Negro Slavery: A Survey of the Supply, Employment, and Control of Negro Labor as Determined by the Plantation Régime*. New York: D. Appleton.

———. 1925. Plantations with Slave Labor and Free. *American Historical Review* 30 (July): 738–53.

———. 1928. The Central Theme in Southern History. *American Historical Review* 24 (October): 30–43.

———. 1929. *Life and Labor in the Old South*. Boston: Little, Brown, and Co.

Piven, Frances Fox, and Richard A. Cloward. 1977. *Poor People's Movements: Why They Succeed, How They Fail*. New York: Pantheon Books.

Pollard, David. 1966a. Meredith Plans to Return Soon from New York. *Commercial Appeal*, June 9, 1.

———. 1966b. King Leads Nine-Mile Trek, Leaves to Plan March. *Commercial Appeal*, June 10, 8.

Pomerantz, Gary. 1996a. *Where Peachtree Meets Sweet Auburn*. New York: Scribner.

———. 1996b. Streets to Success: The Two Atlantas—Black and White. *Atlanta Journal/Atlanta Constitution*, April 21, M1.

Pool Area Ransacked by Negroes in Eutaw. 1969. *Birmingham Post-Herald*, August 12, 1–2.

Porter, Joanell, and Roy C. Nehrt. 1977. *Nonpublic School Statistics, 1976–1977*. Advance Report. Washington, D.C.: U.S. Department of Health, Education, and Welfare.

Porter, Walter K., Jr. 1962. Ten Years of Weed Control in Cotton. In *Summary-Proceedings 1962 Beltwide Cotton Production—Mechanization Conference*, 6–8. Memphis, Tenn.: National Cotton Council.

Powdermaker, Hortense. 1939. *After Freedom: A Cultural Study in the Deep South*. New York: Viking Press.

———. 1966. *Stranger and Friend: The Way of an Anthropologist*. New York: W. W. Norton.

Powledge, Fred. 1964. Dr. King to Renew Southern Protests. *New York Times*, November 5, 1, 33.

Prattville Chamber of Commerce. 1991. *Prattville the Birthplace of Industry in Alabama: 1991 Membership Directory and Newcomer's Guide*. Prattville, Ala.: Prattville Chamber of Commerce.

Pred, Allan. 1963. Business Thoroughfares as Expressions of Urban Negro Culture. *Economic Geography* 39 (July): 217–33.

Price, Margaret. 1959. *The Negro and the Ballot in the South*. Atlanta: Southern Regional Council.

Program for Pickers. 1936. *Time*, March 23, 60.

Prunty, Merle C., Jr. 1955. The Renaissance of the Southern Plantation. *Geographical Review* 45 (October): 459–91.

———. 1962. Deltapine: Field Laboratory for the Neoplantation Occupance Type. In *Festschrift: Clarence F. Jones*, edited by Merle C. Prunty Jr., 151–72. Evanston, Ill., Northwestern University, Department of Geography. Studies in Geography No. 6. Evanston: Department of Geography, Northwestern University.

Prunty, Merle C., Jr., and Charles S. Aiken. 1972. The Demise of the Piedmont Cotton Region. *Annals of the Association of American Geographers* 62 (June): 283–306.

Pyron, Darden Asbury. 1991. *Southern Daughter: The Life of Margaret Mitchell*. New York: Oxford University Press.

Raines, Howell. 1983. *My Soul Is Rested: Movement Days in the Deep South Remembered*. New York: Viking Penguin.

Range, Willard. 1954. *A Century of Georgia Agriculture, 1850–1950*. Athens: University of Georgia Press.

Ransom, Roger L., and Richard Sutch. 1977. *One Kind of Freedom: The Economic Consequences of Emancipation*. New York: Cambridge University Press.

Raper, Arthur F. 1936. *Preface to Peasantry: A Tale of Two Black Belt Counties*. Chapel Hill: University of North Carolina Press.

———. 1943. *Tenants of the Almighty*. New York: Macmillan.

Raper, Arthur F., and Ira De A. Reid. 1941. *Sharecroppers All*. Chapel Hill: University of North Carolina Press.

Read, Frank T. 1975. Judicial Evolution of the Law of School Integration since *Brown v. Board of Education*. *Law and Contemporary Problems* 39 (Winter): 7–49.

Rector, Kyle T. 1995. *Shaw v. Reno* and North Carolina's Twelfth Congressional District: Testing the Constitutionality of a Majority-Minority District. M.S. thesis, Department of Geography, University of Tennessee, Knoxville.

Reed, Roy. 1965a. Alabama Police Use Gas and Clubs to Rout Negroes. *New York Times*, March 8, 1, 20.

———. 1965b. Freedom March Begins at Selma; Troops on Guard. *New York Times*, March 22, 1, 26.

———. 1965c. Dr. King Cheered. *New York Times*, March 26, 1, 22.

Rehder, John. 1978. Diagnostic Landscape Traits of Sugar Plantations in Southern Louisiana. In *Man and Environment in the Lower Mississippi River Valley*, vol. 19 of *Geoscience and Man*, edited by Sam B. Hilliard, 135–50. Baton Rouge: School of Geoscience, Louisiana State University.

Reid, Joseph D. 1973. Sharecropping as an Understandable Market Response: The Post-Bellum South. *Journal of Economic History* 33 (March): 106–30.

Reid, Whitelaw. 1965. *After the War: A Tour of the Southern States, 1865–1866*. New York: Harper and Row. Reprint.

Report to the South. 1947. *Dixie: The Magazine of Southern Progress* (Winter 1947–48): 4–27.

Rhodes, Carroll. 1987. Enforcing the Voting Rights Act in Mississippi through Litigation. *Mississippi Law Journal* 57 (August): 705–37.

Rice, Bradley Robert. 1971. *Progressive Cities: The Commission Government in America, 1901–1920*. Austin: University of Texas Press.

Rice, Roger L. 1968. Residential Segregation by Law, 1910–1917. *Journal of Southern History* 24 (May): 179–98.

Roberts, Gene. 1966a. Delta Area of Mississippi in Turmoil. *New York Times*, February 7, 1, 23.

———. 1966b. Mississippi Reduces Police Protection for Marchers. *New York Times*, June 17, 1, 33.

Roper, Daniel Clifford. 1903. Census Office Cotton Report and the Significant Development of the Cottonseed-Oil Industry. *South Atlantic Quarterly* 2 (July): 239–40.

Roper, John Herbert. 1984. *U. B. Phillips: A Southern Mind*. Macon, Ga.: Mercer University Press.

Rose, Harold M. 1965. The All-Negro Town: Its Evolution and Function. *Geographical Review* 55 (July): 362–81.

———. 1970. The Development of an Urban Sub-system: The Case of the Negro Ghetto. *Annals of the Association of American Geographers* 60 (December): 1–17.

———. 1971. *The Black Ghetto: A Spatial Behavior Perspective*. New York: McGraw-Hill.

———, ed. 1972. *Geography of the Ghetto: Perception, Problems, and Alternatives*. Perspectives in Geography No. 2. DeKalb, Ill.: Northern Illinois University Press.

Rosengarten, Theodore. 1989. *All God's Dangers: The Life of Nate Shaw*. New York: Vintage Books.

Rowan, Richard L. 1970. The Negro in the Textile Industry. In *Negro Employment in Southern Industry*. Vol. 4, *Studies of Negro Employment*, by Herbert R. Northrap, and Richard L. Rowan. Philadelphia: Wharton School of Finance and Commerce, University of Pennsylvania.

Rozier, John. 1982. *Black Boss: Political Revolution in a Georgia County*. Athens: University of Georgia Press.

———, ed. 1988. *The Granite Farm Letters: The Civil War Correspondence of Edgeworth and Sallie Bird*. Athens: University of Georgia Press.

Rubin, Morton. 1951. *Plantation County*. Chapel Hill: University of North Carolina Press.

Rudwick, Elliott M. 1968. Preface to the 2d ed. of *After Freedom: A Cultural Study in the Deep South*, by Hortense Powdermaker. New York: Russell & Russell.

Russell, James M. 1982. Elites and Municipal Politics and Government in Atlanta, 1847–1990. In *Toward a New South?: Studies in Post–Civil War Southern Communities*, edited by Orville Vernon Burton and Robert C. McMath Jr., 37–70. Westport, Conn.: Greenwood Press.

———. 1988. *Atlanta, 1847–1890: City Building in the Old South and the New*. Baton Rouge: Louisiana State University Press.

Russell, John. 1913. *The Free Negro in Virginia, 1619–1865*. Baltimore: Johns Hopkins University Press.

Russell, Lester F. 1977. *Profile of a Black Heritage*. Franklin Square, N.Y.: Graphicopy.

Salamon, Lester M. 1973. Leadership and Modernization: The Emerging Black Political Elite in the American South. *Journal of Politics* 35 (August): 615–46.

Sauer, Carl. 1941. Forward to Historical Geography. *Annals of the Association of American Geographers* 31 (March): 1–24.

Scarborough, Dorothy. 1923. *In the Land of Cotton*. New York: Macmillan.

Scher, Richard, and James Button. 1984. Voting Rights Act: Implementation and Impact. In *Implementation of Civil Rights Policy*, edited by Charles S. Bullock III and Charles M. Lamb, 20–53. Monterey, Calif.: Brooks/Cole.

Schlesinger, Arthur M., Jr. 1978. *Robert Kennedy and His Times*. Boston: Houghton Mifflin.

Scott, Emmett J. 1920. *Negro Migration during the War*. New York: Oxford University Press.

"Secret" Crisis in the Delta. 1966. *Newsweek*, March 7, 28–29.

Sellers, Cleveland. 1973. *The River of No Return: The Autobiography of a Black Militant and the Life and Death of SNCC*. New York: William Morrow.

Shaw, Lucille Ivey. 1988. Porterdale. In *History of Newton County, Georgia*, 419–25. Covington: Newtown County Historical Society.

Shlomowitz, Ralph. 1982. The Squad System on Postbellum Cotton Plantations. In *Toward a New South?: Studies in Post–Civil War Communities*, edited by Orville Vernon Burton and Robert C. McMath Jr., 265–80. Westport, Conn.: Greenwood Press.

Shugg, Roger W. 1939. *Origins of Class Struggle in Louisiana: A Social History of White Farmers and Laborers during Slavery and After, 1840–1875*. Baton Rouge: Louisiana State University Press.

Sikora, Frank. 1969. Greene Negro Leaders Rap Abernathy-led Pool March. *Birmingham News*, August 12, 1–2.

Siler, Julia. 1991. "Environmental Racism": It Could Be a Messy Fight. *Newsweek*, May 20, 116.

Silk, Mark. 1989. Even Today, *Gone with the Wind* Evokes Atlanta's Double Image. *Atlanta Journal/Atlanta Constitution*, December 17, H1-H2.

Silver, James. W. 1964. *Mississippi: The Closed Society*. New York: Harcourt, Brace & World.

Sinclair, Ward. 1986. Black Farmers: A Dying Minority. *Washington Post*, Feb. 18, A1, A4.

Singal, Daniel Joseph. 1982. *The War Within: From Victorian to Modernist Thought in the South, 1919–1945*. Chapel Hill: University of North Carolina Press.

Skinner, Frank. 1976. Paul Bryant, Jr.'s Group Wins Dog Track Franchise. *Birmingham Post-Herald*, December 20, A1.

Smith, Algernon L. 1952. *Continental Gin Company and Its Fifty-Two Years of Service*. Birmingham, Ala.: Birmingham Publishing Co.

Smith, Frank E. 1964. *Congressman from Mississippi*. New York: Pantheon Books.

Smith, Gita M. 1996. Out of the Money: Alabama County Isn't Reaping the Prosperity It Expected along with Its Dog Track. *Atlanta Journal/Atlanta Constitution*, April 14, D12.

Smith, J. Russell, and M. Ogden Phillips. 1925. *North America: Its People and the Resources, Development, and Prospects of the Continent as the Home of Man*. New York: Harcourt Brace.

Smith, John David, and John C. Inscoe. 1990. *Ulrich Bonnell Phillips: A Southern Historian and His Critics*. New York: Greenwood Press.

Smith, T. Lynn. 1974. *Studies of the Great Rural Tap Roots of Urban Poverty in the United States*. New York: Carlton Press.

Smith, Vern E. 1990. Old Enemies, New Questions: Fresh Leads Reopen the Medgar Evers Case. *Newsweek*, July 23, 24.

Somers, Robert. 1871. *The Southern States since the War, 1870–1871*. New York: Macmillan.

Sonnichsen, C. L. 1969. The Sharecropper Novel in the Southwest. *Agricultural History* 43 (April): 249–58.

Sorensen, Theodore C. 1965. *Kennedy*. New York: Harper & Row.

Soule, Andrew M. 1920. Fighting the Boll Weevil. *Georgia Agricultural Quarterly* 13 (January): 51–52.

A South Carolinian. 1877. South Carolina Society. *Atlantic Monthly* 39 (June): 670–84.

South Enters New Era. 1960. *Southern Patriot* 18 (April): 1, 4.

Southern Education Reporting Service. 1961–67. *A Statistical Summary, State by State of Segregation-Desegregation Activity Affecting Southern Schools from 1954 to Present, Together with Pertinent Data on Enrollment, Teachers, Colleges, Litigation, and Legislation*. Nashville, Tenn.: Southern Education Reporting Service.

Southern Governors' Association. 1986. *Cornerstone of Competition*. Washington, D.C.: Southern Governor's Association.

Southern Growth Policies Board. 1986. *Halfway Home and a Long Way to Go*. Research Triangle, N.C.: Southern Growth Policies Board.

Southern Regional Council. 1963. Public Education in Mississippi. Report L-45 (August 19). Atlanta. 6 pp. Mimeograph.

Southern Regional Council, Voter Education Project. 1966. *Voter Registration in the South, Summer 1966*. Atlanta: Southern Regional Council.

The South's Poor. 1967. *Commercial Appeal*, October 9, 6.

Spillman, W. J. 1919. The Agricultural Ladder. *American Economic Review Supplement* 9 (March): 170–79.

Stampp, Kenneth M. 1956. *The Peculiar Institution: Slavery in the Ante-Bellum South*. New York: Alfred A. Knopf.

Steinbeck, John. 1939. *The Grapes of Wrath*. New York: Viking Press.

Stephenson, Wendell Holmes. 1938. *Isaac Franklin: Slave Trader and Planter of the Old South*. Baton Rouge: Louisiana State University Press.

Stevens, Carol. 1964. Selma: A Lonely Outpost. *Southern Patriot* 22 (October): 1.

Stewart, Joseph, Jr., and James F. Sheffield Jr. 1987. Does Interest Group Litigation Matter? The Case of Black Political Mobilization in Mississippi. *Journal of Politics* 49 (August): 780–98.

Stone, Alfred Holt. 1908. *Studies in the American Race Problem*. New York: Doubleday, Page.

Stoney, George C. 1941. No Room in Green Pastures. *Survey Graphic* 30 (January): 14–20.

Stott, William. 1973. *Documentary Expression and Thirties America*. New York: Oxford University Press.

Street, James H. 1957. *The New Revolution in the Cotton Economy: Mechanization and Its Consequences.* Chapel Hill: University of North Carolina Press.

Street, William B. 1966. Sniper Halts Meredith with Shotgun Blasts. *Commercial Appeal*, June 7, 1, 4.

Stuart, Reginald. 1982. Segregated Academies Look to Congress for Tax Relief. *New York Times*, February 2, A18.

———. 1983. Businesses Said to Have Barred New Plants in Largely Black Communities. *New York Times*, February 15, A14.

Student Nonviolent Coordinating Committee. 1963. Operation Mississippi (September 13). SCLC Papers, subgroup E, ser. I, subser. 3, box 141.

Sutherland, Elizabeth, ed. 1965. *Letters from Mississippi.* New York: McGraw-Hill.

Taylor, Paul S. 1938a. Power Farming and Labor Displacement in the Cotton Belt, 1937. Part 1, Northwest Texas. *Monthly Labor Review* 46 (March): 595–607.

———. 1938b. Power Farming and Labor Displacement. Part 2, Southwestern Oklahoma and Mississippi Delta. *Monthly Labor Review* 46 (April): 852–67.

Taylor, Paul S. 1954a. Plantation Agriculture in the United States: Seventeenth to Twentieth Centuries. *Land Economics* 30 (May): 141–52.

———. 1954b. Plantation Labor before the Civil War. *Agricultural History* 30 (January): 1–21.

Thatcher, Gary. 1977. Dixie's Boom: "Good-Life" Incentives Lure Industry South. *Atlanta Journal/Atlanta Constitution*, March 6, 1A, 16A.

This Is Still God's Country. 1922. *Madison (Ga.) Madisonian*, December 22, 6.

Thomas, Norman. 1934. *The Plight of the Share-Cropper.* New York: League for Industrial Democracy.

Thompson, Edgar T. 1940. The Natural History of Agricultural Labor in the South. In *Plantation Societies, Race Relations, and the South: The Regimentation of Populations: Selected Papers of Edgar T. Thompson.* 1975. Durham, N.C.: Duke University Press.

Thompson, Tracy. 1988. Feds, State Take Opposite Sides in Meriwether School Merger Fracas. *Atlanta Journal/Atlanta Constitution*, July 17, E1, E10.

Tilly, Charles. 1968. Race and Migration to the American City. In *The Metropolitan Enigma: Inquiries into the Nature and Dimensions of America's "Urban Crisis,"* edited by James Q. Wilson. Cambridge: Harvard University Press.

Tompkins, Daniel A. 1899. *Cotton Mill, Commercial Features: A Textbook for the Use of Textile Schools and Investors.* Charlotte, N.C.: Published by the author.

———. 1901. *Cotton and Cotton Oil.* 2 vols. Charlotte, N.C.: Published by the author.

———. 1903. *History of Mecklenburg County and the City of Charlotte.* 2 vols. Charlotte, N.C.: Observer Printing House.

Tompkins, Stephen G. 1993. Army Spied on Blacks for Generations. *Commercial Appeal*, March 21, 1.

Trovaioli, August P., and Roulhac B. Toledano. 1972. *William Aiken Walker: Southern Genre Painter.* Baton Rouge: Louisiana State University Press.

Tufs, R. B. 1897. Map of Morgan County, Georgia. ¾″ to one mile. Atlanta: published privately.

Tumin, Melvin M., et al. 1958. *Desegregation: Resistance and Readiness*. Princeton: Princeton University Press.

Turque, Bill, and Howard Manly. 1990. Selma's Unfinished Business. *Newsweek*, February 26, 44.

Twelve Southerners. 1930. *I'll Take My Stand: The South and the Agrarian Tradition*. New York: Harper.

U.S. Bureau of the Census. 1901–90. *Cotton Production in the United States* (title varies). Washington, D.C.: GPO.

———. 1901. *Twelfth Census of the United States Taken in the Year 1900*. Vol. 1, pt. 1. *Population*. Washington, D.C.: GPO.

———. 1902. *Twelfth Census of the United States Taken in the Year 1900*. Vol. 3, *Agriculture*. Pt. 1, *The Southern States*. Washington, D.C.: GPO.

———. 1906. *Cotton Production and Statistics of Cottonseed Products: 1905*. Bulletin 40. Washington, D.C.: GPO.

———. 1912. *Thirteenth Census of the United States Taken in the Year 1910*. Vol. 1, *Reports by States with Statistics for Counties, Cities, and Other Civil Divisions: Alabama-Montana*. Washington, D.C.: GPO.

———. 1913a. *Thirteenth Census of the United States Taken in the Year 1910*. Vol. 6, *Agriculture 1909 and 1910: Alabama-Montana*. Washington, D.C.: GPO.

———. 1913b. *Thirteenth Census of the United States Taken in the Year 1910*. Vol. 8, *Manufactures 1909, General Report and Analysis*. Washington, D.C.: GPO.

———. 1914. *Thirteenth Census of the United States Taken in the Year 1910*. Vol. 5, *Agriculture 1909 and 1910, General Report and Analysis*. Washington, D.C.: GPO.

———. 1916. *Plantation Farming in the United States*. Washington, D.C.: GPO.

———. 1932a. *Fifteenth Census of the United States: 1930*. Vol. 2, *Population*. Pt. 1, *Reports by States, Alabama-Missouri*. Washington, D.C.: GPO.

———. 1932b. *Fifteenth Census of the United States: 1930*. Vol. 2, *Agriculture*. Pt. 2, *The Southern States*.

———. 1943. *Sixteenth Census of the United States: 1940. Characteristics of the Population*. Pt. 1. *United States Summary and Alabama–District of Columbia*. Washington, D.C.: GPO.

———. 1946a. *Cotton Ginning Machinery and Equipment in the United States, 1945*. Washington, D.C.: GPO.

———. 1946b. *United States Census of Agriculture: 1945*. Vol. 1, pt. 22, *Statistics for Counties, Mississippi*. Washington, D.C.: GPO.

———. 1947. *Census of Agriculture, 1945. General Report, Statistics by Subjects, Color, and Tenure of Farm Operator*. Vol. 2, chap. 3. Washington, D.C.: GPO.

———. [1948]. *Special Study: Plantations Based upon Tabulations from the Sixteenth Census of the United States, 1940*. Washington, D.C.: GPO.

———. 1952. *Census of Population, 1950*. Vol. 2, pt. 2, *Characteristics of the Population. Alabama*. Washington, D.C.: GPO.

———. 1962. *U.S. Census of Agriculture, 1959.* Vol. 2, chap. 10, *Statistics by Subjects, Color, Race, and Tenure of Farm Operators.* Washington, D.C.: GPO.

———. 1963a. *Census of Population, 1960.* Vol. 1, *Characteristics of the Population.* Pts. 2, 12, 26, *Alabama, Georgia, Mississippi.* Washington, D.C.: GPO.

———. 1963b. *United States Census of Housing, 1960.* Vol. 1, pt. 2, *States and Small Areas, Michigan–New Hampshire.* Washington, D.C.: GPO.

———. 1967. *1964 United States Census of Agriculture.* Vol. 1, Part 33. *Mississippi.* Washington, D. C.: GPO.

———. 1975. *Historical Statistics of the United States from Colonial Times to 1970.* Washington, D.C.: GPO.

———. 1981. *Census of Population and Housing, 1980. Advance Reports, Final Population and Housing Counts. Alabama, Georgia, Mississippi.* Washington, D.C.: GPO.

———. 1982. *1980 Census of Population and Housing, User's Guide, Glossary.* Pt. B. Washington, D.C.: GPO.

———. 1983a. *Census of Housing, 1980. Characteristics of Housing Units, Details of Housing Units.* Chap. B, pt. 12, *Georgia*; pt. 26, *Mississippi.* Washington, D.C.: GPO.

———. 1983b. *Census of Population, 1980.* Vol. 1, *General Population Characteristics.* Pt. 2, *Alabama*; pt. 12, *Georgia*; pt. 26, *Mississippi.* Washington, D.C.: GPO.

———. 1983c. *Census of Population, 1980.* Vol. 1, chap. C, pt. 26, *Characteristics of the Population, General Social and Economic Characteristics. Mississippi.* Washington, D.C.: GPO.

———. 1991. *1990 Census of Population and Housing.* 1990 CPH-1–26. *Summary, Population and Housing Characteristics, Mississippi.* Washington, D.C.: GPO.

———. 1992a. *1990 Census of Population and Housing. Summary, Social, Economic, and Housing Characteristics of the United States.* Washington, D.C.: GPO.

———. 1992b. *U.S. Statistical Abstract.* Washington, D.C.: GPO.

———. 1993. *1990 Census of Population. Social and Economic Characteristics, Alabama.* 1990 CP-2-2. 1990 CP-2-12; CP-2-26. Washington, D.C.: GPO.

U.S. Census Office. 1864. *Population of the United States in 1860.* Washington, D.C. GPO.

———. 1865. *Manufactures of the United States in 1860.* Washington, D.C.: GPO.

———. 1884. *Special Report on Cotton Production in the United States in 1880.* 2 pts. Washington, D.C.: GPO.

U.S. Commission on Civil Rights. 1959. *Report of the United States Commission on Civil Rights, 1959.* Washington, D.C.: GPO.

———. 1964a. *1963 Staff Report, Public Education.* Washington, D.C.: GPO.

———. 1964b. *1964 Staff Report, Public Education.* Washington, D.C.: GPO.

———. 1966. *Survey of School Desegregation in the Southern and Border States, 1965–1966.* Washington, D.C.: GPO.

———. 1967. *Southern School Desegregation, 1966–1967.* Washington, D.C.: GPO.

———. 1968a. *Cycle to Nowhere.* Clearinghouse Publication 14. Washington, D.C.: GPO.

———. 1968b. *Political Participation: A Study of the Participation by Negroes in the Electoral and Political Processes in Ten Southern States since Passage of the Voting Rights Act of 1965*. Washington, D.C.: GPO.

———. 1983. *Fifteen Years Ago . . . Rural Alabama Revisited*. Clearinghouse Publication 82. Washington, D.C.: GPO.

U.S. Comptroller General. 1969. *Review of Economic Opportunity Programs Pursuant to Title II of the 1967 Amendments to the Economic Opportunity Act of 1964.* U.S. Congress. Joint Senate Committee on Labor and Public Welfare and the House Coimmittee on Education and Labor. 91st Cong., 1st sess. Washington, D.C.: GPO.

U.S. Department of Agriculture. Office of Farm Management. 1918. 1919. *Atlas of American Agriculture*. Pt. 5, *The Crops*. Sec. A, *Cotton*. Pt. 9, *Rural Population and Organizations*. Sec. 1, *Rural Population*. Washington, D.C.: GPO.

U.S. Department of Agriculture. 1922. *Yearbook, 1921*. Washington, D.C.: GPO.

———. 1930. *Yearbook of Agriculture, 1930*. Washington, D.C.: GPO.

U.S. Department of Agriculture. Farmers Home Administration. 1980. *This is FmHA*. Washington, D.C.: FmHA.

———. 1982. *FmHA Instruction, Rural Rental Housing Loan Policies, Procedures, and Authorizations*. Title 7, chap. 18, pt. 1944, subpt. E. Washington, D.C.: FmHA.

———. 1985. 7CFR pts. 1804 and 1924, Revision and Redesignation of pt. 1804, subpt. D. Planning and Performing Site Development Work. *Federal Register*. Vol. 50, 815–34.

———. 1986. Bolivar, Humphreys, Sunflower, and Tunica Counties, Mississippi Farmers Home Administration offices. Subdivision records.

U.S. Department of Commerce. Bureau of Economic Analysis. 1991. *Local Area Personal Income, 1984–1989*. Vol. 4, *Southeast Region*. Washington, D.C.: GPO.

U.S. Department of Housing and Urban Development. Jackson, Mississippi, Area Office. 1986. Directory of Public Housing in Mississippi.

U.S. Department of Justice. 1958, 1959, 1960, 1961, 1962, 1963, 1964, 1965, 1966, 1967, 1968. *Annual Report of the Attorney General of the United States*. Washington, D.C.: U.S. Department of Justice.

U.S. Executive Office of the President. 1937. President's Committee on Farm Tenancy. *Farm Tenancy: Report of the President's Committee*. Washington, D.C.: GPO.

U.S. Executive Office of the President. Commodity Services Administration. 1968, 1969, 1970, 1971, 1972, 1973, 1974, 1975, 1976. *Federal Outlays in Mississippi*. Springfield, Va.: National Technical Information Service.

———. 1977, 1978, 1979, 1980, 1981. *Geographical Distribution of Federal Funds in Mississippi*. Springfield, Va.: National Technical Information Service.

U.S. Geological Survey. 1977. Helena, Arkansas, Mississippi, and Tennessee. 1:250,000. Washington, D.C. Rev. ed.

U.S. House. 1941. Select Committee to Investigate the Interstate Migration of Destitute Citizens. *Interstate Migration Report*. 77th Cong., 1st sess., no. 369.

———. 1967. Committee on Un-American Activities. *The Present-Day Ku Klux Klan Movement.* 90th Cong., 1st sess.

U.S. Senate. 1967. Committee on Labor and Public Welfare, Subcommittee on Employment, Manpower, and Poverty. *Examination of the War on Poverty: Staff and Consultants Reports.* Vol. 4, *Community Action Program: Interpretation and Analysis.* Vol. 8, *Community Action Program: State Technical Assistance Agencies.* 90th Cong., 1st Sess.

———. 1976. Select Committee to Study Governmental Operations with Respect to Intelligence Activities. *Hearings.* Vol. 6, *Federal Bureau of Investigation.* 94th Cong, 1st sess, November 18–19; December 2–3, 1975.

Vance, Rupert B. 1929. *Human Factors in Cotton Culture: A Study in the Social Geography of the American South.* Chapel Hill: University of North Carolina Press.

———. 1935. *Human Geography of the South: A Study in Regional Resources and Human Adequacy.* 2d ed. Chapel Hill: University of North Carolina Press.

Van Deburg, William L. 1992. *New Day in Babylon: The Black Power Movement and American Culture, 1965–1975.* Chicago: University of Chicago Press.

Vesey, Susannah. 1993. A Literary Life: Andrew Lyle Mixed His Love of Worlds and Faith in the Land to Become a Modern Southern Classic. *Atlanta Journal/Atlanta Constitution,* February 28, M1, M4.

Vollers, Maryanne. 1995. *Ghosts of Mississippi: The Murder of Medgar Evers, the Trials of Byron De La Beckwith, and the Haunting of the New South.* Boston: Little, Brown.

Wagley, Charles. 1960. Plantation America: A Culture Sphere. In *Caribbean Studies: A Symposium,* edited by Vera Rubin, 3–13. Seattle: University of Washington Press.

Walker, Tom. 1989. Storied Sun Belt Prosperity Has Bypassed Much of the Rural South. *Atlanta Journal/Atlanta Constitution,* March 12, 4S.

Washington, Booker T. 1928. *Up from Slavery: An Autobiography.* Boston: Houghton Mifflin.

Watters, Pat, and Reese Cleghorn. 1967. *Climbing Jacob's Ladder: The Arrival of Negroes in Southern Politics.* New York: Harcourt, Brace & World.

Wayne, Michael. 1983. *The Reshaping of Plantation Society: The Natchez District, 1860–1880.* Baton Rouge: Louisiana State University Press.

Webb, Constance. 1968. *Richard Wright: A Biography.* New York: Putnam.

Welch, Frank J. 1943. *The Plantation Land Tenure System in Mississippi.* Bulletin 385. State College: Mississippi State College, Agricultural Experiment Station.

Whalen, Charles, and Barbara Whalen. 1985. *The Longest Debate: A Legislative History of the 1964 Civil Rights Act.* Washington, D.C.: Seven Locks Press.

Wheeler, C. W. 1927. Boll Weevil Found in Wilkes. *Washington (Ga.) News-Reporter,* July 1, 1.

White, Betsy. 1989a. After Three Years, Lack of Funds Plagues OBE. *Atlanta Constitution,* August 13, A1, A8.

———. 1989b. Meriwether Split over Plan to Merge Two Schools: County Turns Down $6.5 Million from QBE. *Atlanta Journal/Atlanta Constitution*, October 15, D1, D6.

White, Morris. 1951. *Cotton Production Practices in the Piedmont Area of Alabama*. Circular 102. Auburn: Alabama Polytechnic Institute, Agricultural Experiment Station.

White, Steven A. 1994. Experts Say Redistributing Added to GPO Surge. *New York Times*, November 14, 16.

Whitehead, Don. 1970. *Attack on Terror: The FBI against the Ku Klux Klan in Mississippi*. New York: Funk and Wagnalls.

Whitfield, Stephen J. 1988. *A Death in the Delta: The Story of Emmett Till*. New York: Free Press.

Whitt, Richard. 1989. Prisons Become Boom Industry in Georgia. *Atlanta Journal/Atlanta Constitution*, October 8, B1.

Wicker, Tom. 1963. Johnson Bids Congress Enact Civil Rights Bill with Speed. *New York Times*, November 28, 1, 20.

Wiener, Jonathan M. 1978. *Social Origins of the New South Alabama, 1860–1885*. Baton Rouge: Louisiana State University Press.

Wilhoit, Francis M. 1973. *The Politics of Massive Resistance*. New York: George Braziller.

Wilkinson, J. Harvie, III. 1979. *From Brown to Bakke: The Supreme Court and School Integration, 1954–1978*. New York: Oxford University Press.

Williams, Juan. 1988. *Eyes on the Prize: America's Civil Rights Years, 1954–1965*. New York: Penguin Books.

Williams, Raymond. 1976. *Key Words: A Vocabulary of Culture and Society*. New York: Oxford University Press.

Williams, Vinnie. 1967. Hancock Serves as School Lab. *Atlanta Journal*, April 30, 1.

Wirt, Frederick M. 1970. *Politics of Southern Equality: Law and Social Change in a Mississippi County*. Chicago: Aldine Publishing Co.

Wisdom, John Minor. 1975. Random Remarks on the Role of Social Sciences in the Judicial Decision-Making Process in School Desegregation Cases. *Law and Contemporary Problems* 39 (Winter): 134–49.

Woodson, Carter Godwin. 1930. *The Rural Negro*. Washington, D.C.: Association for the Study of Negro Life and History.

Woodward, C. Vann. 1951. *Origins of the New South, 1877–1913*. Vol. 9 of *A History of the South*, edited by Wendell Holmes Stephenson and E. Merton Coulter. Baton Rouge: Louisiana State University Press.

Woofter, Thomas Jackson, Jr. 1920. *Negro Migration: Changes in Rural Organization and Population of the Cotton Belt*. New York: W. D. Gray.

———. 1928. *Negro Problems in Cities*. 1968 reprint. New York: Negro Universities Press.

Woofter, Thomas Jackson, Jr., et al. 1936. *Landlord and Tenant on the Cotton Plantation*. Research Monograph 5. Washington D.C.: Division of Social Research, Works Progress Administration.

Wooten, James T. 1970. Private Schools Thrive in South, but Finances Restrict Quality. *New York Times*, February 1, L1, L34.

———. 1988. Rural Georgia Is Rich and Rural Georgia Is Poor. *Atlanta Journal/Atlanta Constitution*, June 5, 2C.

Works Progress Administration. Federal Writers Project. Slave Narratives. Published as *The American Slave: A Composite Autobiography*. 1972 and other years. Edited by George P. Rawick. Westport, Conn.: Greenwood Press.

Wright, Gavin. 1986. *Old South, New South: Revolutions in the Southern Economy since the Civil War*. New York: Basic Books.

Wright, Richard, and Edwin Rosskam. 1941. *Twelve Million Black Voices: A Folk History of the Negro in the United States*. New York: Viking Press.

Wyatt-Brown, Bertram. 1994. *The House of Percy: Honor, Melancholy, and Imagination in a Southern Family*. New York: Oxford University Press.

Wynes, Charles E. 1967. *Forgotten Voices: Dissenting Southerners in an Age of Conformity*. Baton Rouge: Louisiana State University Press.

Yarmolinsky, Adam. 1969. The Beginnings of OEO. In *On Fighting Poverty: Perspectives from Experience*, edited by James L. Sundquist, 34–51. New York: Basic Books.

The Youth of the Rural Organizing and Cultural Center. 1991. *Minds Stayed on Freedom: The Civil Rights Struggle in the Rural South: An Oral History*. Boulder, Colo.: Westview Press.

Zinn, Howard. 1965. *SNCC: The New Abolitionists*. 2d ed. Boston: Beacon Press.

Index

Page references to figures are indicated by *f*; to tables, by *t*; to maps, by *m*

OTHER BOOKS IN THE SERIES

LIBRARY OF CONGRESS CATALOGING-IN-PUBLICATION DATA

Aiken, Charles S. (Charles Shelton), 1938–
    The cotton plantation South since the Civil War / Charles S. Aiken. — 1st ed.
        p.   cm. — (Creating the North American landscape)
    Includes bibliographical references and index.
    ISBN 0-8018-5679-5 (alk. paper)
        1. Plantation life—Southern States—History—19th century.   2. Plantation life—Southern
    States—History—20th century.   3. Cotton growing—Southern States—History—19th
    century.   4. Cotton growing—Southern States—History—20th century.   5. Landscape—
    Southern States—history—19th century.   6. Landscape—Southern States—History—20th
    century.   7. Afro-Americans—Civil rights—Southern States—History—19th century.
    8. Afro-Americans—Civil rights—Southern States—History—20th century.   9. Southern
    States—Historical geography.   I. Title.   II. Series.
    F215.A4   1998
    975'.04—dc21   97-30162
                                                                                                                    CIP